AF806616

Strüngmann Forum Reports

Series Editor

Julia R. Lupp, Ernst Strüngmann Forum, Frankfurt Institute for Advanced Studies, Frankfurt, Germany

This series publishes the work of the Ernst Strüngmann Forum, an independent academic initiative dedicated to advancing knowledge in basic science.

The Ernst Strüngmann Forum provides a collaborative environment for invited experts to address contemporary research challenges. Guided by an international advisory board, topics chosen are of high-priority interest to various areas of research, requiring multidisciplinary input to expand understanding of the issues. Past topics include digital ethology, emergent brain dynamics, deliberate ignorance, psychiatric genetics, agrobiodiversity, and computational psychiatry. The insights and future research strategies that emerge from a topic are disseminated through the Strüngmann Forum Report series.

Established in 2006, the Ernst Strüngmann Forum has gained a reputation as a place where intellectual "dead ends" are overcome, where novel ways of conceptualizing issues are articulated, and where new collaborations are established. For further information, see https://esforum.de

This is an Open Access book series.

Dost Öngür • Judith M. Ford
Editors

Metabolic Neuropsychiatry

Springer

Editors
Dost Öngür
McLean Hospital/Harvard Medical School
Boston, MA, USA

Judith M. Ford
University of California San Francisco and the San Francisco VA Medical Center
San Francisco, CA, USA

ISSN 3059-3786 ISSN 3059-3794 (electronic)
Strüngmann Forum Reports
ISBN 978-3-032-05629-0 ISBN 978-3-032-05630-6 (eBook)
https://doi.org/10.1007/978-3-032-05630-6

This work was supported by the Ernst Strüngmann Foundation.

© Frankfurt Institute for Advanced Studies (FIAS) 2025. This book is an open access publication.

Open Access This book is licensed under the terms of the Creative Commons Attribution-NonCommercial-NoDerivatives 4.0 International License (http://creativecommons.org/licenses/by-nc-nd/4.0/), which permits any noncommercial use, sharing, distribution and reproduction in any medium or format, as long as you give appropriate credit to the original author(s) and the source, provide a link to the Creative Commons license and indicate if you modified the licensed material. You do not have permission under this license to share adapted material derived from this book or parts of it.
The images or other third party material in this book are included in the book's Creative Commons license, unless indicated otherwise in a credit line to the material. If material is not included in the book's Creative Commons license and your intended use is not permitted by statutory regulation or exceeds the permitted use, you will need to obtain permission directly from the copyright holder.
This work is subject to copyright. All commercial rights are reserved by the author(s), whether the whole or part of the material is concerned, specifically the rights of translation, reprinting, reuse of illustrations, recitation, broadcasting, reproduction on microfilms or in any other physical way, and transmission or information storage and retrieval, electronic adaptation, computer software, or by similar or dissimilar methodology now known or hereafter developed. Regarding these commercial rights a non-exclusive license has been granted to the publisher.
The use of general descriptive names, registered names, trademarks, service marks, etc. in this publication does not imply, even in the absence of a specific statement, that such names are exempt from the relevant protective laws and regulations and therefore free for general use.
The publisher, the authors and the editors are safe to assume that the advice and information in this book are believed to be true and accurate at the date of publication. Neither the publisher nor the authors or the editors give a warranty, expressed or implied, with respect to the material contained herein or for any errors or omissions that may have been made. The publisher remains neutral with regard to jurisdictional claims in published maps and institutional affiliations.

This Springer imprint is published by the registered company Springer Nature Switzerland AG
The registered company address is: Gewerbestrasse 11, 6330 Cham, Switzerland

If disposing of this product, please recycle the paper.

Preface

Science is a highly specialized enterprise—one that enables areas of inquiry to be minutely pursued, establishes working paradigms and normative standards, and supports rigor in experimental research. Yet all too often, researchers encounter problems that fall outside the scope of their immediate discipline and to progress, new insights and conceptual frameworks are needed.

The Ernst Strüngmann Forum was established in 2006 to address these types of issues. Founded on the tenets of scientific independence and open discourse, we provide a platform for experts to scrutinize topics that necessitate the input from multiple disciplines. At a central gathering, or Forum, existing perspectives are questioned, knowledge gaps are exposed, and strategies are proposed to fill such gaps. To ensure broad access to the insights that emerge, results are disseminated through the Strüngmann Forum Report series.

This volume is the culmination of a proposal initiated by Dost Öngür, put forth to advance the emerging field of metabolic neuropsychiatry. Together with Judith Ford, they sought to convene "leading scientists and clinicians in the field of brain metabolism and its clinical applications in neuropsychiatric disorders to examine the existing strands of progress in understanding how molecular and cellular metabolic abnormalities may impact brain function and how this process may be associated with the emergence of brain disorders" (pers. comm.). Are the metabolic abnormalities seen in neuropsychiatric disorders restricted to the brain or are they distributed throughout the body? Do factors that contribute to the emergence of abnormal brain function also contribute to a higher burden of medical conditions (e.g., diabetes, obesity, and hypertension) in patients with neuropsychiatric disorders?

After approval, Kim Q. Do, Lilianne R. Mujica-Parodi, Zoltan Sarnyai, and Rachel Upthegrove joined us on the organizing committee to set up the discussion framework and select invitees. Four primary areas were defined to focus the discussion:

- Role of metabolism in brain function
- Metabolic abnormalities in neuropsychiatric disorders
- Systemic metabolic aspects of neuropsychiatric disorders
- Metabolism-based therapies

The committee posed initial questions for consideration and identified topics to be addressed by invited papers. Then, from May 12–17, 2024, participants gathered in Frankfurt for lively debate. This volume synthesizes the ideas and perspectives that emerged. Organized around the main thematic areas (detailed in Chapter 1), each section contains the finalized papers written for this Forum as well as summary reports of the group discussions.

Engaging in this type of discourse necessitates an ability to question one's own perspectives while probing those of others. It is an intense experience, and I would like to thank each person who participated for their time, effort, and willingness to contribute. A special word of gratitude goes to the members of the organizing committee as well as to the authors of the background papers. In addition, I would like to highlight the efforts of the group moderators—Zoltan Sarnyai, Kim Do, Rachel Upthegrove, and Lily Mujica-Parodi—and rapporteurs—Lindy Rae, Sabina

Berretta, Toby Pillinger, and Corey Weistuch: to enable and guide a lively debate and transform it into a coherent, multiauthor report is never a simple matter. Finally, I wish to extend my appreciation to Dost Öngür and Judith Ford, whose leadership and expertise were essential at every stage of this project.

Good science requires collaboration and steadfast support. The work of the Ernst Strüngmann Forum is sustained by a number of essential partnerships, which I gratefully acknowledge: We are able to serve science and society due to the generous support of the Ernst Strüngmann Foundation, established by Dr. Andreas and Dr. Thomas Strüngmann in honor of their father. Members of the scientific advisory board guide the review process of proposals. The Deutsche Forschungsgemeinschaft provided supplemental funding for this topic, and the Ernst Strüngmann Institute graciously shared its vibrant setting with us during the Forum.

It is never easy to push the boundaries of knowledge, as long-held views can be difficult to set aside. Yet once these limitations are recognized, the process of developing strategies to move forward becomes an invigorating endeavor. Such was the spirit of everyone involved in the 37th Ernst Strüngmann Forum. We hope this volume will spur further action and guide the development of better treatments to improve the quality of life for people living with mental illnesses.

Julia R. Lupp, Director, Ernst Strüngmann Forum
Frankfurt Institute for Advanced Studies
Ruth-Moufang-Str. 1, 60438 Frankfurt am Main, Germany
https://esforum.de/

Contents

Preface v
Julia R. Lupp

List of Contributors ix

1 **Metabolic Neuropsychiatry** 1
Dost Öngür

Role of Metabolism in Brain Function

2 **Integrating Brain Metabolism Research into Big Data Human Neuroscience** 15
Manu S. Goyal

3 **Brain Energy Constraints and Vulnerability** 29
Caroline D. Rae, L. Felipe Barros, Alexander Behnke, Manu S. Goyal, Suzana Herculano-Houzel, Daria Peleg-Raibstein, Martin Picard, Douglas Rothman, Anthony C. Vernon, and Zoltan Sarnyai

4 **Brain Energy Production: Vulnerability and Pathophysiology** 57
Caroline D. Rae, L. Felipe Barros, Alexander Behnke, Manu S. Goyal, Suzana Herculano-Houzel, Daria Peleg-Raibstein, Martin Picard, Douglas Rothman, Anthony C. Vernon, and Zoltan Sarnyai

Metabolic Abnormalities in Neuropsychiatric Disorders

5 **What Is the Best Evidence for Bioenergetic Abnormalities in Severe Mental Illness?** 83
Hunter Eby, Ali Sajid Imami, William G. Ryan V, Smita Sahay, John Vergis, Jennifer H. Nguyen, Taylen O. Arvay, Priyanka Pulvender, Nicholas Henkel, and Robert E. McCullumsmith

6 **Role of Metabolism in the Emergence of Neuropsychiatric Disorders** 111
Corey Weistuch, Anthony G. Chesebro, Botond B. Antal, and Lilianne R. Mujica-Parodi

7 **Systemic Mitochondrial Alterations in Alzheimer Disease Dementia** 135
Allie Amick and Anthony J. A. Molina

8 **Metabolic Abnormalities in Psychiatric Disorders: A Transdiagnostic, Whole-Body Approach** 155
Sabina Berretta, Ana C. Andreazza, Dorit Ben-Shachar, Javier Gilbert-Jaramillo, Jill R. Glausier, Margaret Hahn, Iris-Tatjana Kolassa, R. Nehir Mavioglu, Anthony J. A. Molina, Martin Picard, Zoltan Sarnyai, Robert E. McCullumsmith, Johann Steiner, and Kim Q. Do

Systemic Metabolic Aspects of Neuropsychiatric Disorders

9 Links between Inflammation and Metabolism in Periphery and Brain: Lessons from Epidemiological Research on Depression **195**
Yuri Milaneschi

10 Metabolism across the Lifespan: Implications for Body and Mind **223**
Benjamin I. Perry, Bodyl A. Brand, Janna N. de Boer, Peter Swann, Toby Pillinger, Lena Pielke, Suchitra Varadarajan, Iain Campbell, and Iris-Tatjana Kolassa

11 Severe Mental Illness as a Multisystem Metabolic Disorder: From Brain to Body and Back Again **245**
Toby Pillinger, Hannelore Ehrenreich, Zachary Freyberg, Margaret Hahn, Matthias Mack, Yuri Milaneschi, Benjamin I. Perry, and Rachel Upthegrove

Metabolism-Based Therapies

12 Best Practices for Randomized Controlled Trials in Metabolic Neuropsychiatry Studies **263**
Melanie M. Wall and Martina Pavlicova

13 Disassociating Causality in Complex Interacting Systems **285**
David Hofmann, Simon Carter, Helmut H. Strey, and Lilianne R. Mujica-Parodi

14 Designing Innovative Interventions for Metabolic Neuropsychiatry **305**
Corey Weistuch, Virginie-Anne Chouinard, Sharmili Edwin Thanarajah, Peter Falkai, David Hofmann, Dost Öngür, Martin P. Paulus, Melanie M. Wall, and Lilianne R. Mujica-Parodi

Subject Index **325**

Strüngmann Forum Reports **331**

List of Contributors

Amick, Allie Sticht Center for Healthy Aging and Alzheimer's Prevention, Dept. of Internal Medicine, Wake Forest Baptist Medical Center, Winston-Salem, NC 27157, USA

Andreazza, Ana C. Dept. of Pharmacology and Toxicology, and Psychiatry and Mitochondrial Innovation Initiative, MITO2i, University of Toronto, Toronto, ON M5S 1A8, Canada

Antal, Botond B. Dept. of Biomedical Engineering and Laufer Center for Physical and Quantitative Biology, Renaissance School of Medicine, State University of New York at Stony Brook, Stony Brook, NY 11794; Martinos Center for Biomedical Imaging, Massachusetts General Hospital, Boston, MA 02114, USA

Arvay, Taylen O. Dept. of Neurosciences, University of Toledo College of Medicine, Toledo, OH 43614, USA

Barros, L. Felipe Centro de Estudios Científicos-CECs and Universidad San Sebastián, Facultad de Medicina, Valdivia, Chile

Behnke, Alexander Dept. of Clinical and Biological Psychology, Ulm University; and German Center for Mental Health (DZPG) at the Mannheim-Heidelberg-Ulm partner site, 89081 Ulm, Germany

Ben-Shachar, Dorit Lab of Psychobiology, Dept. of Neuroscience, Bruce and Ruth Rappaport Faculty of Medicine Faculty of Medicine, Technion-Israel Institute of Technology, Haifa 31096, Israel

Berretta, Sabina Basic Neuroscience Division, McLean Hospital, Belmont, MA 02478; Harvard Medical School, Boston, MA 02115; Stanley Center for Psychiatric Research at Broad Institute, Cambridge, MA 02142, USA

Brand, Bodyl A. Dept. of Psychiatry, University of Oxford, Oxford OX3 7JX, UK

Campbell, Iain Centre for Clinical Brain Sciences, University of Edinburgh, Edinburgh EH8 9YL, UK

Carter, Simon Applied Mathematics and Statistics, State University of New York at Stony Brook, Stony Brook, NY 11794, USA

Chesebro, Anthony G. Dept. of Biomedical Engineering and Laufer Center for Physical and Quantitative Biology, Renaissance School of Medicine, State University of New York at Stony Brook, Stony Brook, NY 11794; Martinos Center for Biomedical Imaging, Massachusetts General Hospital, Boston, MA 02114, USA

Chouinard, Virginie-Anne Psychotic Disorders Division, McLean Hospital/ Harvard Medical School, Belmont, MA 02478, USA

de Boer, Janna N. Faculty of Medical Sciences, University Medical Center Groningen, University of Groningen, 9712 CP Groningen, The Netherlands

Do, Kim Q. Center for Psychiatric Neuroscience, Dept. of Psychiatry, Lausanne University Hospital and University of Lausanne, 1005 Lausanne, Switzerland

Edwin Thanarajah, Sharmili Dept. of Psychiatry, Psychotherapy and Psychosomatics, University Hospital Frankfurt – Goethe University, 60528 Frankfurt am Main, Germany

Eby, Hunter Dept. of Neurosciences and Psychiatry, University of Toledo College of Medicine, Toledo, OH 43614, USA

Ehrenreich, Hannelore Dept. of Clinical Neuroscience, University of Göttingen and Max Planck Institute for Multidisciplinary Sciences, 37073 Göttingen, Germany

Falkai, Peter Dept. of Psychiatry and Psychotherapy, LMU University Hospital, LMU Munich; and Max-Planck-Institute for Psychiatry, Munich 80336, Germany

Ford, Judith M. Dept. of Psychology, UCSF Weill Institute for Neurosciences, San Francisco CA 94158, USA

Freyberg, Zachary Dept. of Psychiatry, University of Pittsburgh, Pittsburgh, PA 15213, USA

Gilbert-Jaramillo, Javier University of Oxford, John Radcliffe Hospital, Nuffield Department of Clinical Neurosciences, Oxford, OX3 9DU, UK

Glausier, Jill R. Dept. of Psychiatry, University of Pittsburgh, Pittsburgh, PA 15213, USA

Goyal, Manu S. Mallinckrodt Institute of Radiology, Washington University School of Medicine in St Louis, St. Louis, MO 63124, USA

Hahn, Margaret Dept. of Psychiatry, University of Toronto, Toronto, ON M4M 1K3, Canada

Henkel, Nicholas Dept. of Neurosciences and Psychiatry, University of Toledo College of Medicine, Toledo, OH 43614, USA

Herculano-Houzel, Suzana Dept. of Psychology, Vanderbilt University, Nashville, TN 37220, USA

Hofmann, David Dept. of Biomedical Engineering and Laufer Center for Physical and Quantitative Biology, Renaissance School of Medicine, State University of New York at Stony Brook, Stony Brook, NY 11794, USA

Imami, Ali Sajid Dept. of Neurosciences and Psychiatry, University of Toledo College of Medicine, Toledo, OH 43614, USA

Kolassa, Iris-Tatjana Dept. of Clinical and Biological Psychology, Institute of Psychology and Education, Ulm University; German Center for Mental Health (DZPG) at the Mannheim-Heidelberg-Ulm partner site; German Center for Child and Adolescent Health (DZKJ) at the Ulm partner site, 89081 Ulm, Germany

Mack, Matthias Dept. of Clinical and Biological Psychology, Ulm University; German Center for Mental Health (DZPG) at the Mannheim-Heidelberg-Ulm partner site, 89081 Ulm, Germany

Mavioglu, R. Nehir Dept. of Clinical and Biological Psychology, Institute of Psychology and Education, Ulm University; German Center for Mental Health (DZPG) at the Mannheim-Heidelberg-Ulm partner site; German Center for Child and Adolescent Health (DZKJ) at the Ulm partner site, 89081 Ulm, Germany

McCullumsmith, Robert E. Dept. of Neurosciences and Psychiatry, University of Toledo College of Medicine, Toledo, OH, 43614; Neurosciences Institute, ProMedica, Toledo, OH 43614, USA

Milaneschi, Yuri Dept. of Psychiatry, Amsterdam UMC, Vrije Universiteit Amsterdam, Amsterdam 1081HJ, The Netherlands

Molina, Anthony J. A. School of Medicine, University of California San Diego; Stein Institute for Research on Aging, Division of Geriatrics, Gerontology, and Palliative Care, and Stein Institute for Research on Aging and the Center for Healthy Aging, La Jolla, CA 92093, USA

Mujica-Parodi, Lilianne R. Dept. of Biomedical Engineering and Laufer Center for Physical and Quantitative Biology, Renaissance School of Medicine, State University of New York at Stony Brook, Stony Brook, NY 11794; Martinos Center for Biomedical Imaging, Massachusetts General Hospital, Boston, MA 02114, USA

Nguyen, Jennifer H. Dept. of Neurosciences and Psychiatry, University of Toledo College of Medicine, Toledo, OH 43614, USA

Öngür, Dost McLean Hospital/Harvard Medical School, Belmont, MA 02478, USA

Paulus, Martin P. Laureate Institute for Brain Research, Tulsa, OK 74136-3326, USA

Pavlicova, Martina Dept. of Biostatistics, Mailman School of Public Health, Columbia University Irving Medical Center, New York, NY 10032, USA

Peleg-Raibstein, Daria Dept. of Health Sciences and Technology, Swiss Federal Institute of Technology / ETH Zürich, 8603 Schwerzenbach, Switzerland

Perry, Benjamin I. Institute for Mental Health, University of Birmingham, Birmingham B15 2TT, UK

Picard, Martin Dept. of Psychiatry, Division of Behavioral Medicine, Columbia University Irving Medical Center; Dept. of Neurology, H. Houston Merritt Center for Neuromuscular and Mitochondrial Disorders, Columbia University Translational Neuroscience Initiative, Columbia University Irving Medical Center, New York; New York State Psychiatric Institute, New York; and Robert N Butler Columbia Aging Center, Columbia University Mailman School of Public Health, New York, 10032, USA

Pielke, Lena Dept. of Clinical and Biological Psychology, Ulm University; German Center for Mental Health (DZPG) at the Mannheim-Heidelberg-Ulm partner site, 89081 Ulm, Germany

Pillinger, Toby Institute of Psychiatry, Psychology and Neuroscience, Kings College London, London WC2R 2LS, UK

Pulvender, Priyanka Dept. of Neurosciences and Psychiatry, University of Toledo College of Medicine, Toledo, OH 43614, USA

Rae, Caroline D. Neuroscience Research Australia and School of Psychology, University of New South Wales, Randwick NSW 2031, Australia

Rothman, Douglas Magnetic Resonance Research Center (MRRC), Yale School of Medicine, New Haven, CT 06510, USA

Ryan V, William G. Dept. of Neurosciences and Psychiatry, University of Toledo College of Medicine, Toledo, OH 43614, USA

Sahay, Smita Dept. of Neurosciences and Psychiatry, University of Toledo College of Medicine, Toledo, OH 43614, USA

Sarnyai, Zoltan College of Public Health, Medical, and Vet Sciences, James Cook University, James Cook Dr, Townsville, 4811, QLD, Australia

Steiner, Johann Dept. of Psychiatry and Laboratory of Translational Psychiatry, Otto-von-Guericke-University Magdeburg; German Center for Mental Health (DZPG), Partner Site Halle-Jena-Magdeburg, Magdeburg, Germany, 39120 Magdeburg, Germany

Strey, Helmut H. Dept. of Biomedical Engineering and Laufer Center for Physical and Quantitative Biology, Renaissance School of Medicine, State University of New York at Stony Brook, Stony Brook, NY 11794, USA

Swann, Peter Dept. of Psychiatry, University of Cambridge, Cambridge CB2 0SZ, UK

Upthegrove, Rachel Dept. of Psychiatry, University of Oxford, Oxford OX3 7JX, UK

Varadarajan, Suchitra Dept. of Clinical and Biological Psychology, Ulm University; German Center for Mental Health (DZPG) at the Mannheim-Heidelberg-Ulm partner site, 89081 Ulm, Germany

Vergis, John Dept. of Neurosciences and Psychiatry, University of Toledo College of Medicine, Toledo, OH 43614, USA

Vernon, Anthony C. Dept. of Basic and Clinical Neuroscience, Institute of Psychiatry, Psychology and Neuroscience, Maurice Wohl Clinical Neuroscience Institute; and the MRC Centre for Neurodevelopmental Disorders, King's College London, London WC2R 2LS, UK

Wall, Melanie M. Dept. of Biostatistics, Mailman School of Public Health; Dept. of Psychiatry, Vegelos College of Physicians and Surgeons, Columbia University Irving Medical Center, New York, NY 10032, USA

Weistuch, Corey Dept. of Medical Physics, Memorial Sloan Kettering Cancer Center, New York, NY 10065, USA

1

Metabolic Neuropsychiatry

Dost Öngür

1 Introduction

A major and growing unmet need in psychiatry is the development of more efficacious and tolerable treatments to improve both longevity and quality of life for people with mental illnesses. Many real-world outcomes for patients have stagnated or worsened over the past several decades. Existing challenges include the absence of robust research paradigms (Paulus and Thompson 2019) and biomarkers (Abi-Dargham et al. 2023) to guide development of improved therapies. Moreover, many existing treatments contribute to already elevated rates of cardiovascular disease and dysglycemia among individuals with serious mental illness (Freyberg et al. 2017; Pillinger et al. 2019, 2020). This situation may be changing, however, with new treatments leveraging novel mechanisms (Kaul et al. 2024) as well as the repurposing of long-available substances as therapies (Bogenschutz et al. 2022). In addition to development of small molecules as new medications, innovative non-pharmacological treatments that impact brain metabolism, which has long been implicated in the pathophysiology of mental illnesses, are also generating growing interest, such as the ketogenic diet (Needham et al. 2023; Palmer 2017; Sethi et al. 2024). Within this emerging field of "metabolic neuropsychiatry," interest in treatment interventions has arguably outstripped clinical and translational research progress. To date, research in this domain is scattered and relatively modest in comparison to other areas of interest within psychiatry. Thus, a research agenda that stimulates concerted efforts to deepen our understanding of brain metabolism and guide the design of improved treatments is needed. This introductory chapter presents an overview of such a research agenda and describe the structure of this volume.

The need to consolidate efforts that underpin such a research agenda led us to propose the topic for this Ernst Strüngmann Forum. At the focal meeting, held from May 12–17, 2024, experts in biochemistry, neuroscience, neurology, and psychiatry

---

Dost Öngür (✉)
McLean Hospital/Harvard Medical School, Boston, MA, USA
Email: dongur@mgb.org

© Frankfurt Institute for Advanced Studies (FIAS) 2025

D. Öngür, J. M. Ford (eds.), *Metabolic Neuropsychiatry*, Strüngmann Forum Reports,
https://doi.org/10.1007/978-3-032-05630-6_1

met to identify (a) basic research progress needed to support translation of findings, (b) tractable clinical interventions aligned with the basic research findings, and (c) strategies to be deployed to test these interventions. The resulting discourse led to many insights and recommendations for future research and treatment development, as well as a high-level summary of these discussions, from which we draw extensively in this introductory chapter (Andreazza et al. 2025). This volume presents in-depth coverage of the topics covered at the Forum, provides an interdisciplinary perspective on energy metabolism in the brain and body in neuropsychiatric disorders, and highlights future research directions. We believe that it will be of interest to students and investigators at all career stages across a range of disciplines, and we hope that it provides impetus for development of better treatments for people suffering from neuropsychiatric disorders.

The basis for excitement in this emerging field comes from our deepening insights into the intricate relationship between brain metabolism and healthy brain function. The origins of brain metabolism research date back to the mid-twentieth century when several principles were established: (a) the brain does not have extensive energy reserves in storage; (b) it is dependent on a constant supply of glucose, its primary energy fuel, and oxygen, requisite to efficiently metabolize glucose, from the bloodstream; and (c) energy supply is modulated by brain activity levels (Shulman et al. 2002). There has been elaboration on these principles over the past thirty years, with discovery of alternative fuel sources, emerging roles for glial cells, and an understanding of the complex balance between supply and consumption during brain activation. In fact, the transient imbalance between supply and demand underlies the blood-oxygen-level-dependent (BOLD) signal (Theriault et al. 2023; Thulborn et al. 1982), which is the biophysical basis of functional MRI, a ubiquitous research technique in modern human neuroscience (Belliveau et al. 1991; Ogawa et al. 1992).

Along with progress in the neuroscience of brain metabolism, a growing literature of patient-oriented research has uncovered significant aberrations in markers of energy metabolism and closely related biochemical processes, including mitochondrial function and redox biology, in the brains of individuals with psychiatric disorders (Henkel et al. 2022; Stein et al. 2023; Townsend et al. 2023). This has led to the suggestion that bioenergetic abnormalities underlie many brain disorders. Currently, however, we do not have a good understanding of how putative abnormalities in energy supply and utilization over time influence neuronal/glial function, information processing, and ultimately symptom formation in pathological states.

2 Energy in the Brain

The role of energy metabolism in supporting healthy brain function has been reviewed in detail elsewhere (Bonvento and Bolanos 2021; Dienel 2019; DiNuzzo et al. 2024; Magistretti and Allaman 2018; Picard and McEwen 2018; Rae et al. 2024; Rothman et al. 2003, 2022; Shulman et al. 2004). The majority of the brain's energy budget (~70–80%) is spent on neuronal electrochemical activity that is the basis of ongoing neural communication, with a small increase for focused cognitive activity

(Harris et al. 2012). This is in contrast to muscle tissue, which has very low energy expenditure rates at rest but experiences many-fold increases with physical activity. These data indicate that maintaining basic brain functions is energetically expensive and imply that the brain is primarily designed for what has been called "resting-state neuronal activity" (Raichle 2015).

The tight correlation between energy supply and brain activity, along with the absence of major energy reserves within the brain itself, makes brain tissue unusually sensitive to bioenergetic disruptions (Trigo et al. 2022). This is evident in cases of recovery from cardiac arrest, when brain death can occur even though the rest of the body emerges without significant damage, and in liver or kidney failure that can lead to reversible encephalopathy in the absence of permanent neuronal damage (Vamadevan et al. 2019). These concepts have been popularized by the notion of the "selfish brain," which places the brain as the final consumer that is critically dependent on energy flows from the rest of the body (Peters 2011). The brain is, after all, an exceptionally catabolic organ with one of the highest oxygen consumption rates on a *per mass* basis (Andreazza et al. 2010).

But what are the consequences of bioenergetic failure? For example, how might brain function be impacted if ATP synthesis rates are reduced, if there is an impairment in the switch from OxPhos to glycolysis as required during development, or if redox balance is disrupted? Would there be changes in membrane potentials, synaptic activity, action potential generation, circuit oscillations, or coordination across large-scale neural networks (Kula et al. 2024)? Would different brain systems (e.g., cortex, subcortical nuclei, or the cerebellum) be affected differently? Would other brain processes such as neurogenesis and neuroplasticity be affected in the long term? And over time, how might such bioenergetic compromise relate to different neuropsychiatric phenotypes emerging at different developmental stages, such as autism, epilepsy, psychotic disorders, or neurodegenerative disorders? These questions remain largely unanswered, but intriguing research has shown that metabolic status indeed controls action potential firing of specific neuron types in the brain as well as activation states of glial cells, and that both neurons and astrocytes can rapidly modulate energy metabolism in response to activity status (Barros et al. 2023; Yellen 2018).

2.1 Role of Brain Energy Supply in Psychiatric Disorders

As with healthy brain bioenergetics, the topic of bioenergetic abnormalities in psychiatric disorders has been reviewed in great detail elsewhere (Gonçalves et al. 2015; Henkel et al. 2022; Stein et al. 2023; Steullet et al. 2016). Convergent findings from a substantial literature that includes postmortem brain and patient-derived neuronal cell line research demonstrate broad bioenergetic alterations in psychiatric disorders, including major depressive disorder, bipolar disorder, and schizophrenia (for in-depth reviews, see Kim et al. 2019; Ni et al. 2022; Sullivan et al. 2018). *In vivo* studies suggest various energy-metabolism-related markers also demonstrate abnormalities in psychiatric disorders (e.g., Townsend et al. 2023).

The findings of molecular and cellular bioenergetic abnormalities in psychiatric disorders raise the temptation to ask whether these may represent causative factors for mental illness. However, this question may not be productive; unidirectional causal models are unlikely to adequately explain the evolution or failure of multi-factorial complex systems such as the brain (Fried and Robinaugh 2020). It is more realistic to expect that the field will continue to explore multiple diverse risk factors and pathophysiologic mechanisms, rather than simply converging upon a small number of causal pathways. Instead of debating whether metabolic abnormalities are causal in the pathophysiology of psychiatric disorders, we simply point out that they are one hallmark of mental illness, analogous to the widely discussed "hallmarks of cancer" (Hanahan and Weinberg 2011).

2.2 Energy in Body and Brain

Although it is shielded from many peripheral biochemical factors by the blood-brain barrier, the brain does not operate in isolation. There are multiple bidirectional interactions between the brain and periphery that provide complex control mechanisms to modulate the biochemistry of our internal milieu (Le Thuc and Garcia-Caceres 2024). This raises two questions: Are the brain bioenergetic abnormalities seen in psychiatric disorders accompanied by parallel changes in the rest of the body? If so, what is the relationship between the two?

The resounding answer to the first question is yes; a large literature demonstrates a range of peripheral metabolic abnormalities in psychiatric disorders, including some that suggest metabolic disturbance predates psychiatric diagnosis and may be a risk indicator for it (Freyberg et al. 2024; Milaneschi et al. 2020; Perry et al. 2020, 2021a; Pillinger et al. 2019). For this reason, the term "metabolic syndrome," a term originally used in the field of internal medicine, now features prominently in any discussion of clinical care for psychiatric disorders, including in children and adolescents (Cortese et al. 2024; Henderson et al. 2015; Vancampfort et al. 2014; Zachos et al. 2024). Put another way, it is highly possible, though not proven, that the same underlying metabolic abnormalities contribute to risk for psychiatric disorders when manifest in the brain and to risk for metabolic syndrome when manifest in the rest of the body (Agarwal et al. 2020; Lee et al. 2024; Perry et al. 2021b; Prestwood et al. 2021).

3 Treatment Interventions

Energy metabolism is a critical process for developing and maintaining healthy brain function, and there are significant abnormalities in this process in the brain and body in individuals with psychiatric disorders. These suppositions have received attention and support in recent years from the success of the ketogenic diet in improving the condition of some patients with mood and psychotic disorders (Freyberg et al. 2024; Needham et al. 2023; Sarnyai and Palmer 2020; Sethi et al.

2024). Therefore, a more detailed and systematic approach to expand our armamentarium of metabolic treatments is warranted.

In service of this goal, it is useful to first consider how energy metabolism is embedded in the broader network of physiologic processes. The brain modulates metabolism through control circuits that sense change and elicit responses to maintain critical parameters within healthy ranges (Tingley et al. 2021). It is likely that these control circuits create complex, brain region-specific oscillatory patterns that respond to perturbations in the environment and the needs of the organism. Dysregulation in such control circuits specific to maintaining the internal milieu (Mujica-Parodi and Strey 2020) would ultimately generate abnormal metabolic setpoints and the metabolic syndrome. Brain areas expressing high levels of neuromodulator chemicals, which have evolved more recently in humans, have higher energy requirements and mitochondria specialized for energy transformation (Castrillon et al. 2023; Mosharov et al. 2025). In people with bioenergetic dysfunction, one possible scenario is that the brain's response to metabolic perturbation is triggered correctly but the correction then overshoots the normal range, leading to uncontrolled oscillatory behavior. Viewed through this lens, the task of treating bioenergetic dysfunction involves not raising or lowering the value of a specific parameter, but rather resetting the relevant control circuits.

Many physiologic interventions can be used to modulate metabolic control circuits. Each of these interventions will have different effect profiles and time constants. Thus, some may be more appropriately seen as "probes" of the system, which cause shorter-lived deviations from a setpoint with rapid return to baseline. These probes, therefore, reveal the magnitude of a system's response and any potential oscillatory behavior, but they do not cause long-lasting change. Other interventions may cause longer-term deviations and reset the system's properties, thereby functioning to "treat" the abnormality.

In the metabolic intervention space, some open questions include: Does early biological or clinical response to an intervention predict long-term change and, if so, for which interventions? How effective might it be to combine interventions with theoretical synergies? Will metabolic interventions be universally effective, or will we need to personalize interventions based on patient characteristics (e.g., difficult to treat symptoms)? If the latter, what psychosocial, biological, or other parameters do we need to know about the system to select the right treatment? Finally, is it possible to stratify patient populations using baseline biological measures to optimize outcomes? Since there is substantial heterogeneity in psychiatric disorders and metabolic alterations are transdiagnostic, the engagement of metabolic pathways may be responsible for the expression of specific clinical manifestations potentially shared across disorders.

4 Structure of this Volume

The remaining chapters are organized into four sections that reflect the narrative arc of this introductory chapter. Many of the points raised here are greatly expanded in the rest of the book to provide much deeper insights regarding profitable directions for the field to take. The chapters in the first section review the basics of energy metabolism in healthy brain function:

- Chapter 2 Integrating Brain Metabolism Research into Big Data Human Neuroscience, by Manu S. Goyal
- Chapter 3 Brain Energy Constraints and Vulnerability, by Caroline D. Rae, L. Felipe Barros, Alexander Behnke, Manu S. Goyal, Suzana Herculano-Houzel, Daria Peleg-Raibstein, Martin Picard, Douglas Rothman, Anthony C. Vernon, and Zoltan Sarnyai
- Chapter 4 Brain Energy Production: Vulnerability and Pathophysiology, by Caroline D. Rae, L. Felipe Barros, Alexander Behnke, Manu S. Goyal, Suzana Herculano-Houzel, Daria Peleg-Raibstein, Martin Picard, Douglas Rothman, Anthony C. Vernon, and Zoltan Sarnyai

The second section provides a survey of metabolic abnormalities in neuropsychiatric disorders:

- Chapter 5 What Is the Best Evidence for Bioenergetic Abnormalities in Severe Mental Illness? by Hunter Eby, Ali Sajid Imami, William G. Ryan V, Smita Sahay, John Vergis, Jennifer H. Nguyen, Taylen O. Arvay, Priyanka Pulvender, Nicholas Henkel, Robert E. McCullumsmith
- Chapter 6 Role of Metabolism in the Emergence of Neuropsychiatric Disorders, by Corey Weistuch, Anthony G. Chesebro, Botond B. Antal, and Lilianne R. Mujica-Parodi
- Chapter 7 Systemic Mitochondrial Alterations in Alzheimer's Disease Dementia, by Allie Amick and Anthony J. A. Molina
- Chapter 8 Metabolic Abnormalities in Psychiatric Disorders: A Transdiagnostic, Whole-Body Approach, by Sabina Berretta, Ana C. Andreazza, Dorit Ben-Shachar, Javier Gilbert-Jaramillo, Jill R. Glausier, Iris-Tatjana Kolassa, R. Nehir Mavioglu, Robert E. McCullumsmith, Anthony J. A. Molina, Johann Steiner, and Kim Q. Do

The third section focuses on systemic metabolic aspects of neuropsychiatric disorders:

- Chapter 9 Links between Inflammation and Metabolism in Periphery and Brain: Lessons from Epidemiological Research on Depression, by Yuri Milaneschi
- Chapter 10 Metabolic Measures across the Lifespan: Implications for Body and Mind, by Benjamin I. Perry, Bodyl A. Brand, Janna N. de Boer, Peter Swan, Toby Pillinger, Lena Pielke, Suchitra Varadarajan, Iain Campbell, Iris-Tatjana Kolassa
- Chapter 11 Serious Mental Illness As a Multisystem Metabolic Disorder, by Toby Pillinger, Hannelore Ehrenreich, Zachary Freyberg, Margaret Hahn, Matthias Mack, Yuri Milaneschi, Benjamin I. Perry, and Rachel Upthegrove

The fourth and final section concerns therapies targeting metabolic function:

Chapter 12 Methodological Best Practices for Planning, Conducting, Analyzing, and Reporting on Randomized Controlled Trials in Metabolic Neuropsychology Studies, by Melanie M. Wall and Martina Pavlicova

Chapter 13 Disassociating Causality in Complex Interacting Systems, by David Hofmann, Simon Carter, Helmut H. Strey, and Lilianne R. Mujica-Parodi

Chapter 14 Designing Innovative Interventions for Metabolic Psychiatry, by Corey Weistuch, Virginie-Anne Chouinard, Sharmili Edwin Thanarajah, Peter Falkai, David Hofmann, Dost Öngür, Martin P. Paulus, Melanie M. Wall, and Lilianne R. Mujica-Parodi

5 Conclusions and Future Directions

Based on emergent insights from this project, the following areas for future work in the field are highlighted:

- Long-term mechanistic studies in cell and animal models are needed to determine causal pathways linking metabolic changes in the body and brain to neural activity patterns, and ultimately to symptom formation.
- We need to understand better the complex and mutually reinforcing relationships between peripheral insulin signaling, immune activation/neuroinflammation, and brain function in model systems.
- We need to identify and assess *in vivo* metabolic measures as biomarkers for use in human studies.
- Measures thus identified should be studied longitudinally in health and disease, especially during perturbation/provocation in controlled environments (i.e., in an experimental medicine framework). To ensure understanding of the behavior of the system, such studies should first be conducted in healthy individuals before moving on to patient groups.
- Longitudinal studies across diagnoses and illness stages are needed to map trajectories of metabolic dysfunction, including prior to symptom onset.
- We need studies to assess whether patient stratification based on metabolic phenotypes have diagnostic or prognostic utility, and whether it can enable selection of individualized interventions.

If metabolic interventions are salutary for mental illness, psychiatry stands to benefit from extensive experience in other fields of medicine and to rapidly adapt interventions for our patient population, a point previously made by Manji et al. (2012). This phenomenon is currently unfolding in studies of the ketogenic diet, where we have a century of experience using this diet to treat seizure disorders. Initial promising serendipitous observations were followed by small clinical studies and ultimately large well-controlled and adequately powered clinical trials. Though far from a cure, the consensus is that over 50% of patients with treatment-resistant epilepsy experience a reduction in seizure frequency on the ketogenic diet. This literature has culminated in meta-analyses, umbrella reviews (Diez-Arroyo et al. 2024), and FDA approval for the ketogenic diet in childhood epilepsy.

The large unmet need in psychiatry leads to pressure on clinicians and researchers to find efficacious and well-tolerated interventions, setting up unrealistic expectations and ultimately disappointment and loss of trust by the public if treatments are not as good as hoped (Hyman 2012). While it is important not to promise miraculous recovery to severely ill patients as the expected outcome because of this background, it is likely that new metabolic treatments offer new hope for improvement for many patients. Holding as our therapeutic goal the improvement of metabolic health rather than symptom suppression could thus deliver an empowering message for our patients and their families (Picard 2022).

References

Abi-Dargham A, Moeller S, Ali F, et al. (2023) Candidate Biomarkers in Psychiatric Disorders: State of the Field. World Psychiatry 22 (2):236–262. https://doi.org/10.1002/wps.21078

Agarwal SM, Caravaggio F, Costa-Dookhan KA, et al. (2020) Brain Insulin Action in Schizophrenia: Something Borrowed and Something New. Neuropharmacology 163:107633. https://doi.org/10.1016/j.neuropharm.2019.05.010

Andreazza, AC, Barros LF, Behnke A et al. (2025) Brain and Body Energy Metabolism and Potential for Treatment of Psychiatric Disorders. Nat Mental Health https://doi.org/10.1038/s44220-025-00422-6

Andreazza AC, Shao L, Wang JF, and Young LT (2010) Mitochondrial Complex I Activity and Oxidative Damage to Mitochondrial Proteins in the Prefrontal Cortex of Patients with Bipolar Disorder. Arch Gen Psychiatry 67 (4):360–368. https://doi.org/10.1001/archgenpsychiatry.2010.22

Barros LF, Ruminot I, Sotelo-Hitschfeld T, Lerchundi R, and Fernandez-Moncada I (2023) Metabolic Recruitment in Brain Tissue. Annu Rev Physiol 85:115–135. https://doi.org/10.1146/annurev-physiol-021422-091035

Belliveau JW, Kennedy DN, Jr., McKinstry RC, et al. (1991) Functional Mapping of the Human Visual Cortex by Magnetic Resonance Imaging. Science 254 (5032):716–719. https://doi.org/10.1126/science.1948051

Bogenschutz MP, Ross S, Bhatt S, et al. (2022) Percentage of Heavy Drinking Days Following Psilocybin-Assisted Psychotherapy vs Placebo in the Treatment of Adult Patients with Alcohol Use Disorder: A Randomized Clinical Trial. JAMA Psychiatry 79 (10):953–962. https://doi.org/10.1001/jamapsychiatry.2022.2096

Bonvento G, and Bolanos JP (2021) Astrocyte-Neuron Metabolic Cooperation Shapes Brain Activity. Cell Metab 33 (8):1546–1564. https://doi.org/10.1016/j.cmet.2021.07.006

Castrillon G, Epp S, Bose A, et al. (2023) An Energy Costly Architecture of Neuromodulators for Human Brain Evolution and Cognition. Sci Adv 9 (50):eadi7632. https://doi.org/10.1126/sciadv.adi7632

Cortese S, Purper-Ouakil D, Apter A, et al. (2024) Psychopharmacology in Children and Adolescents: Unmet Needs and Opportunities. Lancet Psychiatry 11 (2):143–154. https://doi.org/10.1016/S2215-0366(23)00345-0

Dienel GA (2019) Brain Glucose Metabolism: Integration of Energetics with Function. Physiol Rev 99 (1):949–1045. https://doi.org/10.1152/physrev.00062.2017

Diez-Arroyo C, Garcia-Garcia M, Soto-Mendez MJ, et al. (2024) Effect of the Ketogenic Diet as a Treatment for Refractory Epilepsy in Children and Adolescents: A Systematic Review of Reviews. Nutr Rev 82 (4):487–502. https://doi.org/10.1093/nutrit/nuad071

DiNuzzo M, Dienel GA, Behar KL, et al. (2024) Neurovascular Coupling Is Optimized to Compensate for the Increase in Proton Production from Nonoxidative Glycolysis and Glycogenolysis during Brain Activation and Maintain Homeostasis of pH, pCO_2, and pO_2. J Neurochem 168 (5):632–662. https://doi.org/10.1111/jnc.15839
Freyberg Z, Andreazza AC, McClung CA, and Phillips ML (2024) Linking Mitochondrial Dysfunction, Neurotransmitter, Neural Network Abnormalities and Mania: Elucidating Neurobiological Mechanisms of the Therapeutic Effect of the Ketogenic Diet in Bipolar Disorder. Biol Psychiatry Cogn Neurosci Neuroimaging 10 (3):267–277. https://doi.org/10.1016/j.bpsc.2024.07.011
Freyberg Z, Aslanoglou D, Shah R, and Ballon JS (2017) Intrinsic and Antipsychotic Drug-Induced Metabolic Dysfunction in Schizophrenia. Front Neurosci 11:432. https://doi.org/10.3389/fnins.2017.00432
Fried EI, and Robinaugh DJ (2020) Systems All the Way Down: Embracing Complexity in Mental Health Research. BMC Med 18 (1):205. https://doi.org/10.1186/s12916-020-01668-w
Gonçalves VF, Andreazza AC, and Kennedy JL (2015) Mitochondrial Dysfunction in Schizophrenia: an Evolutionary Perspective. Hum Genet 134 (1):13–21. https://doi.org/10.1007/s00439-014-1491-8
Hanahan D, and Weinberg RA (2011) Hallmarks of Cancer: The Next Generation. Cell 144 (5):646–674. https://doi.org/10.1016/j.cell.2011.02.013
Harris JJ, Jolivet R, and Attwell D (2012) Synaptic Energy Use and Supply. Neuron 75 (5):762–777. https://doi.org/10.1016/j.neuron.2012.08.019
Henderson DC, Vincenzi B, Andrea NV, Ulloa M, and Copeland PM (2015) Pathophysiological Mechanisms of Increased Cardiometabolic Risk in People with Schizophrenia and Other Severe Mental Illnesses. Lancet Psychiatry 2 (5):452–464. https://doi.org/10.1016/S2215-0366(15)00115-7
Henkel ND, Wu X, O'Donovan SM, et al. (2022) Schizophrenia: A Disorder of Broken Brain Bioenergetics. Mol Psychiatr 27 (5):2393–2404. https://doi.org/10.1038/s41380-022-01494-x
Hyman SE (2012) Revolution Stalled. Science translational medicine 4 (155):155cm111. https://doi.org/10.1126/scitranslmed.3003142
Kaul I, Sawchak S, Walling DP, et al. (2024) Efficacy and Safety of Xanomeline-Trospium Chloride in Schizophrenia: A Randomized Clinical Trial. JAMA Psychiatry 81 (8):749–756. https://doi.org/10.1001/jamapsychiatry.2024.0785
Kim Y, Vadodaria KC, Lenkei Z, et al. (2019) Mitochondria, Metabolism, and Redox Mechanisms in Psychiatric Disorders. Antioxid Redox Signal 31 (4):275–317. https://doi.org/10.1089/ars.2018.7606
Kula B, Antal B, Weistuch C, et al. (2024) D-β-Hydroxybutyrate Stabilizes the Hippocampal CA3-CA1 Circuit during Acute Insulin Resistance. PNAS Nexus 3 (5):196. https://doi.org/10.1093/pnasnexus/pgae196
Le Thuc O, and Garcia-Caceres C (2024) Obesity-Induced Inflammation: Connecting the Periphery to the Brain. Nat Metab 6 (7):1237–1252. https://doi.org/10.1038/s42255-024-01079-8
Lee J, Xue X, Au E, et al. (2024) Glucose Dysregulation in Antipsychotic-Naive First-Episode Psychosis: In Silico Exploration of Gene Expression Signatures. Transl Psychiatry 14 (1):19. https://doi.org/10.1038/s41398-023-02716-8
Magistretti PJ, and Allaman I (2018) Lactate in the Brain: From Metabolic End-Product to Signalling Molecule. Nat Rev Neurosci 19 (4):235 249. https://doi.org/10.1038/nrn.2018.19
Manji H, Kato T, Di Prospero NA, et al. (2012) Impaired Mitochondrial Function in Psychiatric Disorders. Nat Rev Neurosci 13 (5):293–307. https://doi.org/10.1038/nrn3229
Milaneschi Y, Lamers F, Berk M, and Penninx BWJH (2020) Depression Heterogeneity and Its Biological Underpinnings: Toward Immunometabolic Depression. Biol Psychiatry 88 (5):369–380. https://doi.org/10.1016/j.biopsych.2020.01.014

Mosharov E, Rosenberg A, Monzel A, et al. (2025) A Human Brain Map of Mitochondrial Respiratory Capacity and Diversity. Nature. https://doi.org/10.1038/s41586-025-08740-6
Mujica-Parodi LR, and Strey HH (2020) Making Sense of Computational Psychiatry. Int J Neuropsychopharmacol 23 (5):339–347. https://doi.org/10.1093/ijnp/pyaa013
Needham N, Campbell IH, Grossi H, et al. (2023) Pilot Study of a Ketogenic Diet in Bipolar Disorder. BJPsych Open 9 (6):e176. https://doi.org/10.1192/bjo.2023.568
Ni P, Ma Y, and Chung S (2022) Mitochondrial Dysfunction in Psychiatric Disorders. Schizophr Res 273:62–77. https://doi.org/10.1016/j.schres.2022.08.027
Ogawa S, Tank DW, Menon R, et al. (1992) Intrinsic Signal Changes Accompanying Sensory Stimulation: Functional Brain Mapping with Magnetic Resonance Imaging. PNAS 89 (13):5951–5955. https://doi.org/10.1073/pnas.89.13.5951
Palmer CM (2017) Ketogenic Diet in the Treatment of Schizoaffective Disorder: Two Case Studies. Schizophr Res 189:208–209. https://doi.org/10.1016/j.schres.2017.01.053
Paulus MP, and Thompson WK (2019) The Challenges and Opportunities of Small Effects: The New Normal in Academic Psychiatry. JAMA Psychiatry 76 (4):353–354. https://doi.org/10.1001/jamapsychiatry.2018.4540
Perry BI, Burgess S, Jones HJ, et al. (2021a) The Potential Shared Role of Inflammation in Insulin Resistance and Schizophrenia: A Bidirectional Two-Sample Mendelian Randomization Study. PLoS Med 18:e1003455. https://doi.org/10.1371/journal.pmed.1003455
Perry BI, Khandaker GM, Marwaha S, et al. (2020) Insulin Resistance and Obesity, and Their Association with Depression in Relatively Young People: Findings from a Large UK Birth Cohort. Psychol Med 50 (4):556–565. https://doi.org/10.1017/S0033291719000308
Perry BI, Stochl J, Upthegrove R, et al. (2021b) Longitudinal Trends in Childhood Insulin Levels and Body Mass Index and Associations with Risks of Psychosis and Depression in Young Adults. JAMA Psychiatry 78:416–425. https://doi.org/10.1001/jamapsychiatry.2020.4180
Peters A (2011) The Selfish Brain: Competition for Energy Resources. Am J Hum Biol 23 (1):29–34. https://doi.org/10.1002/ajhb.21106
Picard M (2022) Why Do We Care More About Disease Than Health? Phenomics 2 (3):145–155. https://doi.org/10.1007/s43657-021-00037-8
Picard M, and McEwen BS (2018) Psychological Stress and Mitochondria: A Systematic Review. Psychosom Med 80 (2):141–153. https://doi.org/10.1097/PSY.0000000000000545
Pillinger T, D'Ambrosio E, McCutcheon R, and Howes OD (2019) Is Psychosis a Multisystem Disorder? A Meta-Review of Central Nervous System, Immune, Cardiometabolic, and Endocrine Alterations in First-Episode Psychosis and Perspective on Potential Models. Mol Psychiatr 24 (6):776–794. https://doi.org/10.1038/s41380-018-0058-9
Pillinger T, McCutcheon RA, Vano L, et al. (2020) Comparative Effects of 18 Antipsychotics on Metabolic Function in Patients with Schizophrenia, Predictors of Metabolic Dysregulation, and Association with Psychopathology: A Systematic Review and Network Meta-Analysis. Lancet Psychiatry 7 (1):64–77. https://doi.org/10.1016/s2215-0366(19)30416-x
Prestwood TR, Asgariroozbehani R, Wu S, et al. (2021) Roles of Inflammation in Intrinsic Pathophysiology and Antipsychotic Drug-Induced Metabolic Disturbances of Schizophrenia. Behav Brain Res 402:113101. https://doi.org/10.1016/j.bbr.2020.113101
Rae CD, Baur JA, Borges K, et al. (2024) Brain Energy Metabolism: A Roadmap for Future Research. J Neurochem 168 (5):910–954. https://doi.org/10.1111/jnc.16032
Raichle ME (2015) The Restless Brain: How Intrinsic Activity Organizes Brain Function. Philos Trans R Soc Lond B Biol Sci 370 (1668). https://doi.org/10.1098/rstb.2014.0172
Rothman DL, Behar KL, Hyder F, and Shulman RG (2003) *In Vivo* NMR Studies of the Glutamate Neurotransmitter Flux and Neuroenergetics: Implications for Brain Function. Annu Rev Physiol 65:401–427. https://doi.org/10.1146/annurev.physiol.65.092101.142131

Rothman DL, Dienel GA, Behar KL, et al. (2022) Glucose Sparing by Glycogenolysis (GSG) Determines the Relationship between Brain Metabolism and Neurotransmission. J Cereb Blood Flow Metab 42 (5):844–860. https://doi.org/10.1177/0271678X211064399

Sarnyai Z, and Palmer CM (2020) Ketogenic Therapy in Serious Mental Illness: Emerging Evidence. Int J Neuropsychopharmacol 23 (7):434–439. https://doi.org/10.1093/ijnp/pyaa036

Sethi S, Wakeham D, Ketter T, et al. (2024) Ketogenic Diet Intervention on Metabolic and Psychiatric Health in Bipolar and Schizophrenia: A Pilot Trial. Psychiatry Res 335:115866. https://doi.org/10.1016/j.psychres.2024.115866

Shulman RG, Hyder F, and Rothman DL (2002) Biophysical Basis of Brain Activity: Implications for Neuroimaging. Q Rev Biophys 35 (3):287–325. https://doi.org/10.1017/s0033583502003803

Shulman RG, Rothman DL, Behar KL, and Hyder F (2004) Energetic Basis of Brain Activity: Implications for Neuroimaging. Trends Neurosci 27 (8):489–495. https://doi.org/10.1016/j.tins.2004.06.005

Stein A, Zhu C, Du F, and Öngür D (2023) Magnetic Resonance Spectroscopy Studies of Brain Energy Metabolism in Schizophrenia: Progression from Prodrome to Chronic Psychosis. Curr Psychiatry Rep 25 (11):659–669. https://doi.org/10.1007/s11920-023-01457-1

Steullet P, Cabungcal JH, Monin A, et al. (2016) Redox Dysregulation, Neuroinflammation, and NMDA Receptor Hypofunction: A "Central Hub" in Schizophrenia Pathophysiology? Schizophr Res 176 (1):41–51. https://doi.org/10.1016/j.schres.2014.06.021

Sullivan CR, O'Donovan SM, McCullumsmith RE, and Ramsey A (2018) Defects in Bioenergetic Coupling in Schizophrenia. Biol Psychiatry 83 (9):739–750. https://doi.org/10.1016/j.biopsych.2017.10.014

Theriault JE, Shaffer C, Dienel GA, et al. (2023) A Functional Account of Stimulation-Based Aerobic Glycolysis and Its Role in Interpreting BOLD Signal Intensity Increases in Neuroimaging Experiments. Neurosci Biobehav Rev 153:105373. https://doi.org/10.1016/j.neubiorev.2023.105373

Thulborn KR, Waterton JC, Matthews PM, and Radda GK (1982) Oxygenation Dependence of the Transverse Relaxation Time of Water Protons in Whole Blood at High Field. Biochim Biophys Acta 714 (2):265–270. https://doi.org/10.1016/0304-4165(82)90333-6

Tingley D, McClain K, Kaya E, Carpenter J, and Buzsaki G (2021) A Metabolic Function of the Hippocampal Sharp Wave-Ripple. Nature 597 (7874):82–86. https://doi.org/10.1038/s41586-021-03811-w

Townsend L, Pillinger T, Selvaggi P, et al. (2023) Brain Glucose Metabolism in Schizophrenia: A Systematic Review and Meta-Analysis of ^{18}FDG-PET Studies in Schizophrenia. Psychol Med 53 (11):4880–4897. https://doi.org/10.1017/s003329172200174x

Trigo D, Avelar C, Fernandes M, Sa J, and da Cruz ESO (2022) Mitochondria, Energy, and Metabolism in Neuronal Health and Disease. FEBS Lett 596 (9):1095–1110. https://doi.org/10.1002/1873-3468.14298

Vamadevan T, Howlett D, and Filyridou M (2019) Imaging Appearances of Toxic and Acquired Metabolic Encephalopathic Disorders. Br J Hosp Med (Lond) 80 (7):372–376. https://doi.org/10.12968/hmed.2019.80.7.372

Vancampfort D, Correll CU, Wampers M, et al. (2014) Metabolic Syndrome and Metabolic Abnormalities in Patients with Major Depressive Disorder: A Meta-Analysis of Prevalences and Moderating Variables. Psychol Med 44 (10):2017–2028. https://doi.org/10.1017/S0033291713002778

Yellen G (2018) Fueling Thought: Management of Glycolysis and Oxidative Phosphorylation in Neuronal Metabolism. J Cell Biol 217 (7):2235–2246. https://doi.org/10.1083/jcb.201803152

Zachos KA, Choi J, Godin O, et al. (2024) Mitochondrial Biomarkers and Metabolic Syndrome in Bipolar Disorder. Psychiatry Res 339:116063. https://doi.org/10.1016/j.psychres.2024.116063

Open Access This chapter is licensed under the terms of the Creative Commons Attribution-NonCommercial-NoDerivatives 4.0 International License (http://creativecommons.org/licenses/by-nc-nd/4.0/), which permits any noncommercial use, sharing, distribution and reproduction in any medium or format, as long as you give appropriate credit to the original author(s) and the source, provide a link to the Creative Commons license and indicate if you modified the licensed material. You do not have permission under this license to share adapted material derived from this chapter or parts of it.

The images or other third party material in this chapter are included in the chapter's Creative Commons license, unless indicated otherwise in a credit line to the material. If material is not included in the chapter's Creative Commons license and your intended use is not permitted by statutory regulation or exceeds the permitted use, you will need to obtain permission directly from the copyright holder.

Role of Metabolism in Brain Function

2

Integrating Brain Metabolism Research into Big Data Human Neuroscience

Manu S. Goyal

Abstract Discovering and determining links between brain metabolism and psychiatric illnesses is a challenge. This chapter suggests approaches that integrate brain metabolism research into large-scale human neuroscience efforts. Examples are provided as well as avenues for future investigations. It is hoped that such efforts, combined with rigorous basic science investigations and theory, will ultimately help to translate brain metabolism research to the benefit of patients with mental illness.

Keywords Brain metabolism, mental illness, human neuroscience

1 Introduction

When asked to be a part of the Ernst Strüngmann Forum on "Metabolic Neuropsychiatry," I was struck by the enormous complexity of the problem at hand. How does one begin to tackle this topic? Addressing the relationships between metabolism and neuropsychiatric conditions requires that one understand how metabolism and brain cellular activity are linked, how cellular activity relates to brain function, and how brain function relates to cognition, behavior, and mental illness. The complexity of each of these topics is independently very high; understanding how they influence one another increases this complexity substantially, particularly when scaling up from cellular metabolism to whole-brain function and behavior. Many, including the attendees at this forum, are investigating these topics deeply and rigorously, but the challenge of putting together actionable theories that can reliably impact human mental health remains.

Manu S. Goyal (✉)
Mallinckrodt Institute of Radiology, Washington University School of Medicine in St. Louis, St. Louis, MO 63124, USA
Email: goyalm@wustl.edu

© Frankfurt Institute for Advanced Studies (FIAS) 2025

D. Öngür, J. M. Ford (eds.), *Metabolic Neuropsychiatry*, Strüngmann Forum Reports,
https://doi.org/10.1007/978-3-032-05630-6_2

Much of this challenge is conceptual and addressed by other chapters in this volume. Here, I focus on translational difficulties and potential approaches to surmount them. One difficulty involves translating basic brain metabolism research to clinical neuroscience. This challenge is certainly not unique to "metabolic neuropsychiatry." For example, despite extensive basic science investigations into neuroprotection, efforts to translate this science to stroke patients have thus far largely failed (Kidwell et al. 2001; O'Collins et al. 2011). The potential reasons for this failure have been outlined by numerous articles (see, e.g., Chamorro et al. 2021; Gladstone et al. 2002). Among the proposed reasons, statistical limitations in preclinical research, differences between the brains of humans and other animals, and clinical trial design have been noted, though many other factors are likely also relevant (Chamorro et al. 2021; Dronne et al. 2008; Gladstone et al. 2002; Xu and Pan 2013). It is instructive to contrast this to the remarkable success of thrombectomy trials in stroke. Here, the basic science rapidly involved human investigations that included widely accessible neuroimaging methods appropriate for clinical trials (Albers et al. 2018; Goyal et al. 2016; Nogueira et al. 2018; Smith and Furlan 2016). Ultimately, this not only led to a much better understanding of the pathophysiology of stroke but also to treatments that have revolutionized acute stroke care and outcomes. Clinically accessible perfusion imaging was a component in several of these trials, remains a widely used technique, and is useful for prognosis determination and the detection of potentially treatable strokes (Bathla et al. 2022; Peerlings et al. 2023).

The second challenge is methodological in nature, with regard to measuring human brain metabolism. Measuring brain metabolism, being closely linked to brain function, is vulnerable to various methodological "conveniences." For example, some animal and human studies require sedation, which by definition affects brain function, and thus may have a significant effect upon metabolism (Foster et al. 1987; Guedj et al. 2022; Hypponen et al. 2022). It is plausible that dietary changes such as fasting and differences in caffeine intake—both common in human imaging studies—might also influence patterns of blood flow and glucose utilization (Park et al. 2014). Even wearing a mask during the COVID-19 pandemic had measurable changes on functional neuroimaging studies (Fischer et al. 2021; Law et al. 2022, 2021; Scholkmann et al. 2021). Moreover, many methods of quantifying brain metabolism rely on tractable but variably complex models (Carson et al. 2023); these models may themselves include unmeasured parameters that have to be estimated and accounted for, resulting in risk of error propagation and bias. Though carefully designed and well-controlled experiments aim to address many of these challenges, internal validity does not often translate to external validity, particularly when addressing diseases as heterogeneous as those found in psychiatry. Perfusion imaging in stroke, for example, in part surmounted difficulties in producing consistent quantitative results across scanners and sites by normalizing cerebral blood flow values to the contralateral hemisphere, which became the standard method for several successful clinical trials (Campbell et al. 2011; Vagal et al. 2019).

The third challenge involves the sheer complexity of the brain's cellular and subcellular composition. Efforts such as those from the Allen Institute are demonstrating that the brain has numerous different types of neurons, astrocytes, oligodendrocytes, microglia, pericytes, and other cells, and that they differ further in terms of maturity and activity (Jorstad et al. 2023; Langlieb et al. 2023; Siletti et al. 2023;

Shi et al. 2023; Weninger and Arlotta 2023). Moreover, synapses and their underlying subcellular substrates are also highly heterogeneous (Sun and Schuman 2023). Following the discovery that a great deal of mRNA to protein translation occurs locally near synapses, rather than just in the cell body, recent investigations of "synaptosomes" suggest that even such local synaptic translation can be heterogeneous (Hafner et al. 2019; van Oostrum et al. 2023). It is thus very likely that metabolism also differs according to cell type and synapse and their proportion within different parcels of brain tissue, thereby challenging the generalizability of many smaller-scale cellular studies of brain metabolism.

A general theme across these challenges is that they have to do with scale and complexity. Changes in scale and complexity are apparent when moving from rodent to human brains, from relatively small carefully controlled experiments to large multicenter studies, or from investigating a handful of cells to a massive number of cells. Added to this is the enormous complexity of mental illnesses, which overlap in their features and manifest heterogeneously. To address these challenges, approaches that integrate brain metabolism research into "big data" clinical neuroscience might help to improve translational success as well as to reveal emergent phenomena that arise from complex and large-scale physiology. Though not a comprehensive review, below I delineate four potential approaches.

2 Develop More Readily Accessible PET Imaging of Human Brain Metabolism for Large-Scale Studies

Methods to assess human brain metabolism *in vivo* have and continue to evolve. To illustrate this, it helps to discuss how one might measure oxygen consumption in the human brain. Decades ago, methods to measure whole-brain oxygen metabolism involved placing arterial and venous catheters to assess blood flowing into and leaving the brain. The Kety-Schmidt method then applied the Fick Principle to calculate whole-brain blood flow and the metabolic rates of oxygen and glucose consumption (Kety and Schmidt 1948; Lassen 1959). Advantages of this technique were that it relied on a small set of basic assumptions. However, the Kety-Schmidt method is invasive, with risks to the research participant or patient, and it relies on skilled operators to perform it. In addition, this technique does not provide regional estimates.

Subsequent developments in positron emission tomography (PET) imaging provided an opportunity to use ^{15}O radiotracers to measure regional oxygen metabolism in the brain. Models were constructed that relate observed positron emission counts within a region of interest to its metabolic rate of oxygen consumption (An et al. 2021; Fan et al. 2020; Mintun et al. 1984, 2001). Since these models require an arterial input function, an arterial catheter was typically still needed and often placed in the radial artery. An influential model also required independent measurements of cerebral blood flow and volume (Mintun et al. 1984). One limitation of this technique on older PET scanners was that the total effective radiation dose from a single measurement of brain oxygen metabolism was around 20 mSv, which is 40% of the annual occupational limit in the United States, and thus limits repetitions and other scans. In addition, ^{15}O has a very short half-life, requiring an on-site, experienced cyclotron staff to produce and supply the ^{15}O gases and water rapidly and safely. Accordingly, this technique remains limited to specialized centers. Nonetheless, a randomized clinical trial (the Carotid Occlusion Surgery Study) applied a simplified version of this technique to study whether neurosurgical bypass was effective or not in preventing stroke in high-risk carotid occlusion patients as selected by ^{15}O PET imaging (Powers et al. 2011). The trial was terminated early due to futility, suggesting that bypass surgery may not be beneficial for preventing stroke in carotid occlusion patients.

Due to the resources and technical requirements of this method, clinical trials and other large-scale studies employing ^{15}O PET have been rare. Multiple major recent developments in PET imaging might now help to overcome these obstacles. First, modern PET scanners, including some being built specific to brain imaging, have much better spatial resolution and sensitivity, which leads to several advantages (Li et al. 2024; Prenosil et al. 2022; van Sluis et al. 2020). The arterial input function can now be directly extracted from the imaging itself, thus avoiding the need for an arterial catheter (though arterial blood sampling may still be needed to assess metabolites for other radiotracers) (Volpi et al. 2023). Partial volume effects become less of a concern, limiting the need for sophisticated, computationally expensive algorithms to perform partial volume correction.

Importantly, due to the better sensitivity of modern PET scanners, the dose of radiotracer needed can be reduced substantially while maintaining adequate signal-to-noise; in our experience, we can successfully obtain adequate ^{15}O imaging with a third or less of the typical total dose used on older PET scanners (Lee et al. 2024). Notably, efforts to reduce the dose of an amyloid tracer have found even more remarkable results (Young et al. 2024). Second, there are many ongoing advances in PET data acquisition and analysis, such as better attenuation, motion, and scatter correction algorithms. We have found that with these improvements, we can now assess brain metabolism (including cerebral blood volume, flow, oxygen, and glucose metabolism) at a finer spatial scale, within single individuals, and at a fraction of the total radiation exposure (Lee et al. 2024).

As modern PET scanners become more widely available, there is now an opportunity to perform metabolic brain PET imaging at a much larger scale than previously possible. For example, if the radiation dose of ^{18}F-fluorodeoxyglucose (FDG) PET of the brain can be lowered substantially to a resulting effective dose of 1 mSv or less, it might be feasible to incorporate this technique more often into large-scale human neuroscience studies. Very low doses might also permit longitudinally repeated assessments and including younger at-risk individuals. Since analysis of metabolic PET imaging is technically challenging, software packages that automate analysis of the PET imaging will be needed to ensure accurate and consistent results. Methods to "harmonize" imaging across sites, scanners, and different radiotracers are now available and demonstrate improved longitudinal consistency (Bourgeat et al. 2022); such approaches have already been successful for amyloid PET imaging, which is widely performed in large-scale studies of Alzheimer disease. Applying the same approach to FDG-PET and potentially other metabolic radiotracers might thus allow for the integration of brain metabolism PET imaging into large-scale human studies in neuropsychiatry, and could be applied to the multiple, smaller extant PET studies already performed.

3 Develop and Validate Accessible MRI Methods to Measure Human Brain Metabolism

Though PET imaging has many advantages when measuring brain metabolism, it also has important limitations. For example, ^{15}O gas production and delivery remains resource intensive and challenging; thus, PET imaging of brain oxygen metabolism may remain difficult to use beyond a few select centers. The radiation dose from PET imaging limits its use in children. PET also has interpretive limitations. For example, when performing FDG-PET alone, one cannot determine how much of a difference in FDG metabolism is due to oxygen-dependent metabolism (i.e., oxidative phosphorylation), glycolysis, and its intermediary pathways versus changes in glycogen metabolism. FDG uptake and phosphorylation are also two independent processes that are collectively counted when measuring positron emission counts; kinetic compartmental modeling can help separate these processes to some extent but depend on the acquisition of an arterial input function and accurate modeling (Gjedde 2014). Finally, FDG is not glucose and thus requires further correction for small but potentially relevant differences between the two molecules (Barrio et al. 2020; Gjedde 2014).

Magnetic resonance imaging (MRI) methods are poised to help overcome some of the limitations in PET-based measurements of brain metabolism. There are now an increasing number of attempts to develop MRI-based methods of measuring brain oxygen metabolism (for recent reviews, see An and Lin 2000; Biondetti et al. 2023; Hyder and Rothman 2017; Rodgers et al. 2016). Each method has its own advantages and shortcomings, but can now be validated against ^{15}O PET, which is facilitated with the development of PET/MRI scanners (Cho et al. 2021; Fan et al. 2020). Validation should entail both regional and whole-brain quantitative estimates. Differences between MRI and PET should be understood; assuming that PET is the "gold standard" may not be appropriate in some cases, and discrepancies between modalities might be due to an error in the PET analysis rather than with MRI. Some groups have even recently advocated a hybrid approach, whereby one uses MRI for certain high reliability measurements and PET for others (Fan et al. 2020; Narciso et al. 2021).

Other MRI techniques, including chemical exchange saturation transfer MRI and MR spectroscopy, are also undergoing major developments, such as the use of native proton and isotope imaging techniques as well as following enrichment with stable isotopes and hyperpolarization to boost signal strength (Hyder and Rothman 2017; Koush et al. 2022; Straathof et al. 2021; van Zijl et al. 2021). These techniques can measure various aspects of the metabolism of glucose and other carbohydrates, as well as amino acids and other compounds. Importantly, unlike PET imaging, MR spectroscopy provides information on the fate of various compounds.

MR spectroscopy, in particular, has already shown enormous value in numerous basic science experiments and is now being increasingly applied to study human brain metabolism in clinical settings. For example, phosphorous MR spectroscopy, which can identify metabolites in the brain related to energy metabolism and phospholipid turnover, has been applied to investigate mental illnesses. A review of phosphorous MR spectroscopy studies in schizophrenia found dozens of such studies, but with rather limited consistency in their results (Yüksel et al. 2015).

Both heterogeneity in experimental design as well as statistical issues were noted as reasons for the absence of definitive findings. Phosphorous MR spectroscopy has also been used to measure the brain's redox state via the NAD+/NADH ratio in schizophrenia, finding that it is reduced in both first-episode and chronic psychosis patients (Stein et al. 2023). As for PET imaging above, developing more readily accessible MR spectroscopy methods and analytical packages will be critical to translating these techniques to large-scale and longitudinal human studies.

4 Relate Brain Metabolism Imaging to Fluid Biomarkers

Alongside various developments in neuroimaging of metabolism, there have been remarkable advances in assaying fluids, such as blood and cerebrospinal fluid (CSF), for numerous proteins, metabolites, transcripts, and other biological factors. One example is the use of proteomics to identify over a thousand proteins within CSF in individuals at genetic risk for early-onset Alzheimer disease. In a recently published study, this demonstrated significant rises in several specific proteins during the stages between onset of amyloid-beta deposition and that of symptomatic disease (Johnson et al. 2023). Interestingly, this analysis identified several proteins specifically related to glycolysis as key factors that changed very soon after amyloid-beta deposition began, and again near the onset of symptomatic disease.

Interpretation of such results, however, is limited. What does an increase in one of these proteins, such as lactate dehydrogenase B (LDHB), mean? One can speculate, but given the variety of cells and brain regions from where LDHB might have been released, it is uncertain how to understand this finding. Moreover, the release of LDHB indicates little about the direction of changes in metabolism. However, by coupling metabolic brain imaging with such CSF proteomic studies, one might be able to better infer where in the brain the metabolic changes are occurring. Also, the direction of change in metabolic flux can be measured. The CSF proteomic results remain informative: they indicate the individuals and time period when these changes might be best addressed with neuroimaging. Furthermore, the proteomic results might identify many other affected proteins, providing more insight into affected cells, subcellular components, and metabolic pathways (van Zalm et al. 2023). As more of these investigations are performed, interpretation of the CSF proteome, along with that from brain and plasma, will likely improve, and thereby provide another avenue to study more widely brain metabolism in various human diseases (Yang et al. 2021).

5 Relate Brain Metabolism to Cellular and Subcellular Studies

A major limitation of the approaches discussed above is that they are generally unable to ascribe metabolic measurements to specific cell types or subcellular components. This is currently true for both fluid biomarkers of metabolism and for neuroimaging measures that typically have a spatial resolution on the order of cubic millimeters to cubic centimeters. As noted above, this is particularly important to resolve given the increasing identification of numerous different cell types, including among neurons and the various glial cells. Moreover, it is now established that many subcellular processes, including metabolism and translation, occur locally near synapses (Sun and Schuman 2023).

Methods to probe metabolism at the cellular and subcellular level *in vivo* in the rodent brain are being developed and show welcome promise (Cao et al. 2017, 2021; Zhang et al. 2019). However, these techniques have their own limitations (Dienel and Rothman 2023); more importantly for the current discussion, studies in the rodent brain aimed at addressing human mental illnesses may have limited external validity. A human neuroscience-based approach to address this problem would be desirable, even if it can only currently provide data for hypothesis generation.

One approach in this regard is to take advantage of big data human neuroscience efforts that are arising from the Allen Institute and other large-scale "brain mapping" efforts (Gabitto et al. 2024; Hawrylycz et al. 2012; Jorstad et al. 2023; Kang et al. 2011; Mosharov et al. 2025; Siletti et al. 2023; van Oostrum et al. 2023). Efforts from the Allen Institute initially included whole genome-scale transcriptomics acquired from hundreds of postmortem brain samples taken from a small group of individuals (Hawrylycz et al. 2012). Since then, their efforts and those of several other groups have expanded in terms of the number of individuals, ages that span across mammalian lifespans, diseases such as Alzheimer disease, and methods being used, including single-cell transcriptomics (Gabitto et al. 2024; Kang et al. 2011; Zhu et al. 2018). There is now a deluge of data available, and the challenge is to devise analyses that make informative use of these data.

One approach to studying brain metabolism using these data is simply to target metabolism-related transcripts and determine (a) their interrelationships, (b) their relationships with meta-features (e.g., brain regions, age, disease state), and (c) their relationships with other transcripts, including those that help to identify different cell types and subcellular components. For example, an initial report on the Seattle Alzheimer's Disease Cell Atlas identified decreased expression of many metabolism-related genes specifically within neurons of the middle temporal gyrus, including those related to cholesterol synthesis, glycolysis, and mitochondrial function, in relation to a model estimate of disease progression (Gabitto et al. 2024). Further analyses of these data suggest that somatostatin interneurons might be more selectively vulnerable in the early stages of preclinical Alzheimer disease.

These data might be even more informative when combined with brain metabolism imaging (and potentially fluid biomarker) data, acquired in a spatiotemporal manner. For example, combining the Seattle Alzheimer's Disease Cell Atlas data with brain metabolism imaging spatiotemporally across the trajectory of Alzheimer disease might help to provide further insight into how the disease affects brain metabolism at each of its stages. Previous work by our group employed a similar approach across the human lifespan using the Allen Human Brain Atlas and the BrainSpan dataset (Goyal et al. 2014). We examined gene expression related to the regional pattern of the glycolytic index, which was devised to be a relative measure of aerobic glycolysis in the human brain using multitracer PET imaging. In contrast to FDG-PET alone, which highly correlated with genes related to mitochondria and energy metabolism, the glycolytic index identified a large set of different genes that were related to neurite and synaptic development. Moreover, the BrainSpan dataset was used to identify transcriptionally neotenous regions of the brain, which corresponded to those regions with the highest glycolytic index in young adults. Collectively these results suggest that aerobic glycolysis in the human brain might be related to development and neotenous processes in the human adult brain (Bauernfeind et al. 2014; Goyal et al. 2014); the reasons for this are currently unclear, though several possibilities exist (see, e.g., Cruz et al. 2022; Harris et al. 2019; Rae et al. 2024; Segarra-Mondejar et al. 2018; Theriault et al. 2023; Zhang et al. 2019).

The external validity of this correlative approach is currently limited. However, it can be useful to generate hypotheses, identify target genes and proteins for further investigation, and act as a component of a larger convergent analysis. For example, the genes identified above, specifically in relation to aerobic glycolysis, have recently been reported to be strongly associated with adolescent cortical development in humans (Lotter et al. 2024). More recently, the general approach of performing spatial (and spatiotemporal) correlations between brain imaging findings and those arising from big data transcriptomic datasets has been codified into a set of tools called "neuromaps" (Markello et al. 2022). This set of tools includes statistical corrections, such as for spatial autocorrelation, and can compare a large variety of brain maps according to different atlases and datasets. As more investigators learn to grapple with these large transcriptional (and other similar "-omics") datasets, perhaps we will discover new ways to integrate them with imaging of brain metabolism. Finally, another approach to linking imaging to specific genes and metabolic pathways is to identify individuals with rare mutations to link their neuroimaging findings with findings from cellular experiments, such as with induced pluripotent stem cells (Reid et al. 2022).

6 Conclusions

With these proposals in hand, what might a future effort aiming to understand brain metabolism in mental illness look like? A large cohort of patients—ranging from those at risk to those experiencing their first symptoms and others in a chronic phase of the illness—would help to evaluate the "trajectory" of metabolic abnormalities related to the illness. Several metabolic measures could be obtained and analyzed in this cohort, including fluid biomarkers from blood and CSF, low-dose radiotracer PET studies on a modern scanner, and brain MRI/MRS. Ideally, this would be done longitudinally, though the long time course of many mental illnesses might make this challenging. Nonetheless, large-scale longitudinal efforts that include metabolism-related fluid biomarkers and brain imaging have been successfully applied to Alzheimer disease, such as in the Alzheimer's Disease Neuroimaging Initiative and Dominantly Inherited Alzheimer Disease Network (McKay et al. 2023; Weber et al. 2021); indeed, these efforts provide a paradigm that can guide similar efforts for mental illness.

In this context, it may also be instructive to consider an expertimental medicine approach where investigators perturb brain metabolism using well-defined interventions such as GLP-1 agonists or a ketogenic diet and use the various big data neuroscience measures proposed above as outcome measures. This framework can be scaled up in adequately powered clinical trials and provide strong evidence for target engagement and validation for the interventions' mechanism of action. Ultimately, this approach can provide a bridge between the knowledge base generated by big data neuroscience and real-world clinical applications.

These efforts may still not adequately address the challenges facing "metabolic neuropsychiatry." Certainly, detailed investigations of various open questions in brain metabolism research are necessary, as thoroughly reviewed recently (Rae et al. 2024), and will require mechanistic studies in animal models as well as "reverse translational" approaches. However, combining animal-based mechanistic investigations with large-scale human studies will help to improve the likelihood of translating basic science findings and avoid repeated failures in future clinical trials. Parenthetically, fostering interest in applying brain metabolism research into big data neuroscience initiatives is equally important to achieve success when proposing and launching these efforts. It is notable that brain metabolism does not currently have a dedicated chapter in two of the top neuroscience textbooks (Kandel et al. 2021; Purves 2018). This Ernst Strüngmann Forum helps to address this gap. Ultimately, increasing future efforts in brain metabolism research—at multiple scales and using a variety of methodologies—will hopefully help to identify new effective treatments for patients with mental illness.

References

Albers GW, Marks MP, Kemp S, et al. (2018) Thrombectomy for Stroke at 6 to 16 Hours with Selection by Perfusion Imaging. N Engl J Med 378 (8):708–718. https://doi.org/10.1056/NEJMoa1713973

An H, Goyal MS, and Powers WJ (2021) ^{15}O PET Imaging: Methods and Applications. In: Franceschi AM, Franceschi D (eds) Hybrid PET/MR Neuroimaging: A Comprehensive Approach. Springer, Cham, pp 197–216. https://doi.org/10.1007/978-3-030-82367-2

An H, and Lin W (2000) Quantitative Measurements of Cerebral Blood Oxygen Saturation Using Magnetic Resonance Imaging. J Cereb Blood Flow Metab 20 (8):1225–1236. https://doi.org/10.1097/00004647-200008000-00008

Barrio JR, Huang SC, Satyamurthy N, et al. (2020) Does 2-FDG PET Accurately Reflect Quantitative *in Vivo* Glucose Utilization? J Nucl Med 61 (6):931–937. https://doi.org/10.2967/jnumed.119.237446

Bathla G, Pillenahalli Maheshwarappa R, Soni N, et al. (2022) CT Perfusion Maps Improve Detection of M2-MCA Occlusions in Acute Ischemic Stroke. J Stroke Cerebrovasc Dis 31 (6):106473. https://doi.org/10.1016/j.jstrokecerebrovasdis.2022.106473

Bauernfeind AL, Barks SK, Duka T, et al. (2014) Aerobic Glycolysis in the Primate Brain: Reconsidering the Implications for Growth and Maintenance. Brain Struct Funct 219 (4):1149–1167. https://doi.org/10.1007/s00429-013-0662-z

Biondetti E, Cho J, and Lee H (2023) Cerebral Oxygen Metabolism from MRI Susceptibility. Neuroimage 276:120189. https://doi.org/10.1016/j.neuroimage.2023.120189

Bourgeat P, Dore V, Burnham SC, et al. (2022) B-Amyloid PET Harmonisation across Longitudinal Studies: Application to AIBL, ADNI and OASIS3. Neuroimage 262:119527. https://doi.org/10.1016/j.neuroimage.2022.119527

Campbell BC, Christensen S, Levi CR, et al. (2011) Cerebral Blood Flow Is the Optimal CT Perfusion Parameter for Assessing Infarct Core. Stroke 42 (12):3435–3440. https://doi.org/10.1161/STROKEAHA.111.618355
Cao R, Li J, Ning B, et al. (2017) Functional and Oxygen-Metabolic Photoacoustic Microscopy of the Awake Mouse Brain. Neuroimage 150:77–87. https://doi.org/10.1016/j.neuroimage.2017.01.049
Cao R, Tran A, Li J, et al. (2021) Hemodynamic and Oxygen-Metabolic Responses of the Awake Mouse Brain to Hypercapnia Revealed by Multi-Parametric Photoacoustic Microscopy. J Cereb Blood Flow Metab 41 (10):2628–2639. https://doi.org/10.1177/0271678X211010352
Carson RE, Naganawa M, and Gallezot J-D (2023) PET Quantification and Kinetic Analysis. In: Cross DJ, Mosci K, Minoshima S (eds) Molecular Imaging of Neurodegenerative Disorders. Springer, Cham, pp 183–194. https://doi.org/10.1007/978-3-031-35098-6_12
Chamorro A, Lo EH, Renu A, van Leyen K, and Lyden PD (2021) The Future of Neuroprotection in Stroke. Journal of Neurochemistry 92 (2):129–135. https://doi.org/10.1136/jnnp-2020-324283
Cho J, Lee J, An H, et al. (2021) Cerebral Oxygen Extraction Fraction (OEF): Comparison of Challenge-Free Gradient Echo QSM+qBOLD (QQ) with ^{15}O PET in Healthy Adults. J Cereb Blood Flow Metab 41 (7):1658–1668. https://doi.org/10.1177/0271678X20973951
Cruz E, Bessieres B, Magistretti P, and Alberini CM (2022) Differential Role of Neuronal Glucose and PFKFB3 in Memory Formation during Development. Glia 70 (11):2207–2231. https://doi.org/10.1002/glia.24248
Dienel GA, and Rothman DL (2023) *In Vivo* Calibration of Genetically Encoded Metabolite Biosensors Must Account for Metabolite Metabolism during Calibration and Cellular Volume. Journal of Neurochemistry 168 (5):506–532. https://doi.org/10.1111/jnc.15775
Dronne MA, Grenier E, Chapuisat G, Hommel M, and Boissel JP (2008) A Modelling Approach to Explore Some Hypotheses of the Failure of Neuroprotective Trials in Ischemic Stroke Patients. Prog Biophys Mol Biol 97 (1):60–78. https://doi.org/10.1016/j.pbiomolbio.2007.10.001
Fan AP, An H, Moradi F, et al. (2020) Quantification of Brain Oxygen Extraction and Metabolism with [^{15}O]-Gas PET: A Technical Review in the Era of PET/MRI. Neuroimage 220:117136. https://doi.org/10.1016/j.neuroimage.2020.117136
Fischer JB, Kobayashi Frisk L, Scholkmann F, et al. (2021) Cerebral and Systemic Physiological Effects of Wearing Face Masks in Young Adults. PNAS 118 (41):e2109111118. https://doi.org/10.1073/pnas.2109111118
Foster NL, VanDerSpek AF, Aldrich MS, et al. (1987) The Effect of Diazepam Sedation on Cerebral Glucose Metabolism in Alzheimer's Disease as Measured Using Positron Emission Tomography. J Cereb Blood Flow Metab 7 (4):415–420. https://doi.org/10.1038/jcbfm.1987.84
Gabitto MI, Travaglini KJ, Rachleff VM, et al. (2024) Integrated Multimodal Cell Atlas of Alzheimer's Disease. Nat Neurosci 27 (12):2366–2383. https://doi.org/10.1038/s41593-024-01774-5
Gjedde A (2014) Positron Emission Tomography of Brain Glucose Metabolism with [18F] Fluorodeoxyglucose in Humans. In: Hirrlinger J, Waagepetersen HS (eds) Brain Energy Metabolism. Springer, New York, pp 341–364. https://doi.org/10.1007/978-1-4939-1059-5_14
Gladstone DJ, Black SE, Hakim AM, Heart, and Stroke Foundation of Ontario Centre of Excellence in Stroke R (2002) Toward Wisdom from Failure: Lessons from Neuroprotective Stroke Trials and New Therapeutic Directions. Stroke 33 (8):2123–2136. https://doi.org/10.1161/01.str.0000025518.34157.51
Goyal MS, Hawrylycz M, Miller JA, Snyder AZ, and Raichle ME (2014) Aerobic Glycolysis in the Human Brain Is Associated with Development and Neotenous Gene Expression. Cell Metab 19 (1):49–57. https://doi.org/10.1016/j.cmet.2013.11.020

Goyal MS, Menon BK, van Zwam WH, et al. (2016) Endovascular Thrombectomy after Large-Vessel Ischaemic Stroke: A Meta-Analysis of Individual Patient Data from Five Randomised Trials. Lancet 387 (10029):1723–1731. https://doi.org/10.1016/S0140-6736(16)00163-X

Guedj E, Varrone A, Boellaard R, et al. (2022) Eanm Procedure Guidelines for Brain PET Imaging Using [^{18}F]FDG, Version 3. Eur J Nucl Med Mol Imaging 49 (2):632–651. https://doi.org/10.1007/s00259-021-05603-w

Hafner AS, Donlin-Asp PG, Leitch B, Herzog E, and Schuman EM (2019) Local Protein Synthesis Is a Ubiquitous Feature of Neuronal Pre- and Postsynaptic Compartments. Science 364 (6441). https://doi.org/10.1126/science.aau3644

Harris RA, Lone A, Lim H, et al. (2019) Aerobic Glycolysis Is Required for Spatial Memory Acquisition but Not Memory Retrieval in Mice. eNeuro 6 (1). https://doi.org/10.1523/ENEURO.0389-18.2019

Hawrylycz MJ, Lein ES, Guillozet-Bongaarts AL, et al. (2012) An Anatomically Comprehensive Atlas of the Adult Human Brain Transcriptome. Nature 489 (7416):391–399. https://doi.org/10.1038/nature11405

Hyder F, and Rothman DL (2017) Advances in Imaging Brain Metabolism. Annu Rev Biomed Eng 19:485–515. https://doi.org/10.1146/annurev-bioeng-071516-044450

Hypponen V, Stenroos P, Nivajarvi R, et al. (2022) Metabolism of Hyperpolarised [1-^{13}C] Pyruvate in Awake and Anaesthetised Rat Brains. NMR Biomed 35 (2):e4635. https://doi.org/10.1002/nbm.4635

Johnson ECB, Bian S, Haque RU, et al. (2023) Cerebrospinal Fluid Proteomics Define the Natural History of Autosomal Dominant Alzheimer's Disease. Nat Med 29 (8):1979–1988. https://doi.org/10.1038/s41591-023-02476-4

Jorstad NL, Close J, Johansen N, et al. (2023) Transcriptomic Cytoarchitecture Reveals Principles of Human Neocortex Organization. Science 382 (6667):eadf6812. https://doi.org/10.1126/science.adf6812

Kandel E, Koester J, Mack S, and Siegelbaum S (2021) Principles of Neural Science, 6th Edn. McGraw Hill, New York

Kang HJ, Kawasawa YI, Cheng F, et al. (2011) Spatio-Temporal Transcriptome of the Human Brain. Nature 478 (7370):483–489. https://doi.org/10.1038/nature10523

Kety SS, and Schmidt CF (1948) The Effects of Altered Arterial Tensions of Carbon Dioxide and Oxygen on Cerebral Blood Flow and Cerebral Oxygen Consumption of Normal Young Men. J Clin Invest 27 (4):484–492. https://doi.org/10.1172/JCI101995

Kidwell CS, Liebeskind DS, Starkman S, and Saver JL (2001) Trends in Acute Ischemic Stroke Trials through the 20th Century. Stroke 32 (6):1349–1359. https://doi.org/10.1161/01.str.32.6.1349

Koush Y, Rothman DL, Behar KL, de Graaf RA, and Hyder F (2022) Human Brain Functional MRS Reveals Interplay of Metabolites Implicated in Neurotransmission and Neuroenergetics. J Cereb Blood Flow Metab 42 (6):911–934. https://doi.org/10.1177/0271678X221076570

Langlieb J, Sachdev NS, Balderrama KS, et al. (2023) The Molecular Cytoarchitecture of the Adult Mouse Brain. Nature 624 (7991):333–342. https://doi.org/10.1038/s41586-023-06818-7

Lassen NA (1959) Cerebral Blood Flow and Oxygen Consumption in Man. Physiol Rev 39 (2):183–238. https://doi.org/10.1152/physrev.1959.39.2.183

Law CSW, Lan PS, and Glover GH (2021) Effect of Wearing a Face Mask on fMRI BOLD Contrast. Neuroimage 229:117752. https://doi.org/10.1016/j.neuroimage.2021.117752

——— (2022) Response to Scholkmann Commentary: "Effect of Wearing a Face Mask on fMRI BOLD Contrast." Neuroimage 246:118773. https://doi.org/10.1016/j.neuroimage.2021.118773

Lee JJ, Metcalf N, Durbin TA, et al. (2024) Multi-Tracer Studies of Brain Oxygen and Glucose Metabolism Using a Time-of-Flight Positron Emission Tomography-Computed Tomography Scanner. J Vis Exp 208:65510. https://doi.org/10.3791/65510

Li H, Badawi RD, Cherry SR, et al. (2024) Performance Characteristics of the Neuroexplorer, a Next-Generation Human Brain PET/CT Imager. J Nucl Med 65 (8):1320–1326. https://doi.org/10.2967/jnumed.124.267767

Lotter LD, Saberi A, Hansen JY, et al. (2024) Human Cortex Development Is Shaped by Molecular and Cellular Brain Systems. Nat Commun 15 (1):7987. https://doi.org/10.1038/s41467-024-52366-7

Markello RD, Hansen JY, Liu ZQ, et al. (2022) Neuromaps: Structural and Functional Interpretation of Brain Maps. Nat Methods 19 (11):1472–1479. https://doi.org/10.1038/s41592-022-01625-w

McKay NS, Gordon BA, Hornbeck RC, et al. (2023) Positron Emission Tomography and Magnetic Resonance Imaging Methods and Datasets within the Dominantly Inherited Alzheimer Network (DIAN). Nat Neurosci 26 (8):1449–1460. https://doi.org/10.1038/s41593-023-01359-8

Mintun MA, Lundstrom BN, Snyder AZ, et al. (2001) Blood Flow and Oxygen Delivery to Human Brain during Functional Activity: Theoretical Modeling and Experimental Data. PNAS 98 (12):6859–6864. https://doi.org/10.1073/pnas.111164398

Mintun MA, Raichle ME, Martin WR, and Herscovitch P (1984) Brain Oxygen Utilization Measured with O-15 Radiotracers and Positron Emission Tomography. J Nucl Med 25 (2):177–187

Mosharov E, Rosenberg A, Monzel A, et al. (2025) A Human Brain Map of Mitochondrial Respiratory Capacity and Diversity. Nature. https://doi.org/10.1038/s41586-025-08740-6

Narciso L, Ssali T, Liu L, et al. (2021) A Noninvasive Method for Quantifying Cerebral Metabolic Rate of Oxygen by Hybrid PET/MRI: Validation in a Porcine Model. J Nucl Med 62 (12):1789–1796. https://doi.org/10.2967/jnumed.120.260521

Nogueira RG, Jadhav AP, Haussen DC, et al. (2018) Thrombectomy 6 to 24 Hours after Stroke with a Mismatch between Deficit and Infarct. N Engl J Med 378 (1):11–21. https://doi.org/10.1056/NEJMoa1706442

O'Collins VE, Macleod MR, Cox SF, et al. (2011) Preclinical Drug Evaluation for Combination Therapy in Acute Stroke Using Systematic Review, Meta-Analysis, and Subsequent Experimental Testing. J Cereb Blood Flow Metab 31 (3):962–975. https://doi.org/10.1038/jcbfm.2010.184

Park CA, Kang CK, Son YD, et al. (2014) The Effects of Caffeine Ingestion on Cortical Areas: Functional Imaging Study. Magn Reson Imaging 32 (4):366–371. https://doi.org/10.1016/j.mri.2013.12.018

Peerlings D, de Jong H, Bennink E, et al. (2023) Spatial CT Perfusion Data Helpful in Automatically Locating Vessel Occlusions for Acute Ischemic Stroke Patients. Front Neurol 14:1136232. https://doi.org/10.3389/fneur.2023.1136232

Powers WJ, Clarke WR, Grubb RL, Jr., et al. (2011) Extracranial-Intracranial Bypass Surgery for Stroke Prevention in Hemodynamic Cerebral Ischemia: The Carotid Occlusion Surgery Study Randomized Trial. JAMA 306 (18):1983–1992. https://doi.org/10.1001/jama.2011.1610

Prenosil GA, Sari H, Furstner M, et al. (2022) Performance Characteristics of the Biograph Vision Quadra PET/CT System with a Long Axial Field of View Using the NEMA NU 2-2018 Standard. J Nucl Med 63 (3):476–484. https://doi.org/10.2967/jnumed.121.261972

Purves D (2018) Neuroscience. 6th Edn. Oxford Univ. Press, New York

Rae CD, Baur JA, Borges K, et al. (2024) Brain Energy Metabolism: A Roadmap for Future Research. Journal of Neurochemistry 168 (5):910–954. https://doi.org/10.1111/jnc.16032

Reid MJ, Rogdaki M, Dutan L, et al. (2022) Cell Line Specific Alterations in Genes Associated with Dopamine Metabolism and Signaling in Midbrain Dopaminergic Neurons Derived from 22q11.2 Deletion Carriers with Elevated Dopamine Synthesis Capacity. Schizophr Res 273:98–106. https://doi.org/10.1016/j.schres.2022.05.010

Rodgers ZB, Detre JA, and Wehrli FW (2016) MRI-Based Methods for Quantification of the Cerebral Metabolic Rate of Oxygen. J Cereb Blood Flow Metab 36 (7):1165–1185. https://doi.org/10.1177/0271678X16643090

Scholkmann F, Fischer JB, Frisk LK, et al. (2021) Influence of Study Design on Effects of Mask Wearing on fMRI BOLD Contrast and Systemic Physiology: A Comment on Law et al. (2021). Neuroimage 244:118549. https://doi.org/10.1016/j.neuroimage.2021.118549

Segarra-Mondejar M, Casellas-Diaz S, Ramiro-Pareta M, et al. (2018) Synaptic Activity-Induced Glycolysis Facilitates Membrane Lipid Provision and Neurite Outgrowth. EMBO J 37 (9):e97368. https://doi.org/10.15252/embj.201797368

Shi H, He Y, Zhou Y, et al. (2023) Spatial Atlas of the Mouse Central Nervous System at Molecular Resolution. Nature 622 (7983):552–561. https://doi.org/10.1038/s41586-023-06569-5

Siletti K, Hodge R, Mossi Albiach A, et al. (2023) Transcriptomic Diversity of Cell Types across the Adult Human Brain. Science 382 (6667):eadd7046. https://doi.org/10.1126/science.add7046

Smith WS, and Furlan AJ (2016) Brief History of Endovascular Acute Ischemic Stroke Treatment. Stroke 47 (2):e23–e26. https://doi.org/10.1161/STROKEAHA.115.010863

Stein A, Zhu C, Du F, and Öngür D (2023) Magnetic Resonance Spectroscopy Studies of Brain Energy Metabolism in Schizophrenia: Progression from Prodrome to Chronic Psychosis. Curr Psychiatry Rep 25 (11):659–669. https://doi.org/10.1007/s11920-023-01457-1

Straathof M, Meerwaldt AE, De Feyter HM, de Graaf RA, and Dijkhuizen RM (2021) Deuterium Metabolic Imaging of the Healthy and Diseased Brain. Neuroscience 474:94–99. https://doi.org/10.1016/j.neuroscience.2021.01.023

Sun C, and Schuman E (2023) A Multi-Omics View of Neuronal Subcellular Protein Synthesis. Curr Opin Neurobiol 80:102705. https://doi.org/10.1016/j.conb.2023.102705

Theriault JE, Shaffer C, Dienel GA, et al. (2023) A Functional Account of Stimulation-Based Aerobic Glycolysis and Its Role in Interpreting BOLD Signal Intensity Increases in Neuroimaging Experiments. Neurosci Biobehav Rev 153:105373. https://doi.org/10.1016/j.neubiorev.2023.105373

Vagal A, Wintermark M, Nael K, et al. (2019) Automated CT Perfusion Imaging for Acute Ischemic Stroke: Pearls and Pitfalls for Real-World Use. Neurology 93 (20):888–898. https://doi.org/10.1212/WNL.0000000000008481

van Oostrum M, Blok TM, Giandomenico SL, et al. (2023) The Proteomic Landscape of Synaptic Diversity across Brain Regions and Cell Types. Cell 186 (24):5411–5427.e5423. https://doi.org/10.1016/j.cell.2023.09.028

van Sluis J, Boellaard R, Somasundaram A, et al. (2020) Image Quality and Semiquantitative Measurements on the Biograph Vision PET/CT System: Initial Experiences and Comparison with the Biograph MCT. J Nucl Med 61 (1):129–135. https://doi.org/10.2967/jnumed.119.227801

van Zalm PW, Ahmed S, Fatou B, et al. (2023) Meta-Analysis of Published Cerebrospinal Fluid Proteomics Data Identifies and Validates Metabolic Enzyme Panel as Alzheimer's Disease Biomarkers. Cell Rep Med 4 (4):101005. https://doi.org/10.1016/j.xcrm.2023.101005

van Zijl PCM, Brindle K, Lu H, et al. (2021) Hyperpolarized MRI, Functional MRI, MR Spectroscopy and CEST to Provide Metabolic Information *in Vivo*. Curr Opin Chem Biol 63:209–218. https://doi.org/10.1016/j.cbpa.2021.06.003

Volpi T, Maccioni L, Colpo M, et al. (2023) An Update on the Use of Image-Derived Input Functions for Human PET Studies: New Hopes or Old Illusions? EJNMMI Res 13 (1):97. https://doi.org/10.1186/s13550-023-01050-w

Weber CJ, Carrillo MC, Jagust W, et al. (2021) The Worldwide Alzheimer's Disease Neuroimaging Initiative: ADNI-3 Updates and Global Perspectives. Alzheimers Dement 7 (1):e12226. https://doi.org/10.1002/trc2.12226

Weninger A, and Arlotta P (2023) A Family Portrait of Human Brain Cells. Science 382 (6667):168–169. https://doi.org/10.1126/science.adk4857

Xu SY, and Pan SY (2013) The Failure of Animal Models of Neuroprotection in Acute Ischemic Stroke to Translate to Clinical Efficacy. Med Sci Monit Basic Res 19:37–45. https://pmc.ncbi.nlm.nih.gov/articles/PMC3638705/

Yang C, Farias FHG, Ibanez L, et al. (2021) Genomic Atlas of the Proteome from Brain, Csf and Plasma Prioritizes Proteins Implicated in Neurological Disorders. Nat Neurosci 24 (9):1302–1312. https://doi.org/10.1038/s41593-021-00886-6

Young P, Heeman F, Axelsson J, et al. (2024) Impact of Simulated Reduced Injected Dose on the Assessment of Amyloid PET Scans. Eur J Nucl Med Mol Imaging 51 (3):734–748. https://doi.org/10.1007/s00259-023-06481-0

Yüksel C, Tegin C, O'Connor L, et al. (2015) Phosphorus Magnetic Resonance Spectroscopy Studies in Schizophrenia. J Psychiatr Res 68:157–166. https://doi.org/10.1016/j.jpsychires.2015.06.014

Zhang L, Shi L, Shen Y, et al. (2019) Spectral Tracing of Deuterium for Imaging Glucose Metabolism. Nat Biomed Eng 3 (5):402–413. https://doi.org/10.1038/s41551-019-0393-4

Zhu Y, Sousa AMM, Gao T, et al. (2018) Spatiotemporal Transcriptomic Divergence across Human and Macaque Brain Development. Science 362 (6420). https://doi.org/10.1126/science.aat8077

Open Access This chapter is licensed under the terms of the Creative Commons Attribution-NonCommercial-NoDerivatives 4.0 International License (http://creativecommons.org/licenses/by-nc-nd/4.0/), which permits any noncommercial use, sharing, distribution and reproduction in any medium or format, as long as you give appropriate credit to the original author(s) and the source, provide a link to the Creative Commons license and indicate if you modified the licensed material. You do not have permission under this license to share adapted material derived from this chapter or parts of it.

The images or other third party material in this chapter are included in the chapter's Creative Commons license, unless indicated otherwise in a credit line to the material. If material is not included in the chapter's Creative Commons license and your intended use is not permitted by statutory regulation or exceeds the permitted use, you will need to obtain permission directly from the copyright holder.

3

Brain Energy Constraints and Vulnerability

Caroline D. Rae, L. Felipe Barros, Alexander Behnke, Manu S. Goyal, Suzana Herculano-Houzel, Daria Peleg-Raibstein, Martin Picard, Douglas Rothman, Anthony C. Vernon, and Zoltan Sarnyai

Abstract As a highly active organ, the awake brain operates at near capacity. It has limited ability to increase delivery of blood, and hence oxygen and glucose, due to restrictions dictated by capillary density and the space within the skull. In addition, the chemoelectric operating environment (homeostasis) of the brain restricts the amount of glucose and oxygen that can be supplied and the level of generated protons (H^+) and CO_2 that can be tolerated without impacting brain functions. This chapter describes the neurochemical basis of how the brain operates carefully within these homeostatic limits: why they exist and what can happen when these limits are infringed.

Keywords Homeostasis, blood supply, glucose, oxygen, lactate

1 Introduction

Impairments in brain energy and neurotransmitter metabolism have been increasingly implicated in the pathogenesis of neurological and psychiatric disorders (Hagihara et al. 2024; Minhas et al. 2024). Furthermore, prevalence of psychiatric disorders has increased among those with impaired whole-body metabolism, referred to as metabolic syndrome, which implies a relationship between brain and body metabolic health (Nousen et al. 2014). Over the last 35 years, the theory of allostasis has been developed to explain the linkage between whole-body metabolic health and mental disorders (McEwen 2003; Sterling 1988). In the allostasis

Caroline D. Rae (✉)
Neuroscience Research Australia and School of Psychology, University of New South Wales, Randwick NSW 2031, Australia
Email: c.rae@unsw.edu.au

Affiliations for the coauthors are available in the List of Contributors.

© Frankfurt Institute for Advanced Studies (FIAS) 2025

D. Öngür, J. M. Ford (eds.), *Metabolic Neuropsychiatry*, Strüngmann Forum Reports,
https://doi.org/10.1007/978-3-032-05630-6_3

Group photos (top left to bottom right) Caroline Rae, L. Felipe Barros, Daria Peleg-Raibstein, Douglas Rothman, Martin Picard, Manu Goyal, Zoltan Sarnyai, Alexander Behnke, Anthony Vernon, Suzana Herculano-Houzel, Manu Goyal, Douglas Rothman, Martin Picard, Anthony Vernon, Suzana Herculano-Houzel, Daria Peleg-Raibstein, Caroline Rae, L. Felipe Barros (stability through change) theory of disease, the brain alters whole-body energy metabolism and immune system activity, via neurotransmission and endocrine factors, to a new homeostatic state to respond to immediate and anticipated environmental stressors, referred to as an allostatic load. In contrast, earlier homeostatic theories proposed that there was only one preferable state that the brain and body tried to maintain when challenged. Ideally, the new state is optimal for meeting the allostatic load. However, when the allostatic load exceeds the range of body metabolic and immune system homeostatic mechanisms, a dysfunctional state can ensue. Excessive allostatic load has also been proposed to lead to dysfunctional brain metabolic and immune states that contribute to the pathogenesis of chronic mental disorders (McEwen and Gianaros 2011). In this chapter, we describe our efforts to adapt allostasis theories of mental health dysfunction to incorporate the unique properties of brain energy and neurotransmitter metabolism and constrained energy supply.

1.1 Allostasis Theory: A Framework to Understand Brain Health Deterioration

We start with a brief review of how a branch of allostasis theory, referred to as the stress resilience and vulnerability model, explains how allostatic load can overwhelm the capacity of whole-body energy metabolism to support appropriate allostatic adaptation. In allostasis theory, the energy-consuming processes are divided into vital and nonvital components (Bobba-Alves et al. 2022). As shown in Figure 3.1, stress, due to environmental and internal factors, puts an additional energy load on the organism, referred to as an allostatic load. Finally, there is a limit on the capacity of the organism to support the energy budget, which when exceeded due to an excessive allostatic load, leads to improper adaptation and dysfunctional homeostatic states. In this chapter, the key processes that define energy in terms of these components are summarized as follows:

- Vital life-sustaining processes that are needed for survival and cannot be interrupted. These include organ-level physiological processes (e.g., pulmonary, cardiac and hepatic function) needed for the survival of the organism as well as intercellular and subcellular processes needed for organ function. For the brain, the main vital energy costs are associated with neuronal signaling (primarily the costs of K^+ and glutamate as well as GABA neurotransmitter uptake by neurons and glia during and after depolarization) to support mental processes (e.g., cognition, sensory perception and interpretation, and motor control).
- Processes needed for growth, maintenance, and repair (GMR). These processes are not acutely needed for survival but if chronically reduced will lead to deterioration of the organism as well as improper developmental trajectories. After early childhood, these processes in the brain are largely related to synaptic pruning and remodeling as opposed to increasing cellular number or turnover.

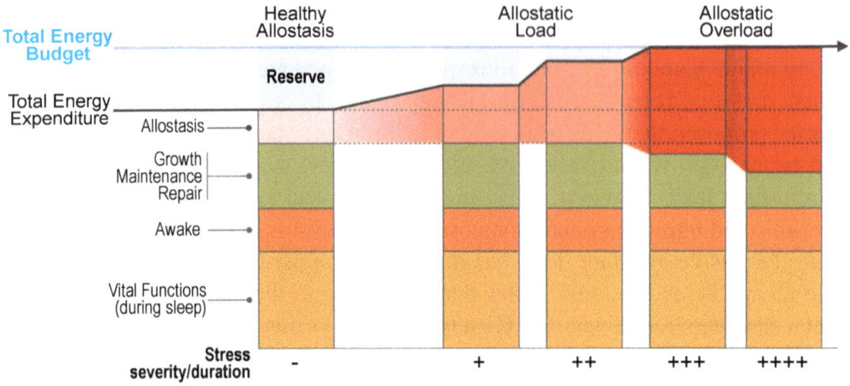

Figure 3.1 Schematic of the stress resilience and vulnerability energy budget model: increasing allostatic load in excess of the limit on total energy expenditure leads to a state of allostatic overload, which can lead to reduced growth, maintenance, and repair (GMR). With increasing allostatic load, the body and its component organs, including the brain, undergo allostasis to new adaptive states. However, when the extra energy costs of allostatic load exceed the total capacity of energy metabolism, insufficient energy is available for non-immediate GMR functions. Extended allostatic overload will lead to deterioration of organismic function, due to a reduction in the energy available for GMR, and potentially chronic disease. For the brain, the primary vital energy costs are the processes required to support neuronal signaling (after Bobba-Alves et al. 2022 and used with permission from from Elsevier).

- Allostasis and stress-related load. When the organism experiences environmental or social stressors, which can be internally generated, increased energy expenditure is required to adapt to the new homeostatic state. As shown in Figure 3.1, progressive increase of allostatic load can lead to energy demands that exceed the organism's adaptability.

1.2 Susceptibility to Allostatic Overload

Over the last 25 years, our understanding of the brain energy budget has dramatically changed. In the late 1980s and early 1990s, it was widely believed that the energy costs associated with the neuronal signaling required to support the vital processes of cognition and other mental functions were on the order of 5–10% of the total energy budget (Raichle 2011). The remaining 90–95% was assigned to processes independent of mental functions, based on assumptions in cognitive psychology that were used to interpret early functional neuroimaging studies (Morcom and Fletcher 2007; Shulman and Rothman 2019). Starting in the late 1990s, however, the development of direct measurements of total neuronal signaling in combination with energetics—such as ^{13}C magnetic resonance spectroscopy (MRS) (Gruetter et al. 2001; Rothman et al. 2011; Sibson et al. 1998), multiunit recordings of neuronal ensembles (Smith et al. 2002), and theoretical models (Attwell and Laughlin 2001; Yu et al. 2018)—showed that the large majority of brain energy metabolism (approximately 85–90%, even in the "resting" awake state) supported neuronal signaling, either directly or indirectly. The remaining 10–15% was devoted to GMR processes (Attwell and Laughlin 2001; Yu et al. 2018).

An additional critical discovery was that the energy required to support enhanced neuronal signaling associated with sensory stimulation and cognitive and other behavioral tasks usually increases regional brain energy metabolism by 5–10%. Only under intense sensory stimulation of sensory cortices are increases on the order of even 30% achieved (DiNuzzo et al. 2024; Shulman and Rothman 1998). The high energy consumption of the brain during rest underpins the activity we see in resting-state magnetic resonance imaging (MRI) (Raichle et al. 2001). Today, measurements of networks of neuronal signaling in the resting state inferred from the coupling of neuronal signaling to energy metabolism via the BOLD effect (Chen et al. 2022) are the dominant application of functional MRI in humans and increasingly in preclinical models (Finn 2021). Similar conclusions have been reached based on the relative EEG amplitude versus different states of consciousness and activation, and single-cell recording measurements of ensemble spiking rate (Hyder et al. 2013; Yu et al. 2018).

Despite this large revision of the brain energy budget and its accounting for a disproportionate (20%) fraction of whole-body energy consumption, brain energy budget models have not considered whether supply of oxygen, glucose, and other biophysical limitations (e.g., mitochondrial density) has an impact on the energy budget (e.g., Attwell and Laughlin 2001; Howarth et al. 2012; Karbowski 2011). Instead, limitations on the energy budget have been implicated in the pathogenesis of psychiatric and neurological disease, reflecting pathological processes such as loss of vascular and mitochondrial capacity in severe dementia (Bhatia et al. 2022). Those considerations aside, the size of the total brain energy budget has been assumed to be able to expand to meet the specific energy demands of the organism due to evolutionary and developmental factors or even short-term adaptation (Herculano-Houzel and Rothman 2022).

The lack of consideration of supply limitations in energy budget models of the brain arises from studies of another unique brain property: neurovascular coupling (NVC). Unlike other organs, for which the ratio of blood flow and oxygen metabolism are close to proportional (see Figure 3.1), for brain energy needs above the awake resting state, there is a disproportionate increase in blood flow versus oxygen metabolism of 2.3-fold and, under specific conditions, up to three- to fourfold (Buxton 2021; DiNuzzo et al. 2024; Herculano-Houzel and Rothman 2022). Traditionally, NVC has been interpreted as a safety factor; at maximum metabolic demand, the brain can be supplied by blood flow with threefold more oxygen and ninefold more glucose than energetic needs (Leithner and Royl 2014).

Over the last decade, however, experimental and theoretical studies of brain oxygen and glucose transport and metabolism—conducted under both resting awake and stimulated conditions, from isolated tissue slices to awake mice and humans—have reached a different conclusion. The large excess of oxygen and glucose supply to awake and especially to activated brain regions is due to the restricted transport of glucose and diffusion of oxygen across the blood-brain barrier (BBB). This ensures activity-independent homeostasis of fuel levels (glucose, oxygen) *and* waste products of energy metabolism (CO_2, protons, lactate, heat). Recent papers summarize supporting findings and outline a new paradigm of supply and homeostatic constraints that restrict the brain's energy budget (Buxton 2021; DiNuzzo et al. 2024; Herculano-Houzel and Rothman 2022).

Below, we summarize how the brain's energy budget limits, due to homeostatic and supply constraints and high functional energy requirements, may make the brain highly vulnerable to increases in allostatic load. We compare the brain and muscle energy budgets, highlighting that the brain's tighter homeostatic constraints on fuel and waste products may explain why the maximum oxidative rate of adenosine triphosphate (ATP) production in the brain is about three times less than that of muscle, despite similar blood flow. We review evidence showing that critical brain functions at the synaptic level, such as rapid K^+ clearance during and after depolarization, rely heavily on tight temporal coordination with metabolic ATP production, fuel delivery, and waste clearance. We conclude by describing an adaptation to the model presented in Figure 3.1, which takes into account the unique energy budget and supply and homeostatic limitations of the brain, and how it leads to vulnerable periods for the development of dysfunctional mental states during brain development and aging. Evidence supporting this new model and proposals for studies to test it in animal models and humans are discussed.

2 Brain versus Muscle Energy Budgets

To illustrate the uniqueness of the brain energy budget, let us consider the energy budget of skeletal muscle. At rest, muscle energy consumption is far from capacity, with 15- and sixfold increases in blood flow and oxygen consumption (which matches capillary delivery), respectively, under maximal aerobic conditions in active adult humans (Joyner and Casey 2015). These increases, shown in Figure 3.2, occur when muscle goes from an inactive resting state to performing aerobic exercise via repeated contractions. Elite athletes show an even higher fractional increase (Laughlin 1987; Saltin 2007). The greater fractional increase in metabolic rate in muscle is due to the rapid increase in oxygen extraction fraction with workload, starting at approximately 25% at rest, and with a near-constant, high oxygen extraction fraction around 90% above low to moderate workloads (Hargreaves and Spriet 2020). The take-home message is that muscle has a great deal of reserve capacity to increase output, metabolism, blood flow, and hence functional activity. Figure 3.2 shows the breakdown of the muscle energy budget into vital and GMR processes. At rest in the muscle, almost all the energy budget supports GMR processes due to the lack of contractile work. In contrast, at maximum aerobic exercise, the majority of the energy budget supports the vital function of contraction.

Most ATP production in muscles (and the brain) comes from mitochondrial substrate oxidation, but at higher workloads, muscles also rely on non-oxidative glycolysis of glycogen and glucose. This leads to a mismatch between metabolism and vascular coupling, causing the production of lactate and protons. When the vascular system is unable to remove lactate fast enough, muscle pH drops which affects metabolic rate. Eventually, the muscle must rest to restore pH, replenish glycogen, and repair damage. This recovery requires time during which energy is used for GMR processes.

For the brain, the relative contributions of the energy budget components are dramatically different, as shown in Figure 3.2. If functional activity is defined in

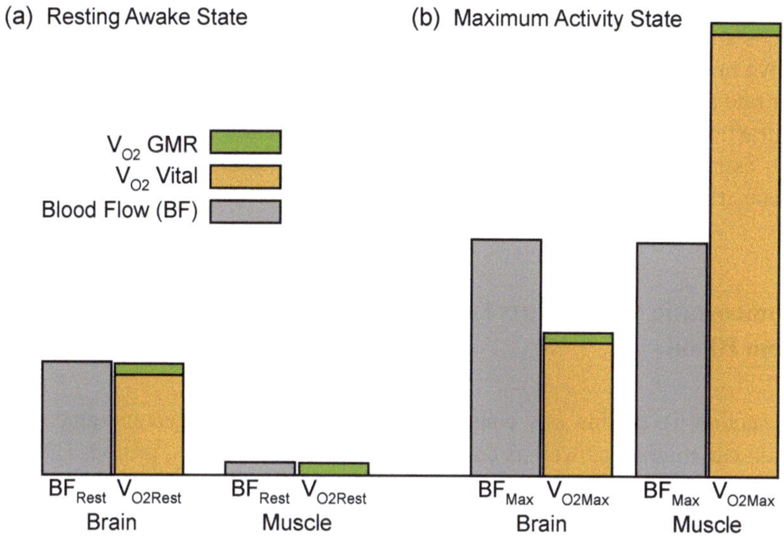

Figure 3.2 Comparison of brain and muscle blood flow (BF) and energy budget for oxidative metabolism (VO_2) under (a) resting awake (Rest) and (b) maximum (Max) activity states. BF and VO_2 are all expressed relative to the values in cerebral cortex in the resting awake state. (a) The majority of the brain energy budget supports vital processes needed to support mental activity (brown), with approximately 10% supporting growth, maintenance, and repair (GMR) processes (green). In contrast, the energy required to support vital processes in muscle are minimal due to no contractile activity, and the energy budget is dominated by GMR processes. (b) Despite up to a twofold increase in BF, the maximum regional increase in oxygen metabolism (VO_{2Max}) is only ~40% due to homeostatic limitations on oxygen transport (see text). In contrast, there is a greater than tenfold increase in muscle BF and VO_{2Max}. Note that despite similar absolute values of BF, muscle extracts threefold more oxygen from the blood due to its higher O_2 capillary diffusivity. As described in the text and (DiNuzzo et al. 2024), the low O_2 extraction fraction is due to brain function requiring homeostatic levels of O_2, CO_2, and pH. Muscle function, in contrast, is tolerant of lower O_2 and higher CO_2 levels.

accordance with the brain energy budget literature as the ATP requirements for all molecular activities that support neuronal signaling (e.g., the sodium potassium ATPase), the brain is undertaking a great deal of functional activity, even in the so-called "resting state," as noted 100 years ago when Hans Berger recorded electrical activity from the resting brain, now known as EEG. Even at "rest," 80–90% of the brain energy budget is dedicated to vital processes that support mental activity. As referred to in Section 1, whole brain oxygen consumption rates can be reduced by only ~10–20% in states such as deep sleep (stage 3–4) (Boyle et al. 1994) and increase at most by 20–30% under very intense physical exercise (Smith and Ainslie 2017), consistent with positron emission tomography and arteriovenous difference studies (Herculano-Houzel and Rothman 2022). Regional gray matter in sensory regions has a somewhat higher capability to increase the cerebral metabolic rate of oxygen ($CMRO_2$) of at most 40% during maximum stimulus intensity (DiNuzzo et al. 2024). Under more natural stimuli, increases on the order of 5–20% are mostly observed. Thus, even in the so-called resting state, the brain is already near the limit of the available energy budget.

The limits to the ability of the brain to support its energy budget relative to muscle is shown clearly in Figure 3.2 by the comparison of the maximum rate of oxidative metabolism per volume in brain ($CMRO_2$) and muscle (VO_2). Despite a similar rate of blood flow per volume, the muscle has an over threefold higher oxygen extraction fraction and therefore metabolic rate. Next we describe how this low oxygen extraction fraction arises due to the brain having to meet stricter homeostatic constraints for proper function compared with exercising muscle.

3 Homeostatic Constraints Limit Oxygen and Glucose Extraction from Blood

In this section we outline why constraints on the levels of oxygen and glucose that the brain can tolerate as well as controls on the amount of CO_2, protons (H^+), and lactate that can be produced without impacting brain function, limit the brain to a narrow range of homeostatic operation.

As shown in Figure 3.2, despite the brain having one of the highest blood flows of any organ, the extraction of oxygen from blood at maximum activity is only 30% (relative to over 90% for muscle and 80% for heart) and only 10% for glucose. Even under resting awake conditions, there is a similar low extraction fraction. The traditional explanation has long been that the high blood flow supply acts as a reserve in case, for some reason, there is a transient impairment of flow or extreme increase in energy needs. Over the last decade and a half, an alternate view has emerged, based on theoretical modeling of experimental studies of brain activation; namely, that the supply of oxygen to the brain is limited by its low capillary density resulting in limited permeability of oxygen through the BBB (Buxton 2010). The exact value of this permeability varies across brain regions and species so that homeostasis of average tissue O_2 levels is maintained independent of activation level (Herculano-Houzel and Rothman 2022). Consistent with this proposal of a requirement for oxygen homeostasis, ultra-high-resolution optical microscopy of brain tissue O_2 in the awake mouse brain has shown that even under intense sensory activation there is at most a 5% increase in regional tissue oxygen partial pressure (Moeini et al. 2019). Similarly, optical and MRS measurements of brain pH during intense sensory stimulation shows only very small changes (Zhu et al. 2018).

Although the brain traditionally is thought to be resistant to all but extreme hypoxia, recent studies in awake animal models and humans have shown that regional electrical activity is highly sensitive to even small variations in oxygen partial pressure (Drew 2022). Furthermore, studies show that even modest reductions of 10% in blood oxygen partial pressure lead to impaired cognitive performance in humans (McMorris et al. 2017; Shaw et al. 2021). The precise impact of small reductions in pO_2 has, however, been hard to determine due to accompanying changes in pCO_2 and other physiological parameters.

In addition to oxidative ATP production, the brain requires glycolytic ATP from glucose and glycogen for rapid ATP-requiring processes (Dienel 2019). Neurotransmission and action potentials result in the rapid (milliseconds) release of neuronal potassium (K^+), most of which is actively taken up by glial cells, astrocytes, and oligodendrocytes (Looser et al. 2024; MacAulay 2020). A fast process, glial K^+ uptake is powered initially by glycolysis, which is fed by glucose and glycogen, producing an increase in tissue lactate and excess H^+ (Barros et al. 2023; DiNuzzo et al. 2024; Fox and Raichle 1986; Fox et al. 1988; Prichard et al. 1991; Rothman et al. 2022; Ruminot et al. 2011). The excess H^+ production can leave the brain to the blood through co-transport with lactate via monocarboxylate transporters (MCT) or be converted in the brain to bicarbonate via carbonic anhydrase activity with a stoichiometry of approximately one bicarbonate and one H^+ to one CO_2 (DiNuzzo et al. 2024; Hladky and Barrand 2019). The increased CO_2 is in rapid exchange with the capillary blood pool due to the high diffusivity and solubility of CO_2, which is 20 times higher than for O_2.

On average, during brain activation, there is a twofold increase in the ratio of incremental total glucose metabolism to oxidative glucose metabolism (Dienel 2019; Madsen et al. 1998; Rothman et al. 2022), often referred to as neurometabolic coupling (NMC). The ratio is similar to the average 2.3-fold increase in cerebral blood flow (CBF) relative to $CMRO_2$ during activation; see NVC in Figure 3.2 (Rothman et al. 2022). Recent studies have used metabolic modeling to explain this similarity. The NVC ratio is almost exactly what is required to manage the increased production of CO_2 and protons to maintain a constant level of tissue pH and CO_2 partial pressure (Buxton 2021, 2024; DiNuzzo et al. 2024).

The importance of maintaining a near constant CO_2 partial pressure and pH has been illustrated in extensive studies, which show that brain electrical activity is extremely sensitive to even small changes in CO_2 partial pressure and pH (Bonnet et al. 1998; Chesler and Kaila 1992; Lee et al. 1996; Schweitzer et al. 2000). Even at the level of the synapse, regulation is highly dependent on small shifts in pH at glial/dendritic junctions (Ransom 2000) and would be overwhelmed by the larger pH decrease due to NMC unless this was offset. The coupling between these phenomena is illustrated and explained further in Figure 3.3.

3.1 Glucose Transport

The oxidation of glucose is the main source of energy for brain tissue, with lactate and ketone bodies playing a minor role in lactation, exercise, and starvation. For glucose, as with oxygen, there is a large apparent excess of supply via blood flow to the brain. This is around ninefold greater than the brain's energetic requirements even during maximum glucose metabolism, outstripping even the threefold higher O_2 delivery during maximum activity.

As in the case of oxygen, where low BBB oxygen permeability restricts metabolism despite high delivery via CBF, maximum glucose metabolism is limited due to restricted BBB capillary transporter density. Glucose is hydrophilic and must therefore enter the brain via the endothelial transporter GLUT1. There is a pronounced standing glucose concentration gradient measured between blood and tissue, which even under resting conditions amounts to 4 mM (i.e., 80% of the maximum possible gradient at euglycemia) (Barros et al. 2005, 2007; de Graaf et al. 2001; Leybaert 2005). Hypoglycemia, even if only mild, perturbs cognition (Lobmann et al. 2000).

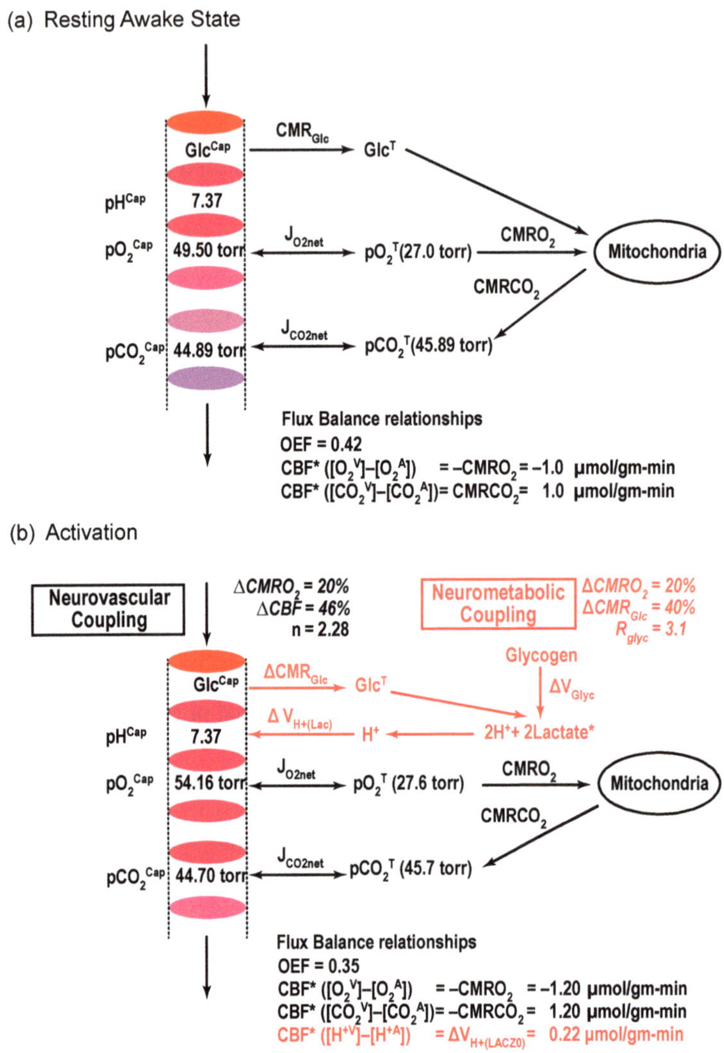

Figure 3.3 Disproportionate increase in blood flow compensates for proton production from non-oxidative glycolysis during functional activation. Neuronal and astrocytic compartment glucose (Glc) and glycogen metabolic fluxes are combined to focus on the relationship between neurometabolic and neurovascular coupling. (a) Resting awake state: Approximately 90 to 95% of Glc transported into the brain is oxidized, leading to an equal flux of CO_2 release into the blood. Cerebral blood flow (CBF) and oxygen metabolism ($CMRO_2$) scale approximately proportionately: $\Delta CBF/CBF = \Delta CMRO_2/CMRO_2$ (b) Activation: There is a large increase in non-oxidative glycolysis of Glc and glycogen to lactate and protons (red). To clear these protons and prevent acidosis, there is a disproportionate increase in the ratio CBF relative to $CMRO_2$, referred to as neurovascular coupling (NVC). Based on homeostatic flux balance modeling, the NVC ratio must increase to 2.3 from 1.0 to maintain pCO_2 and pH homeostasis. Consequently, the oxygen extraction fraction decreases rather than increases with activity (as in muscle and heart), which limits the maximum $CMRO_2$ to only 30–40% above resting awake values. *Abbreviations* pO_2: partial pressure of oxygen; lac: lactate; CMR: cerebral metabolic rate of total (CMR_{Glc}), oxidative ($CMR_{Glc,ox}$), and non-oxidative $CMR_{Glcnonox}$) glucose metabolism; J_{O2}, J_{CO2}, J_{H+}: transmembrane fluxes of O_2, CO_2, and protons; tissue, arteriole, capillary, and venule compartments denoted by superscript T, A, C, and V. Figure from DiNuzzo et al. (2024), used with permission from John Wiley and Sons.

A moderate genetic GLUT1 deficiency causes a severe childhood disease characterized by persistent hypoglycorrhachia, seizures, and developmental delay (De Vivo et al. 1991) and exacerbates vasculo-neuronal dysfunction in Alzheimer disease models (Winkler et al. 2015). It is known that endothelial GLUT1 glucose uptake is stimulated by brain activity (Cura and Carruthers 2012) and exercise (Takimoto and Hamada 2014), whereas GLUT1 expression is suggested to respond to brain energy levels, such as via AMPK signaling (Barnes et al. 2002; Cura and Carruthers 2012; Liemburg-Apers et al. 2016).

The concept that BBB glucose transporter activity can be limiting for metabolism has been questioned based on studies that have shown its maximum activity is on the order of threefold more than resting-awake glucose requirements. As with diffusive oxygen transport to the brain, however, the rate of transport does not just depend on glucose transporter activity but also on the absolute concentration gradient of glucose in the blood and the blood-to-brain glucose gradient across the BBB (Barros et al. 2005, 2007; de Graaf et al. 2001; Leybaert 2005). When these factors are considered, the rate of glucose transport predicted from transporter kinetics and the actual *in vivo* net rate are in good agreement.

3.2 Lactate Levels

The development of improved quantitative MRS methods has shown that even during maximum sensory stimulation, tissue average lactate only increases on the order of 10–30% (Mangia et al. 2007; Maddock et al. 2011), although a larger percentage increase in extracellular lactate occurs (Caesar et al. 2008; Mangia et al. 2009). Multiple roles for lactate have been proposed (Rae et al. 2024), including as a volume neurotransmitter/neuromodulator (Bergersen and Gjedde 2012) and as a key energy source for oxidation by neighboring cells (Pellerin and Magistretti 2012); the latter role, however, remains controversial (Bak and Schousboe 2017; Dienel 2017b). Lactate concentrations in specific brain regions (dorsomedial prefrontal cortex and dorsal anterior cingulate cortex) mediate perceived effort, which can be interpreted to mean that the brain can sense or feel elevated lactate, which signals “energetic stress” (Clairis et al. 2024). High circulating and brain lactate concentrations are associated with less motivation or with aversion to exerting effort (Clairis et al. 2024), as if to conserve energy, which could have implications for psychopathology.

Independent of these alternate roles, tight homeostatic control of lactate levels is important for regulating the NADH redox charge, which is critical for brain function. Thus, effective rapid efflux mechanisms for lactate are of high importance; clearance of lactate, particularly during periods of high activity, is via multiple routes (Hladky and Barrand 2018). Lactate and a proton (H^+) can leave brain cells via MCTs (Halestrap 2013) or leave the local area via gap junctions (Dienel 2017a) and the perivascular space (Ball et al. 2010). MCTs are low-affinity transporters (Halestrap and Wilson 2012; Rae and Bröer 2015). Their expression in the membrane is finely tuned to metabolic activity and must be controlled to maintain intracellular monocarboxylate levels (particularly pyruvate, which must be available to convert to lactate to regenerate NAD^+, thus allowing glycolysis to continue) (Zhang et al. 2020). To maintain this balance, protons also leave the cell via CO_2 (DiNuzzo et al. 2024; Hladky and Barrand 2019).

3.3 Protection against Toxicity from Chronic Elevated Glucose and Oxygen Levels

Based on modeling of NVC, low BBB oxygen diffusivity is not needed to maintain CO_2 and pH homeostasis (Buxton 2021, 2024; DiNuzzo et al. 2024); only the proper ratio of CBF to $CMRO_2$ is required to maintain tissue O_2 at a low, near-constant level. Similarly, brain glucose levels do not impact NMC, other than displacing the use of glycogen (Rothman et al. 2022). Thus BBB transport of glucose and oxygen could be increased to where blood flow is limiting, allowing a threefold higher rate of glucose oxidative metabolism and a ninefold greater rate of non-oxidative glucose metabolism. However, in the undisturbed awake state, brain tissue glucose and oxygen concentrations would be close to arterial values.

We propose that the explanation for low BBB O_2 diffusion and regulated glucose transport capacity is to protect brain tissue from damage resulting from chronic exposure to high levels of glucose and oxygen during nonstimulated resting awake conditions. Chronic hyperglycemia (as occurs in diabetes) and hyperoxia have been extensively shown to lead to damage to brain and peripheral nervous tissue. Biochemical pathways leading to glucose toxicity are non-enzymatic glycation of proteins (Münch et al. 1997), sorbitol formation with concomitant redox imbalance (Varma and Kinoshita 1974), hexosamine metabolism (Kim et al. 2023), and the generation of reactive aldehydes such as methylglyoxal (Rae et al. 1992; Richard 1991). Common to some of these pathways is chronic oxidative stress (Pignalosa et al. 2021; Rossetti et al. 1990). Operating at higher tissue pO_2 levels would be detrimental as ground state molecular O_2 is a diradical and thus a potentially toxic gas (Cobley et al. 2018). Further, oxygen radicals, the production of which is increased at higher pO_2, are used as signaling molecules (Halliwell 1992) so their levels must be tightly controlled.

3.4 Summary

For short-term function and longer-term brain health, the brain's capacity to take up glucose and oxygen from the blood is highly restricted compared to other organs, such as muscle. This is because brain tissue is highly sensitive to O_2, glucose, CO_2, pH, and lactate levels, which need to be maintained under strict homeostatic control. Essentially, the brain faces a trade-off between increasing BBB transport and diffusion and rising toxicity, versus restricting transport and diffusion at the cost of potential local hypoxia and hypoglycemia during activation. The fact that the brain operates under the latter option emphasizes the importance of regulating the cardiovascular system to support proper brain function: from total blood supply rate to the brain, to regulation of microvascular flow to meet local metabolic needs.

4 Synaptic Metabolic Homeostasis for Proper Brain Function

Let us now turn in greater detail to the energy budget required for the brain to support vital mental functions. These ATP-requiring processes have been shown to take place largely in the region of synapses (including surrounding astrocytic processes) and small neuronal/glial circuits associated with neuronal signaling. These processes require constant stable levels of ATP at millisecond timescales, despite rapid changes in energetic requirements. This leads to the requirement for NMC with non-oxidative glucose and glycogen metabolism described above, and the homeostatic constraints it puts on NVC. Loss of metabolic homeostasis at this level, due to impaired energy supply or by-product clearance, will impact all the mechanisms described below, as well as other key synaptic mechanisms (e.g., neurotransmitter and receptor kinetics) and can directly lead to dysfunction. Figure 3.4 shows the metabolic, glutamate, and GABA neurotransmission and recycling, and ion transport pathways that link neuronal and astrocytic metabolism and ion transport.

During synaptic signaling, presynaptic terminals release the amino acid neurotransmitters glutamate and GABA, which lead, respectively, to postsynaptic terminal depolarization and hyperpolarization. Depolarization results in an increased K^+ permeability and release into the interstitial fluid space due to activation of voltage-sensitive K^+ channels as well as the reduced membrane potential to counteract the inside to outside cellular K^+ gradient. To maintain proper synaptic function, the K^+ concentration in the synaptic and interstitial fluid space must be maintained at a near-constant level of ~3–5 mM. After the initial high concentration elevation in the synaptic cleft, glutamate and GABA are returned to micromolar concentrations within 10 ms (Bergles and Jahr 1997; Bergles et al. 1999). To meet these constraints, synapses are surrounded by astrocytes that have a very high capacity for active uptake of glutamate, GABA, and K^+ (Andersen and Schousboe 2023). The primary energy costs for signaling are the cost of astrocytic clearance of K^+, glutamate, and GABA followed by the later transfer of K^+ from the astrocyte and interstitial fluid space back to the neuron (DiNuzzo et al. 2017). Glutamate and GABA are returned to the neuron after being converted to glutamine by the glutamate/GABA/glutamine cycle and associated metabolic pathways, such as anaplerosis and glutamate dehydrogenase (Figure 3.3). While glycolytic ATP production rises transiently during activation, the observation that the ratio between glucose and oxygen consumption approaches 5.5 over time (close to the ratio for complete glucose oxidation of six) shows that in the longer term, the large majority of ATP required to support these clearance and resynthesis processes is mitochondrial (Blazey et al. 2018; Hyder et al. 2016; Madsen et al. 1995).

After vesicular release, glutamate binds multiple classes of postsynaptic glutamate receptors, leading to voltage-sensitive channel Na^+ uptake and K^+ release into the synaptic space, most of which is coupled to the opening of AMPA channels on a timescale below 10 ms. However, long time-constant glutamate receptors, such as NMDA channels, can lead to depolarization being maintained for over 100 ms

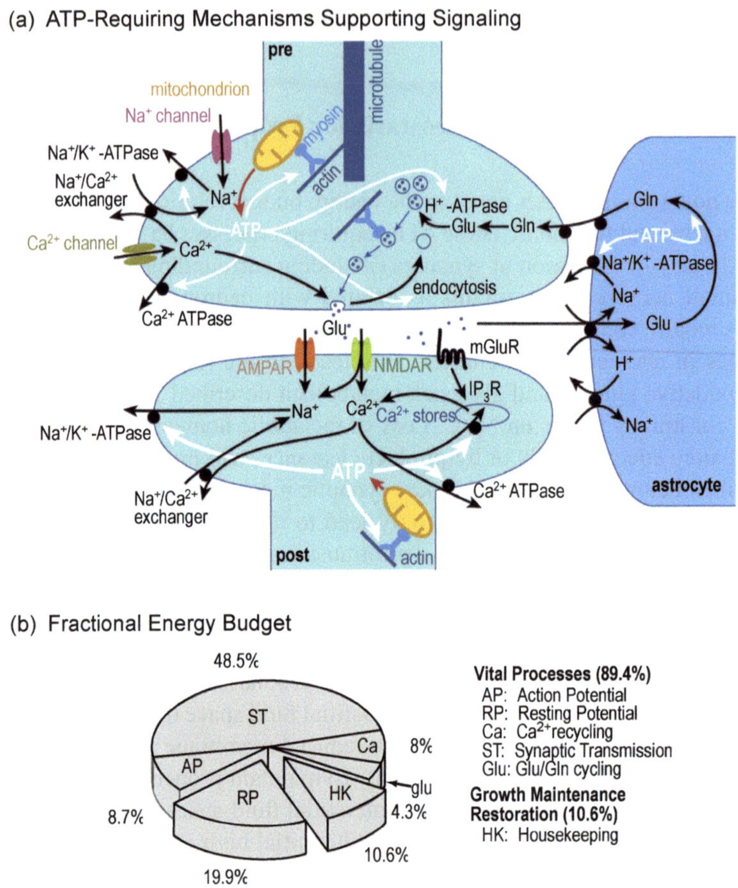

Figure 3.4 ATP-requiring mechanisms at an excitatory cortical synapse and calculated relative contributions to the energy budget in rodent and human cerebral cortex gray matter under resting awake conditions. (a) Schematic of the synaptic ATP-requiring mechanisms associated with excitatory neurotransmission (Harris et al. 2012; http://creativecommons.org/licenses/by/3.0/), showing presynaptic, postsynaptic, and astrocytic components. The presynaptic Na^+/K^+ ATPase sodium pump pumps out Na^+ ions and pumps in K^+ ions to rebalance presynaptic and axonal levels after Na^+ influx and K^+ efflux during depolarization. This ATPase also provides energy for Ca^{2+} removal from the presynapse after receptor-mediated influx associated with depolarization via the Na^+/Ca^{2+} exchanger. It is also used to maintain the resting potential in conjunction with K inward rectifier channels and Na^+ and Cl^- channels. Vacuolar H^+-ATPase pumps glutamate or GABA into vesicles. Other costs include motor proteins (kinesin, dynein, myosin) that move mitochondria and vesicles and vesicle retrieval by dynamin. Postsynaptic: Same mechanisms as above but a higher fraction of the total ATP energy budget due to extended time duration depolarization periods and higher glutamate receptor density. Astrocytes: Na^+/K^+ ATPase provides energy to remove Na^+ associated with glutamate transport from the interstitial fluid to the astrocyte to restore astrocytic ion balance. However, the majority of ATP consumption associated with this ATPase is required to maintain the high astrocytic K^+ permeability and resting potential and maintain interstitial fluid K^+ homeostasis by buffering excess K^+ released from the neuron during depolarization (DiNuzzo et al. 2017; Rothman et al. 2022). Glutamine synthetase uses 1 ATP per glutamate synthesized in the astrocyte from glutamate. (b) Fractional energy budget for cortical gray matter in rodents and humans in the nonstimulated awake state (Yu et al. 2018; reprinted by permission of Sage Publications). Vital processes account for 89.4% of the total energy budget; growth, maintenance, and restoration processes (HK) account for only 10.6% of the energy budget, in contrast to most other organs.

(Lüscher and Malenka 2012). The timescale required to restore neuronal homeostasis completely is on the order of 100 ms or more, because it requires a complete neuronal depolarization and repolarization cycle. When there is a rapid increase in neuronal signaling, restoration of homeostasis may take up to a second while the neuronal Na^+/K^+ ATPase and K^+ inward rectifier (Kir) activity adapts (MacAulay 2020). The energy costs for maintaining interstitial space K^+ and Na^+ homeostasis during this process and restoring astrocytic and neuronal cellular levels accounts for the majority of ATP cost in brain function (Attwell and Laughlin 2001; DiNuzzo et al. 2017). Below, we briefly describe the coordinated neuronal and astrocytic K^+ and Na^+ release and reuptake, how neurons and astrocytes support their ATP needs via non-oxidative glycolysis from astrocytic glycogen, and neuronal glucose oxidation.

Despite full restoration of neuronal ion levels requiring over 100 ms, homeostasis of interstitial K^+ levels is maintained throughout the depolarization/repolarization cycle. The effectiveness of interstitial K^+ homeostasis is demonstrated by recent studies that measured interstitial K^+ levels in awake rats performing an extended locomotion task (Rasmussen et al. 2019). Interstitial K^+ levels rapidly increased at the start of activation, but only by 10–25%, and were maintained at close to this value throughout the activation period. This small increase happens despite the increase in K^+ release flux, which is approximately 24 times higher than the increase in glucose metabolism (DiNuzzo et al. 2017).

Maintaining K^+ and Na^+ homeostasis across the depolarization and repolarization cycle depends on the coordinated activity of neuronal and astrocytic Na^+/K^+ ATPase and Kir channel activity (see Figure 3.4). Although often referred to as low activity (MacAulay 2020), based on mass balance constraints, the neuronal Na^+/K^+ ATPase and Kir channels have sufficient activity to completely restore neuronal (and therefore interstitial space) and astrocytic homeostasis across a full depolarization repolarization cycle (DiNuzzo et al. 2017). However, during the cycle, there are rapidly changing K^+ reuptake and Na^+ release requirements, with the highest occurring during the early part of the depolarization period. In this early period, Na^+ and K^+ homeostasis is maintained by activation of the interstitial K^+ concentration-sensitive astrocytic Na^+/K^+ATPase and Kir channels. The activity of the ATPase and associated Kir channels is rapidly increased by small elevations in interstitial space K^+ (Larsen et al. 2016; MacAulay 2020). During the repolarization cycle, their activity drops parallel to the decrease in interstitial K^+ levels, and the astrocytes switch to net K^+ release; this is taken up by the neuronal Na^+/K^+ATPase and Kir channels until homeostasis of neuronal and astrocytic levels are restored (Figure 3.4).

Maintaining homeostasis of interstitial K^+ concentration during activation is particularly critical due to the dendritic membrane potential, which determines whether a neuron undergoes an action potential being coupled to K^+ permeability and therefore the intracellular to interstitial K^+ gradient. Interstitial K^+ is only ~3–4 mM, and the interstitial fluid volume is only 20%; thus, the rapid K^+ release flux during depolarization can easily collapse the gradient. To maintain homeostasis, released K^+ is rapidly captured by the astrocytic Na^+/K^+ ATPase pump and Kir channels (MacAulay 2020) and spirited away via gap junctions, to exit the astrocytic syncytium elsewhere—a phenomenon termed K^+ buffering (Barros 2022; MacAulay 2020) (Figure 3.4).

To support energetically these transient, very high-energy demands of K^+ uptake released by the neurons, astrocytes will shift to obtain ATP from non-oxidative glycolysis of glycogen—a process referred to as the glycogen shunt (Dienel and Rothman 2019; Schousboe et al. 2010; Shulman et al. 2001; Walls et al. 2009) (Figure 3.3). However, the glycogen shunt is energetically inefficient compared with complete glucose oxidation to ATP, yielding three net ATPs per glucose molecule versus 30. In addition, it results in the production of lactate and protons, both of which must be removed from the brain for proper synaptic function. The requirement of supporting rapid shifts in synaptic energy use while maintaining homeostasis of ions and neurotransmitters puts an extraordinary demand on the coordination of activity between the neuronal and astrocytic metabolic pathways and ion transporters/exchangers such as the Na^+/K^+ ATPase. It has been suggested that high use of glycogen under transient conditions of high activity, as well as studies in animal models, make it an important metabolic target for study in cognitive and psychiatric disorders (Jope and Roh 2006; Rahman et al. 2024).

In contrast to K^+, the high initial influx of Na^+ into the neuron can only be rebalanced by Na^+ transport via neuronal Na^+/K^+ ATPase activity. Due to its lower activity relative to the astrocytes, neuronal Na^+ homeostasis after Na^+ influx is initially maintained primarily by diffusion away from the dendritic spines (MacAulay 2020). The average Na^+ ion entering the dendritic spine will diffuse for several seconds before being pumped out, a time that determines a dendritic shaft segment of 200–300 μm as the "minimum energy unit" of the neuron (i.e., the zone where a given excitatory postsynaptic potential leads to ATP usage). However, once there is a steady-state increase in Na^+/K^+ ATPase activity, based on mass balance constraints, constant average neuronal, astrocytic, and interstitial space Na^+ and K^+ concentrations are maintained with small fluctuations during the depolarization and repolarization cycle, as a consequence of the multiple mechanisms summarized above.

In neurons, the consumption of ATP by the Na^+/K^+ ATPase is exactly matched by ATP production, by mitochondria (95%), and glycolysis (5%); neuronal ATP levels remain constant through neurotransmission and only fall under supraphysiological stimulation (Baeza-Lehnert et al. 2019; Rangaraju et al. 2014). In astrocytes, glycolytic ATP production transiently surpasses consumption, and ATP levels rise (Fernández-Moncada et al. 2018). On a per volume basis, the ATP costs of the astrocytic Na^+/K^+ ATPase and Kir channels is at least as high as neurons when the more than three times lower volume fraction is considered (Barros 2022; Dienel and Rothman 2020).

In many descriptions of the maintenance of interstitial K^+ homeostasis, astrocytic and neuronal Kir channels are given a prominent role. Kir channels, however, are often described as not requiring energy to pump interstitial K^+ into neurons and astrocytes against a K^+ gradient. As Hibino et al. (2010) explain, the energy for the inward flux against a K^+ gradient is due to the membrane potential being hyperpolarized relative to the K^+ Nernst potential. The membrane potential itself is created by the charge imbalance produced by the Na^+/K^+ ATPase activity (3:2 ratio). Thus, Kir channel activity depends directly on ATP consumption. To quote Hibino et al. (2010:292): "These defining characteristics of Kir currents result not from a bending of the rules of biological chemistry by the Kir channel but from asymmetric open channel pore block by intracellular divalent cations and other molecules."

A better understanding of the energetics consequences of Kir channel activity is needed to appreciate fully its functional dependence on ATP synthesis. Consistent with the critical importance of K^+ homeostasis, hypokalemia is an easily identifiable, clinically important but commonly neglected condition in psychiatric patients (Lam et al. 2009).

The glutamatergic synapse is relevant when considering the brain energy budget, as it dominates neuronal signaling in the gray matter: excitatory neurons outnumber inhibitory neurons by a factor of 9 to 1, and ~90% of synapses release glutamate (Lezmy et al. 2021). Alterations in glutamatergic signaling, particularly in responses to glutamate (whether by increased release, decreased uptake, or receptor response), can incur significant energy costs (Attwell and Laughlin 2001; Robinson and Jackson 2016; Yu et al. 2018). Furthermore, direct measurements of glutamatergic neuron signaling by ensemble neuronal recording are proportional to 2-deoxyglucose measurements of glucose metabolism, as is the ratio of the rate of the glutamate/GABA-glutamine cycle and glucose oxidation (Rothman et al. 2022; Yu et al. 2018).

5 Evidence for Vulnerable Periods

In addition to large differences in the energy budget of the mature adult brain compared to other organs (see Figure 3.2 for comparison with muscle), there are also large changes in the brain energy budget from birth to adulthood. Here, we propose and provide evidence (see also Chapter 4) for a model in which critical periods for the development of psychiatric disorders are associated with ages in which there is a reduced gap between maximum oxygen and glucose transport and energy budget requirements for vital and GMR processes. Consequently, if there is an increase in allostatic load (see Figure 3.1), there is less scope to accommodate it. Because the energy cost of the vital processes needed to support mental function cannot be reduced, when the allostatic load increases to where the total energy budget exceeds the maximum limit, GMR processes need to be reduced. If reduced for an extended time, brain development (as well as synaptic turnover and remodeling in adults) would likely be impaired, leading to a potential vulnerability for mental ill health. We start with the example of the brain energy budget during childhood development and aging, in which large shifts in energy needs relative to maximum capacity to supply energy are seen throughout the lifespan.

Figure 3.5a plots the rates of CBF, $CMRO_2$, and CMR_{Glc} from the study by Goyal et al. (2014) across the lifespan. The general shape of these trajectories (the peak in infancy) parallels the lifespan curves for whole-body energy expenditure measured by the doubly labeled water method, where maximal weight-adjusted whole-body energy expenditure is observed between ages 2–10 yr (peak around age 5) (Pontzer et al. 2021). For ease of comparison, child (age 5 to 20 yr), adult (age 21 to 40 yr), and older adult (age 65 to 82 yr) groups are shown as bars. Consistent with other studies in human and rodent brain, there is a rapid increase in all three parameters until age 5, when the average adult values are reached. The increase continues, however, until CBF and CMR_{Glc} are ~twofold; $CMRO_2$ is ~40% above average

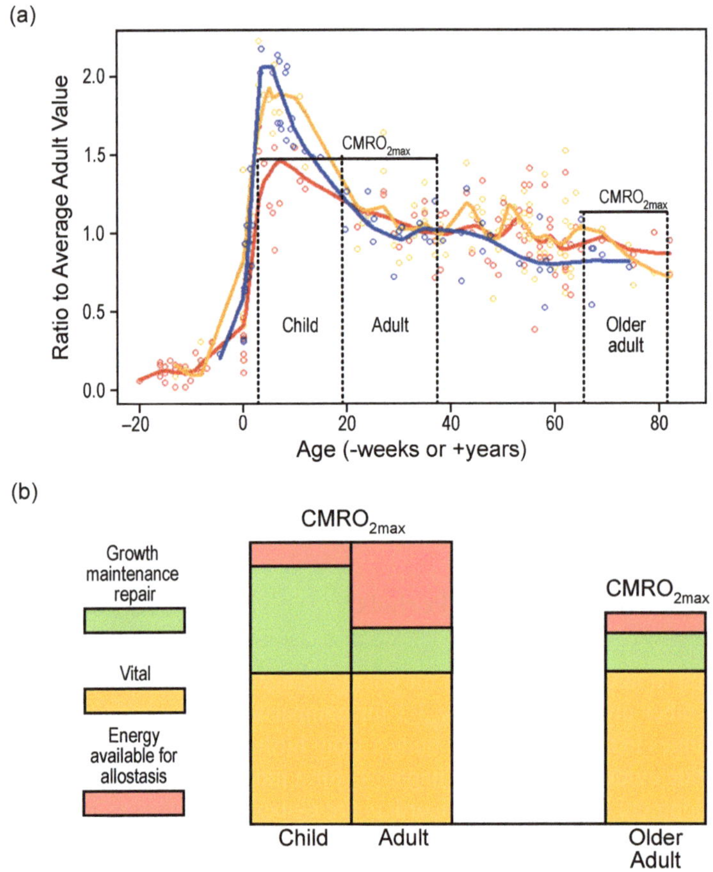

Figure 3.5 Age course of resting awake CBF, $CMRO_2$, CMR_{Glc}, estimated $CMRO_{2max}$, and allostatic energy budgets demonstrate periods of increased vulnerability in children and older adults due to limited energy available to support allostatic loads. (a) Based on the meta-analysis of Goyal et al. (2014; figure used with permission from Elsevier), data is grouped into three demographics (delineated by dashed vertical lines): children (5–20 yr), adults (21–40 yr), and older adults (65–82 yr). Middle-aged adults (41–64 yr) have not been included due to large interindividual variations. The maximum rate of net oxygen transport for each of these groups is indicated by the horizontal lines for the children, adult, and older adult ranges (dashed lines): CBF (yellow), CMR_{Glc} (blue), and $CMRO_2$ (red). All measurements normalized to the average of the adult groups. (b) The energy budget for the age groups is broken up into vital (gold) and GMR processes (green). The energy associated with maximum allostatic stress load (pink) is the gap between maximum supply capacity and the sum of vital and GMR processes. The gap is lowest in children due to increased need for GMR processes during development. In older adults, the gap is reduced relative to young adults, due to the age-dependent reduction in maximum energy supply during the adult age course because of reduced CBF.

adult values by age 10. After this peak, all three parameters decline until adulthood is reached. After the adult range, there is a secondary decline in the older adult group. This is primarily for CBF and CMR_{Glc}, but not for $CMRO_2$. In the unstimulated awake state, the high rate of CBF measured in the child group is supported by the finding of an increased brain artery cross-sectional area in the Circle of Willis in children (Buxton 2010; Guilliams et al. 2021), consistent with the need of this group to sustain up to 50–60% elevated whole brain CBF chronically (Guilliams et al. 2021; Kennedy and Sokoloff 1957). This is similar to the maximal amount of recorded *transient* increase in CBF in adults under conditions of extreme activation (Smith and Ainslie 2017). Also similar to what occurs during brain activation in adults, the relative amount of aerobic glycolysis is higher in development and adolescence (Goyal et al. 2014), paralleling the higher CBF. These lifespan differences highlight the tight relationship between the metabolic capacity of the brain and the limits imposed by its blood supply.

To estimate the gap between unstimulated awake and maximum energy production as a function of age, the maximum value of $CMRO_2$ was calculated for each age group using the relationship reported by Buxton (2010). This value of $CMRO_2$ is determined primarily by capillary density and maximum CBF and has been shown to agree with experimental data (Buxton 2010; Herculano-Houzel and Rothman 2022). For the adult and older adult groups, we determined maximum CBF by multiplying their average values measured in the unstimulated awake state (see Figure 3.2) by the maximum increases for each group reported during stimulation (DiNuzzo et al. 2024). For the child group, maximum $CMRO_2$ was estimated to be similar to the adult maximum, based on findings that capillary density (which is the major determinant of maximum $CMRO_2$) and brain size are close to mature values by age 10 (Buxton 2010).

As shown in the Figure 3.5b, the smallest gap between the estimated maximum $CMRO_2$ and the unstimulated awake value is in the child and older adult groups. These are periods of high vulnerability for developing mental health (childhood) and neurological (aging) disorders; for details, see Chapter 4. We propose that an important factor in this higher vulnerability is this smaller energy reserve, which during periods of extended high neuronal activity, such as occurs during chronic stress, reduces the energy available to support restorative and developmental processes (growth, maintenance, and repair).

6 Conclusions

Because the awake brain operates at near capacity, its ability to increase blood delivery, and hence oxygen and glucose, is limited due to restrictions dictated by capillary density and available space within the skull. To avoid impacting brain functions, homeostasis restricts both the amount of glucose and oxygen that can be supplied and the level of generated protons (H^+) and CO_2 that can be tolerated. Having detailed the neurochemical basis of how the brain operates carefully within these homeostatic limits, we presented a model that suggests when brain energy supply is limited, resource allocation becomes key to longer-term function.

Understanding how brain energy production and utilization are modulated at rest, during activation, and throughout the lifespan is essential before the consequences of exceeding these limits can be appropriately addressed. In Chapter 4, we present examples that describe how infringement takes place and leads to pathophysiology of neuropsychiatric disorders.

References

Andersen JV, and Schousboe A (2023) Milestone Review: Metabolic Dynamics of Glutamate and GABA Mediated Neurotransmission—the Essential Roles of Astrocytes. J Neurochem 166 (2):109–137. https://doi.org/10.1111/jnc.15811

Attwell D, and Laughlin SB (2001) An Energy Budget for Signaling in the Grey Matter of the Brain. J Cereb Blood Flow Metab 21 (10):1133–1145. https://doi.org/10.1097/00004647-200110000-00001

Baeza-Lehnert F, Saab AS, Gutiérrez R, et al. (2019) Non-Canonical Control of Neuronal Energy Status by the Na^+ Pump. Cell Metab 29 (3):668–680.e664. https://doi.org/10.1016/j.cmet.2018.11.005

Bak LK, and Schousboe A (2017) Misconceptions Regarding Basic Thermodynamics and Enzyme Kinetics Have Led to Erroneous Conclusions Regarding the Metabolic Importance of Lactate Dehydrogenase Isoenzyme Expression. J Neurosci Res 95 (11):2098–2102. https://doi.org/10.1002/jnr.23994

Ball KK, Cruz NF, Mrak RE, and Dienel GA (2010) Trafficking of Glucose, Lactate, and Amyloid-Beta from the Inferior Colliculus through Perivascular Routes. J Cereb Blood Flow Metab 30 (1):162–176. https://doi.org/10.1038/jcbfm.2009.206

Barnes K, Ingram JC, Porras OH, et al. (2002) Activation of Glut1 by Metabolic and Osmotic Stress: Potential Involvement of AMP-Activated Protein Kinase (AMPK). J Cell Sci 115 (11):2433–2442. https://doi.org/10.1242/jcs.115.11.2433

Barros LF (2022) How Expensive Is the Astrocyte? J Cereb Blood Flow Metab 42 (5):738–745. https://doi.org/10.1177/0271678X221077343

Barros LF, Bittner CX, Loaiza A, and Porras OH (2007) A Quantitative Overview of Glucose Dynamics in the Gliovascular Unit. Glia 55 (12):1222–1237. https://doi.org/10.1002/glia.20375

Barros LF, Porras OH, and Bittner CX (2005) Why Glucose Transport in the Brain Matters for PET. Trends Neurosci 28 (3):117–119. https://doi.org/10.1016/j.tins.2005.01.002

Barros LF, Ruminot I, Sotelo-Hitschfeld T, Lerchundi R, and Fernandez-Moncada I (2023) Metabolic Recruitment in Brain Tissue. Annu Rev Physiol 85:115–135. https://doi.org/10.1146/annurev-physiol-021422-091035

Bergersen LH, and Gjedde A (2012) Is Lactate a Volume Transmitter of Metabolic States of the Brain? Front Neuroenergetics 4:5. https://doi.org/10.3389/fnene.2012.00005

Bergles DE, Diamond JS, and Jahr CE (1999) Clearance of Glutamate inside the Synapse and Beyond. Curr Opin Neurobiol 9 (3):293–298. https://doi.org/10.1016/S0959-4388(99)80043-9

Bergles DE, and Jahr CE (1997) Synaptic Activation of Glutamate Transporters in Hippocampal Astrocytes. Neuron 19 (6):1297–1308. https://doi.org/10.1016/S0896-6273(00)80420-1

Bhatia S, Rawal R, Sharma P, et al. (2022) Mitochondrial Dysfunction in Alzheimer's Disease: Opportunities for Drug Development. Curr Neuropharmacol 20 (4):675–692. http://dx.doi.org/10.2174/1570159X19666210517114016

Blazey T, Snyder AZ, Goyal MS, Vlassenko AG, and Raichle ME (2018) A Systematic Meta-Analysis of Oxygen-to-Glucose and Oxygen-to-Carbohydrate Ratios in the Resting Human Brain. PLoS One 13 (9):e0204242. https://doi.org/10.1371/journal.pone.0204242

Bobba-Alves N, Juster R-P, and Picard M (2022) The Energetic Cost of Allostasis and Allostatic Load. Psychoneuroendocrinology 146:105951. https://doi.org/10.1016/j.psyneuen.2022.105951

Bonnet U, Wiemann M, and Bingmann D (1998) CO_2/Hco_3^--Withdrawal from the Bath Medium of Hippocampal Slices: Biphasic Effect on Intracellular pH and Bioelectric Activity of CA3-Neurons. Brain Res 796 (1–2):161–170. https://doi.org/10.1016/S0006-8993(98)00341-2

Boyle PJ, Scott JC, Krentz AJ, et al. (1994) Diminished Brain Glucose Metabolism Is a Significant Determinant for Falling Rates of Systemic Glucose Utilization during Sleep in Normal Humans. J Clin Invest 93 (2):529–535. https://doi.org/10.1172/JCI117003
Buxton RB (2010) Interpreting Oxygenation-Based Neuroimaging Signals: The Importance and the Challenge of Understanding Brain Oxygen Metabolism. Front Neuroenergetics 2 (8):1648. https://doi.org/10.3389/fnene.2010.00008
——— (2021) The Thermodynamics of Thinking: Connections between Neural Activity, Energy Metabolism and Blood Flow. Philosophical Transactions of the Royal Society B 376 (1815):20190624. https://doi.org/10.1098/rstb.2019.0624
——— (2024) Thermodynamic Limitations on Brain Oxygen Metabolism: Physiological Implications. J Physiol 602 (4):683–712. https://doi.org/10.1113/JP284358
Caesar K, Hashemi P, Douhou A, et al. (2008) Glutamate Receptor-Dependent Increments in Lactate, Glucose and Oxygen Metabolism Evoked in Rat Cerebellum *in Vivo*. J Physiol 586 (5):1337–1349. https://doi.org/10.1113/jphysiol.2007.144154
Chen JJ, Uthayakumar B, and Hyder F (2022) Mapping Oxidative Metabolism in the Human Brain with Calibrated fMRI in Health and Disease. J Cereb Blood Flow Metab 42 (7):1139–1162. https://doi.org/10.1177/0271678X221077338
Chesler M, and Kaila K (1992) Modulation of pH by Neuronal Activity. Trends Neurosci 15:396–402. https://doi.org/10.1016/0166-2236(92)90191-A
Clairis N, Barakat A, Brochard J, Xin L, and Sandi C (2024) A Neurometabolic Mechanism Involving Dmpfc/Dacc Lactate in Physical Effort-Based Decision-Making. Mol Psychiatr 30 August 2024. https://doi.org/10.1038/s41380-024-02726-y
Cobley JN, Fiorello ML, and Bailey DM (2018) 13 Reasons Why the Brain Is Susceptible to Oxidative Stress. Redox Biol 15:490–503. https://doi.org/10.1016/j.redox.2018.01.008
Cura AJ, and Carruthers A (2012) Amp Kinase Regulation of Sugar Transport in Brain Capillary Endothelial Cells during Acute Metabolic Stress. Am J Physiol Cell Physiol 303 (8):C806–C814. https://doi.org/10.1152/ajpcell.00437.2011
de Graaf RA, Pan JW, Telang F, et al. (2001) Differentiation of Glucose Transport in Human Brain Gray and White Matter. J Cereb Blood Flow Metab 21 (5):483–492. https://doi.org/10.1097/00004647-200105000-00002
De Vivo DC, Trifiletti RR, Jacobson RI, et al. (1991) Defective Glucose Transport across the Blood-Brain Barrier as a Cause of Persistent Hypoglycorrhachia, Seizures, and Developmental Delay. N Engl J Med 325 (10):703–709. https://www.nejm.org/doi/full/10.1056/NEJM199109053251006
Dienel GA (2017a) Fluxes of Lactate into, from, and Among Gap Junction-Coupled Astroglia and Their Interaction with Noradrenaline. In: Vardjan N, Zorec R (eds) Noradrenergic Signaling and Astroglia. Elsevier, London, pp 145–166. https://doi.org/10.3389/fnins.2014.00261
——— (2017b) Lack of Appropriate Stoichiometry: Strong Evidence against an Energetically Important Astrocyte–Neuron Lactate Shuttle in Brain. J Neurosci Res 95 (11):2103–2125. https://doi.org/10.1002/jnr.24015
——— (2019) Brain Glucose Metabolism: Integration of Energetics with Function. Physiol Rev 99 (1):949–1045. https://doi.org/10.1152/physrev.00062.2017
Dienel GA, and Rothman DL (2019) Glycogenolysis in Cerebral Cortex during Sensory Stimulation, Acute Hypoglycemia, and Exercise: Impact on Astrocytic Energetics, Aerobic Glycolysis, and Astrocyte-Neuron Interactions. In: DiNuzzo M, Schousboe A (eds) Brain Glycogen Metabolism. Advances in Neurobiology. Springer, Cham, pp 209–267. https://link.springer.com/chapter/10.1007/978-3-030-27480-1_8
——— (2020) Reevaluation of Astrocyte-Neuron Energy Metabolism with Astrocyte Volume Fraction Correction: Impact on Cellular Glucose Oxidation Rates, Glutamate–Glutamine Cycle Energetics, Glycogen Levels and Utilization Rates vs. Exercising Muscle, and Na^+/K^+ Pumping Rates. Neurochem Res 45 (11):2607–2630. https://doi.org/10.1007/s11064-020-03125-9

DiNuzzo M, Dienel GA, Behar KL, et al. (2024) Neurovascular Coupling Is Optimized to Compensate for the Increase in Proton Production from Nonoxidative Glycolysis and Glycogenolysis during Brain Activation and Maintain Homeostasis of pH, pCO_2, and pO_2. J Neurochem 168 (5):632–662. https://doi.org/10.1111/jnc.15839
DiNuzzo M, Giove F, Maraviglia B, and Mangia S (2017) Computational Flux Balance Analysis Predicts That Stimulation of Energy Metabolism in Astrocytes and Their Metabolic Interactions with Neurons Depend on Uptake of K^+ Rather Than Glutamate. Neurochem Res 42 (1):202–216. https://doi.org/10.1007/s11064-016-2048-0
Drew PJ (2022) Neurovascular Coupling: Motive Unknown. Trends Neurosci 45 (11):809–819. https://doi.org/10.1016/j.tins.2022.08.004
Fernández-Moncada I, Ruminot I, Robles-Maldonado D, et al. (2018) Neuronal Control of Astrocytic Respiration through a Variant of the Crabtree Effect. PNAS 115 (7):1623–1628. https://doi.org/10.1073/pnas.1716469115
Finn ES (2021) Is It Time to Put Rest to Rest? Trends Cogn Sci 25 (12):1021–1032. https://doi.org/10.1016/j.tics.2021.09.005
Fox PT, and Raichle ME (1986) Focal Physiological Uncoupling of Cerebral Blood Flow and Oxidative Metabolism during Somatosensory Stimulation in Human Subjects. PNAS 83 (4):1140–1144. https://doi.org/10.1073/pnas.83.4.1140
Fox PT, Raichle ME, Mintun MA, and Dence C (1988) Non-Oxidative Glucose Consumption during Focal Physiological Neuronal Activity. Science 241 (4864):462–464. https://www.science.org/doi/10.1126/science.3260686
Goyal MS, Hawrylycz M, Miller JA, Snyder AZ, and Raichle ME (2014) Aerobic Glycolysis in the Human Brain Is Associated with Development and Neotenous Gene Expression. Cell Metab 19 (1):49–57. https://doi.org/10.1016/j.cmet.2013.11.020
Gruetter R, Seaquist ER, and Ugurbil K (2001) A Mathematical Model of Compartmentalized Neurotransmitter Metabolism in the Human Brain. Am J Physiol Endocrinol Metab 281 (1):E100–E112. https://doi.org/10.1152/ajpendo.2001.281.1.E100
Guilliams K, Gupta N, Srinivasan S, et al. (2021) MR Imaging Differences in the Circle of Willis between Healthy Children and Adults. AJNR Am J Neuroradiol 42 (11):2062–2069. https://doi.org/10.3174/ajnr.A7290
Hagihara H, Shoji H, Hattori S, et al. (2024) Large-Scale Animal Model Study Uncovers Altered Brain pH and Lactate Levels as a Transdiagnostic Endophenotype of Neuropsychiatric Disorders Involving Cognitive Impairment. eLife 12:RP89376. https://doi.org/10.7554/eLife.89376.3
Halestrap AP (2013) Monocarboxylic Acid Transport. Compr Physiol 3 (4):1611–1643. https://doi.org/10.1002/cphy.c130008
Halestrap AP, and Wilson MC (2012) The Mo nocarboxylate Transporter Family—Role and Regulation. IUBMB Life 64 (2):109–119. https://doi.org/10.1002/iub.572
Halliwell B (1992) Reactive Oxygen Species and the Central Nervous System. J Neurochem 59 (5):1609–1623. https://doi.org/10.1111/j.1471-4159.1992.tb10990.x
Hargreaves M, and Spriet LL (2020) Skeletal Muscle Energy Metabolism during Exercise. Nat Metab 2 (9):817–828. https://doi.org/10.1038/s42255-020-0251-4
Harris JJ, Jolivet R, and Attwell D (2012) Synaptic Energy Use and Supply. Neuron 75 (5):762–777. https://doi.org/10.1016/j.neuron.2012.08.019
Herculano-Houzel S, and Rothman DL (2022) From a Demand-Based to a Supply-Limited Framework of Brain Metabolism. Front Integr Neurosci 16:818685. https://doi.org/10.3389/fnint.2022.818685
Hibino H, Inanobe A, Furutani K, et al. (2010) Inwardly Rectifying Potassium Channels: Their Structure, Function, and Physiological Roles. Physiol Rev 90 (1):291–366. https://doi.org/10.1152/physrev.00021.2009
Hladky SB, and Barrand MA (2018) Elimination of Substances from the Brain Parenchyma: Efflux via Perivascular Pathways and via the Blood–Brain Barrier. Fluids Barriers CNS 15:1–73. https://doi.org/10.1186/s12987-018-0113-6

——— (2019) Metabolite Clearance during Wakefulness and Sleep. In: Landolt H-P, Dijk D-J (eds) Sleep-Wake Neurobiology and Pharmacology, vol 253. Handbook of Experimental Pharmacology. Springer, Cham, pp 385–423. https://link.springer.com/book/10.1007/978-3-030-11272-1

Howarth C, Gleeson P, and Attwell D (2012) Updated Energy Budgets for Neural Computation in the Neocortex and Cerebellum. J Cereb Blood Flow Metab 32 (7):1222–1232. https://doi.org/10.1038/jcbfm.2012.35

Hyder F, Herman P, Bailey CJ, et al. (2016) Uniform Distributions of Glucose Oxidation and Oxygen Extraction in Gray Matter of Normal Human Brain: No Evidence of Regional Differences of Aerobic Glycolysis. J Cereb Blood Flow Metab 36 (5):903–916. https://doi.org/10.1177/0271678X15625349

Hyder F, Rothman DL, and Bennett MR (2013) Cortical Energy Demands of Signaling and Nonsignaling Components in Brain Are Conserved across Mammalian Species and Activity Levels. PNAS 110 (9):3549–3554. https://doi.org/10.1073/pnas.1214912110

Jope RS, and Roh M-S (2006) Glycogen Synthase Kinase-3 (Gsk3) in Psychiatric Diseases and Therapeutic Interventions. Curr Drug Targets 7 (11):1421–1434. http://dx.doi.org/10.2174/138945006778559238

Joyner MJ, and Casey DP (2015) Regulation of Increased Blood Flow (Hyperemia) to Muscles during Exercise: A Hierarchy of Competing Physiological Needs. Physiol Rev 95 (2):549–601. https://doi.org/10.1152/physrev.00035.2013

Karbowski J (2011) Scaling of Brain Metabolism and Blood Flow in Relation to Capillary and Neural Scaling. PLoS One 6 (10):e26709. https://doi.org/10.1371/journal.pone.0026709

Kennedy C, and Sokoloff L (1957) An Adaptation of the Nitrous Oxide Method to the Study of the Cerebral Circulation in Children: Normal Values for Cerebral Blood Flow and Cerebral Metabolic Rate in Childhood. J Clin Invest 36 (7):1130–1137. https://doi.org/10.1172/JCI103509

Kim DY, Park J, and Han I-O (2023) Hexosamine Biosynthetic Pathway and *O*-Glcnac Cycling of Glucose Metabolism in Brain Function and Disease. Am J Physiol Cell Physiol 325 (4):C981–C998. https://doi.org/10.1152/ajpcell.00191.2023

Lam MH-b, Chau SW-h, and Wing Y-k (2009) High Prevalence of Hypokalemia in Acute Psychiatric Inpatients. Gen Hosp Psychiatry 31 (3):262–265. https://doi.org/10.1016/j.genhosppsych.2009.02.007

Larsen BR, Stoica A, and MacAulay N (2016) Managing Brain Extracellular K^+ during Neuronal Activity: The Physiological Role of the Na^+/K^+-ATPase Subunit Isoforms. Front Physiol 7:141. https://doi.org/10.3389/fphys.2016.00141

Laughlin MH (1987) Skeletal Muscle Blood Flow Capacity: Role of Muscle Pump in Exercise Hyperemia. Am J Physiol Heart Circ Physiol 253 (5):H993–H1004. https://doi.org/10.1152/ajpheart.1987.253.5.H993

Lee J, Taira T, Pihlaja P, Ransom BR, and Kaila K (1996) Effects of $C0_2$ on Excitatory Transmission Apparently Caused by Changes in Intracellular pH in the Rat Hippocampal Slice. Brain Res 706 (2):210–216. https://doi.org/10.1016/0006-8993(95)01214-1

Leithner C, and Royl G (2014) The Oxygen Paradox of Neurovascular Coupling. J Cereb Blood Flow Metab 34 (1):19–29. https://doi.org/10.1038/jcbfm.2013.181

Leybaert L (2005) Neurobarrier Coupling in the Brain: A Partner of Neurovascular and Neurometabolic Coupling? J Cereb Blood Flow Metab 25 (1):2–16. https://doi.org/10.1038/sj.jcbfm.9600001

Lezmy J, Harris JJ, and Attwell D (2021) Optimising the Energetic Cost of the Glutamatergic Synapse. Neuropharmacology 197:108727. https://doi.org/10.1016/j.neuropharm.2021.108727

Liemburg-Apers DC, Wagenaars JA, Smeitink JA, Willems PH, and Koopman WJ (2016) Acute Stimulation of Glucose Influx Upon Mitoenergetic Dysfunction Requires LKB1, AMPK, Sirt2 and mTOR–Raptor. J Cell Sci 129 (23):4411–4423. https://doi.org/10.1242/jcs.194480

Lobmann R, Smid HGOM, Pottag G, et al. (2000) Impairment and Recovery of Elementary Cognitive Function Induced by Hypoglycemia in Type-1 Diabetic Patients and Healthy Controls. J Clin Endocrinol Metab 85 (12):4818. https://doi.org/10.1210/jcem.85.12.9999

Looser ZJ, Faik Z, Ravotto L, et al. (2024) Oligodendrocyte–Axon Metabolic Coupling Is Mediated by Extracellular K^+ and Maintains Axonal Health. Nat Neurosci 27 (3):433–448. https://doi.org/10.1038/s41593-023-01558-3

Lüscher C, and Malenka RC (2012) NMDA Receptor-Dependent Long-Term Potentiation and Long-Term Depression (LTP/LTD). Cold Spring Harb Perspect Biol 4 (6):a005710. https://doi.org/10.1101/cshperspect.a005710

MacAulay N (2020) Molecular Mechanisms of K^+ Clearance and Extracellular Space Shrinkage—Glia Cells as the Stars. Glia 68 (11):2192–2211. https://doi.org/10.1002/glia.23824

Maddock RJ, Casazza GA, Buonocore MH, and Tanase C (2011) Vigorous Exercise Increases Brain Lactate and Glx (Glutamate + Glutamine): A Dynamic 1H-MRS Study. Neuroimage 57 (4):1324–1330. https://doi.org/10.1016/j.neuroimage.2011.05.048

Madsen PL, Hasselbalch SG, Hagemann LP, et al. (1995) Persistent Resetting of the Cerebral Oxygen/Glucose Uptake Ratio by Brain Activation: Evidence Obtained with the Kety-Schmidt Technique. J Cereb Blood Flow Metab 15 (3):485–491. https://doi.org/10.1038/jcbfm.1995.60

Madsen PL, Linde R, Hasselbalch SG, Paulson OB, and Lassen NA (1998) Activation-Induced Resetting of Cerebral Oxygen and Glucose Uptake in the Rat. J Cereb Blood Flow Metab 18 (7):742–748. https://doi.org/10.1097/00004647-199807000-00005

Mangia S, Simpson IA, Vannucci SJ, and Carruthers A (2009) The *in Vivo* Neuron-to-Astrocyte Lactate Shuttle in Human Brain: Evidence from Modeling of Measured Lactate Levels during Visual Stimulation. J Neurochem 109 (Suppl 1):55–62. https://doi.org/10.1111/j.1471-4159.2009.06003.x

Mangia S, Tkac I, Gruetter R, et al. (2007) Sustained Neuronal Activation Raises Oxidative Metabolism to a New Steady-State Level: Evidence from ^{1}H NMR Spectroscopy in the Human Visual Cortex. J Cereb Blood Flow Metab 27 (5):1055–1063. https://doi.org/10.1038/sj.jcbfm.9600401

McEwen BS (2003) Mood Disorders and Allostatic Load. Biol Psychiatry 54 (3):200–207. https://doi.org/10.1016/S0006-3223(03)00177-X

McEwen BS, and Gianaros PJ (2011) Stress- and Allostasis-Induced Brain Plasticity. Annu Rev Med 62:431–445. https://doi.org/10.1146/annurev-med-052209-100430

McMorris T, Hale BJ, Barwood M, Costello J, and Corbett J (2017) Effect of Acute Hypoxia on Cognition: A Systematic Review and Meta-Regression Analysis. Neurosci Biobehav Rev 74 (Part A):225–232. https://doi.org/10.1016/j.neubiorev.2017.01.019

Minhas PS, Jones JR, Latif-Hernandez A, et al. (2024) Restoring Hippocampal Glucose Metabolism Rescues Cognition across Alzheimer's Disease Pathologies. Science 385 (6711):eabm6131. https://doi.org/10.1126/science.abm6131

Moeini M, Lu X, Bélanger S, et al. (2019) Cerebral Tissue pO_2 Response to Stimulation Is Preserved with Age in Awake Mice. Neurosci Lett 699:160–166. https://doi.org/10.1016/j.neulet.2019.02.007

Morcom AM, and Fletcher PC (2007) Does the Brain Have a Baseline? Why We Should Be Resisting a Rest. Neuroimage 37 (4):1073–1082. https://doi.org/10.1016/j.neuroimage.2006.09.013

Münch G, Thome J, Foley P, Schinzel R, and Riederer P (1997) Advanced Glycation Endproducts in Ageing and Alzheimer's Disease. Brain Res Rev 23 (1–2):134–143. https://doi.org/10.1016/S0165-0173(96)00016-1

Nousen EK, Franco JG, and Sullivan EL (2014) Unraveling the Mechanisms Responsible for the Comorbidity between Metabolic Syndrome and Mental Health Disorders. Neuroendocrinology 98 (4):254–266. https://doi.org/10.1159/000355632

Pellerin L, and Magistretti PJ (2012) Sweet Sixteen for ANLS. J Cereb Blood Flow Metab 32 (7):1152–1166. https://doi.org/10.1038/jcbfm.2011.149

Pignalosa FC, Desiderio A, Mirra P, et al. (2021) Diabetes and Cognitive Impairment: A Role for Glucotoxicity and Dopaminergic Dysfunction. Int J Mol Sci 22 (22):12366. https://doi.org/10.3390/ijms222212366

Pontzer H, Yamada Y, Sagayama H, et al. (2021) Daily Energy Expenditure through the Human Life Course. Science 373 (6556):808–812. https://doi.org/10.1126/science.abe5017

Prichard J, Rothman D, Novotny E, et al. (1991) Lactate Rise Detected by 1H NMR in Human Visual Cortex Following Physiological Stimulation. PNAS 88 (13):5829–5831. https://doi.org/10.1073/pnas.88.13.5829

Rae CD, Baur JA, Borges K, et al. (2024) Brain Energy Metabolism: A Roadmap for Future Research. J Neurochem 168 (5):910–954. https://doi.org/10.1111/jnc.16032

Rae CD, and Bröer S (2015) Creatine as a Booster for Human Brain Function. How Might It Work? Neurochem Int 89:249–259. http://dx.doi.org/10.1016/j.neuint.2015.08.010

Rae CD, Bubb WA, and Kuchel PW (1992) Aldolase-Catalyzed Diketone Phosphate Formation from Oxoaldehydes. NMR Studies and Metabolic Significance. J Biol Chem 267 (14):9713–9717. https://doi.org/10.1016/S0021-9258(19)50149-1

Rahman MS, Harrison E, Biggs H, et al. (2024) Dynamics of Cognitive Variability with Age and Its Genetic Underpinning in Nihr Bioresource Genes and Cognition Cohort Participants. Nat Med 30:1739–1748. https://doi.org/10.1038/s41591-024-02960-5

Raichle ME (2011) The Restless Brain. Brain Connect 1 (1):3–12. https://doi.org/10.1089/brain.2011.0019

Raichle ME, Macleod KM, Snyder AZ, et al. (2001) A Default Mode of Brain Function. PNAS 98 (2):676–682. https://doi.org/10.1073/pnas.98.2.676

Rangaraju V, Calloway N, and Ryan TA (2014) Activity-Driven Local ATP Synthesis Is Required for Synaptic Function. Cell 156 (4):825–835. https://doi.org/10.1016/j.cell.2013.12.042

Ransom BR (2000) Glial Modulation of Neural Excitability Mediated by Extracellular pH: A Hypothesis Revisited. Prog Brain Res 125:217–228. https://doi.org/10.1016/S0079-6123(00)25012-7

Rasmussen R, Nicholas E, Petersen NC, et al. (2019) Cortex-Wide Changes in Extracellular Potassium Ions Parallel Brain State Transitions in Awake Behaving Mice. Cell Rep 28 (5):1182–1194.e1184. https://doi.org/10.1016/j.celrep.2019.06.082

Richard JP (1991) Kinetic Parameters for the Elimination Reaction Catalyzed by Triosephosphate Isomerase and an Estimation of the Reaction's Physiological Significance. Biochemistry 30 (18):4581–4585. https://doi.org/10.1021/bi00232a031

Robinson MB, and Jackson JG (2016) Astroglial Glutamate Transporters Coordinate Excitatory Signaling and Brain Energetics. Neurochem Int 98:56–71. https://doi.org/10.1016/j.neuint.2016.03.014

Rossetti L, Giaccari A, and DeFronzo RA (1990) Glucose Toxicity. Diabetes Care 13 (6):610–630. https://doi.org/10.2337/diacare.13.6.610

Rothman DL, De Feyter HM, de Graaf RA, Mason GF, and Behar KL (2011) ^{13}C MRS Studies of Neuroenergetics and Neurotransmitter Cycling in Humans. NMR Biomed 24 (8):943–957. https://doi.org/10.1002/nbm.1772

Rothman DL, Dienel GA, Behar KL, et al. (2022) Glucose Sparing by Glycogenolysis (Gsg) Determines the Relationship between Brain Metabolism and Neurotransmission. J Cereb Blood Flow Metab 42 (5):844–860. https://doi.org/10.1177/0271678X211064399

Ruminot I, Gutiérrez R, Peña-Münzenmayer G, et al. (2011) Nbce1 Mediates the Acute Stimulation of Astrocytic Glycolysis by Extracellular K^+. J Neurosci 31 (40):14264–14271. https://doi.org/10.1523/JNEUROSCI.2310-11.2011

Saltin B (2007) Exercise Hyperaemia: Magnitude and Aspects on Regulation in Humans. J Physiol 583 (3):819–823. https://doi.org/10.1113/jphysiol.2007.136309

Schousboe A, Sickmann HM, Walls AB, Bak LK, and Waagepetersen HS (2010) Functional Importance of the Astrocytic Glycogen-Shunt and Glycolysis for Maintenance of an Intact Intra/Extracellular Glutamate Gradient. Neurotox Res 18:94–99. https://doi.org/10.1007/s12640-010-9171-5

Schweitzer JS, Wang H, Xiong Z-Q, and Stringer JL (2000) pH Sensitivity of Non-Synaptic Field Bursts in the Dentate Gyrus. J Neurophysiol 84 (2):927–933. https://doi.org/10.1152/jn.2000.84.2.927

Shaw DM, Cabre G, and Gant N (2021) Hypoxic Hypoxia and Brain Function in Military Aviation: Basic Physiology and Applied Perspectives. Front Physiol 12:665821. https://doi.org/10.3389/fphys.2021.665821

Shulman RG, Hyder F, and Rothman DL (2001) Cerebral Energetics and the Glycogen Shunt: Neurochemical Basis of Functional Imaging. PNAS 98 (11):6417–6422. https://doi.org/10.1073/pnas.101129298

Shulman RG, and Rothman DL (1998) Interpreting Functional Imaging Studies in Terms of Neurotransmitter Cycling. PNAS 95 (20):11993–11998. https://doi.org/10.1073/pnas.95.20.11993

Shulman RG, and Rothman DL (2019) A Non-Cognitive Behavioral Model for Interpreting Functional Neuroimaging Studies. Front Hum Neurosci 13:28. https://doi.org/10.3389/fnhum.2019.00028

Sibson NR, Dhankhar A, Mason GF, et al. (1998) Stoichiometric Coupling of Brain Glucose Metabolism and Glutamatergic Neuronal Activity. PNAS 95 (1):316–321. https://doi.org/10.1073/pnas.95.1.316

Smith AJ, Blumenfeld H, Behar KL, et al. (2002) Cerebral Energetics and Spiking Frequency: The Neurophysiological Basis of fMRI. PNAS 99 (16):10765–10770. https://doi.org/10.1073/pnas.132272199

Smith KJ, and Ainslie PN (2017) Regulation of Cerebral Blood Flow and Metabolism during Exercise. Exp Physiol 102 (11):1356–1371. https://doi.org/10.1113/EP086249

Sterling P (1988) Allostasis: A New Paradigm to Explain Arousal Pathology. In: Fisher S, Reason J (eds) Handbook of Life Stress, Cognition and Health. John Wiley & Sons Ltd, Hoboken, pp 629–649

Takimoto M, and Hamada T (2014) Acute Exercise Increases Brain Region-Specific Expression of MCT1, MCT2, MCT4, GLUT1, and COX IV Proteins. J Appl Physiol 116 (9):1238–1250. https://doi.org/10.1152/japplphysiol.01288.201 3

Varma SD, and Kinoshita JH (1974) Sorbitol Pathway in Diabetic and Galactosemic Rat Lens. Biochim Biophys Acta Gen Subj 338 (2):632–640. https://doi.org/10.1016/0304-4165(74)90326-2

Walls AB, Heimbürger CM, Bouman SD, Schousboe A, and Waagepetersen HS (2009) Robust Glycogen Shunt Activity in Astrocytes: Effects of Glutamatergic and Adrenergic Agents. Neuroscience 158 (1):284–292. https://doi.org/10.1016/j.neuroscience.2008.09.058

Winkler EA, Nishida Y, Sagare AP, et al. (2015) Glut1 Reductions Exacerbate Alzheimer's Disease Vasculo-Neuronal Dysfunction and Degeneration. Nat Neurosci 18 (4):521–530. https://doi.org/10.1038/nn.3966

Yu Y, Herman P, Rothman DL, Agarwal D, and Hyder F (2018) Evaluating the Gray and White Matter Energy Budgets of Human Brain Function. J Cereb Blood Flow Metab 38 (8):1339–1353. https://doi.org/10.1177/0271678X17708691

Zhang B, Jin Q, Xu L, et al. (2020) Cooperative Transport Mechanism of Human Monocarboxylate Transporter 2. Nat Commun 11 (1):2429. https://doi.org/10.1038/s41467-020-16334-1

Zhu X-H, Lee B-Y, and Chen W (2018) Functional Energetic Responses and Individual Variance of the Human Brain Revealed by Quantitative Imaging of Adenosine Triphosphate Production Rates. J Cereb Blood Flow Metab 38 (6):959–972. https://doi.org/10.1177/0271678X18769039

Open Access This chapter is licensed under the terms of the Creative Commons Attribution-NonCommercial-NoDerivatives 4.0 International License (http://creativecommons.org/licenses/by-nc-nd/4.0/), which permits any noncommercial use, sharing, distribution and reproduction in any medium or format, as long as you give appropriate credit to the original author(s) and the source, provide a link to the Creative Commons license and indicate if you modified the licensed material. You do not have permission under this license to share adapted material derived from this chapter or parts of it.

The images or other third party material in this chapter are included in the chapter's Creative Commons license, unless indicated otherwise in a credit line to the material. If material is not included in the chapter's Creative Commons license and your intended use is not permitted by statutory regulation or exceeds the permitted use, you will need to obtain permission directly from the copyright holder.

4

Brain Energy Production

Vulnerability and Pathophysiology

Caroline D. Rae, L. Felipe Barros, Alexander Behnke, Manu S. Goyal, Suzana Herculano-Houzel, Daria Peleg-Raibstein, Martin Picard, Douglas Rothman, Anthony C. Vernon, and Zoltan Sarnyai

Abstract The brain's homeostatic constraints require it to operate optimally within narrow metabolic and physiological limits. When these limits are breached due to internal (genetics, adolescence, aging, reproduction) or external (environment, stress, trauma) factors, there are consequences. Applying the concepts developed in Chapter 3, examples are presented that describe how infringement takes place and lead to pathophysiology. Allostatic changes in demand can overwhelm the brain's energy systems and lead to symptom formation and abnormalities in information processing.

Keywords Homeostasis, allostasis, stress, blood supply

1 Impact of High Allostatic Loads on Energy Competition and Brain Homeostasis

In Chapter 3, we proposed a model based on allostasis theory, which suggests that when brain energy supply is limited, resource allocation becomes key to longer-term function. The key difference between brain metabolism and the rest of the body is that under all awake conditions, the energy consumption by vital processes is close to the maximum capacity of energy supply to brain tissue. According to fundamental biophysical and biological processes, chronic excessive increases in allostatic load, which in the brain will be largely associated with an increase in

---

Caroline D. Rae (✉)
Neuroscience Research Australia and School of Psychology, University of New South Wales, Randwick NSW 2031, Australia
Email: c.rae@unsw.edu.au

Affiliations for the coauthors are available in the List of Contributors.

© Frankfurt Institute for Advanced Studies (FIAS) 2025

D. Öngür, J. M. Ford (eds.), *Metabolic Neuropsychiatry*, Strüngmann Forum Reports,
https://doi.org/10.1007/978-3-032-05630-6_4

the energy allocated to the vital activities required for mental functions, necessarily requires reduction of other energy-requiring activities (Yang et al. 2021). This involves acutely dispensable activities grouped under growth, maintenance, and repair (GMR), which can be paused for variable periods of time but are necessary to the longer-term efficient functioning of various organ systems, including the brain (Bobba-Alves et al. 2022).

To satisfy energy constraints, operation of the brain needs to occur within specific homeostatic state spaces where some configuration of the critical constraints compatible with life is satisfied. The brain regularly moves between different state spaces under healthy conditions; between awake and asleep states, for example, where the relative (and potentially total) energy allocation and the functional outcome between each state is different (Figure 4.1) (Gu et al. 2018). In the case of sleep, there is less allocation to functional activity and more to transiently dispensable maintenance activities, such as memory consolidation, waste clearance, and restoration of brain glycogen levels. Transition between states can be fast (e.g., going to sleep or loss of consciousness due to a sudden drop of blood pressure) or relatively slow, induced, for example by chronic stress.

Dysfunctional mental states can therefore be considered as allostatic shifts in brain homeostasis, where the brain moves from a "normal" homeostatic state to another semipermanent local minimum resulting in altered function. We can consider this as a manifold of brain energy contributions with an overlying temporal component and state alterations driven by intrinsic and extrinsic inputs. The functional consequences are determined by the underlying perturbations and their context, the

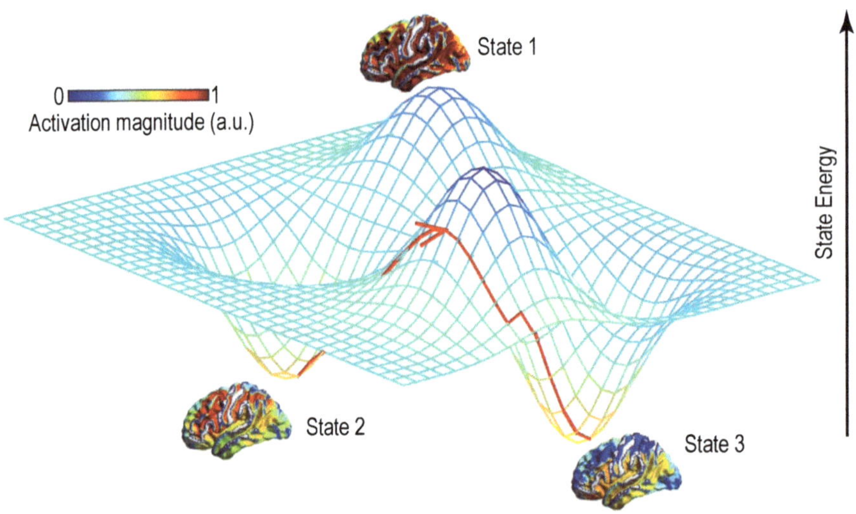

Figure 4.1 Homeostatic minima and maxima of brain function and behavior and their associated energy requirements. The manifold shows the example of three stable (homeostatic) mental states and their associated energy requirements. A homeostatic mental state is a state with stable brain function and behavior. Shifts between homeostatic mental states range from minutes and hours (e.g., the sleep-wake cycle) to years (development). Improper allostatic shifts induced by factors such as long-term stress result in dysfunctional homeostatic mental state spaces with stable average energy costs but dysfunctional function and behavior, such as in psychiatric illness. From Gu et al. (2018) (http://creativecommons.org/licenses/by/4.0/).

brain region in which they occur, as well as factors such as age, sex, brain reserve (Stern et al. 2020), sleep status, and personal history.

From this perspective, state shifting in mental health conditions can be considered as failed allostatic transfers (Bobba-Alves et al. 2022) which result in dysfunctional behavior. Healthy mental state spaces require proper metabolic energy and homeostatic processes to maintain.

We propose that failure in the mechanisms that support energy production and homeostasis can lead to allostatic shifts in mental states, associated with short- and long-term loss of mental health. These shifts can occur during critical periods of development (see Chapter 3, Figure 3.5) due to elevated functional energy needs approaching the maximum capacity of fuel delivery and maintenance of homeostasis; this results in insufficient energy to support developmental processes. Shifts can also occur when stress causes excessive high brain activity, leading to loss of synaptic neurotransmitter and ion homeostasis. The resultant adaptations in synaptic organization, neuronal firing patterns, and metabolism are neuroprotective (e.g., reduction in synaptic density and glutamate signaling in the hippocampus during chronic stress), but they lead to allostatic shifts in mental states. These alterations in the energy budget as well as failure of underlying metabolic homeostatic mechanisms, described in Chapter 3, can lead to harmful allostatic shifts in mental states.

2 Impaired Circuit Functions During Vulnerable Developmental Periods

This internal competition for a limited supply resource has manifold implications for the disruption of critical brain circuits during vulnerable periods in response to high allostatic loads. One potential consequence is that increased activity may need to be focused on a specific functional network, which may limit activity in other networks at any moment in time. In this context, any dysregulation that alters excitability in one specific structure might affect other brain functions through the change in the availability of blood flow to these other regions of the brain. Thus, even specific genetic alterations of excitatory-inhibitory balance in restricted circuits are expected to have brain-wide repercussions that could manifest as neuropsychiatric disorders (Sohal and Rubenstein 2019).

A similar competition for limited resources is predicted to occur in development. In genetically and/or environmentally vulnerable individuals, such as children exposed to maltreatment (Nakama et al. 2023; Schaefer et al. 2022), altered brain organization may impact macroscopic (e.g., regional cortical thickness) and microscopic (e.g., cortical neuronal density and distribution) levels. Studies have found that both excitatory and inhibitory neurons have dendritic arbors of sizes that correlate inversely with their excitability. Those that fire at higher rates, primarily GABAergic neurons, have smaller dendritic arbors, which affects the range of synaptic integration and therefore the function of the circuits to which they belong (Xu and Herculano-Houzel, in preparation). Such a trade-off between neuronal excitability (and thus rate of energy consumption per cell volume) and dendritic arbor size is a predictor of neuronal development in a supply-limited energetic environment. It also highlights the potential for developmental alterations in neuronal circuitry that may underlie neuropsychiatric disorders.

3 Impact of Increased Allostatic Loads on Synaptic and Metabolic Homeostasis

As described in Chapter 3, the large majority of the brain energy budget supports the vital signaling activities of glutamatergic and GABAergic neurons and associated large fluxes of potassium, sodium, and calcium release and reuptake during neuronal depolarization and repolarization. Because of limitations on brain energy capacity, under conditions of excessive allostatic load, the energy available to support these high energy-consuming processes may be overwhelmed, leading to loss of cellular and interstitial fluid homeostasis of the levels of these transmitters and ions. This loss of homeostasis may be exacerbated by extended periods, when insufficient energy is available to support GMR processes required to maintain homeostatic synaptic processes. It has also been shown that neurometabolic coupling and its associated lactate and proton production is a consequence of supply limitations on the generation of oxidative adenosine triphosphate (ATP). A diversion of oxygen and glucose supply to support allostatic loads would, in principle, further increase non-oxidative glycolytic lactate production and decrease pH. In this section, we describe several clinical findings and preclinical studies that exemplify the failure of homeostatic processes, due to supply limitations, to meet allostatic loads. In addition, we summarize studies that show how increased allostatic load directly increases the brain energy load as well as impairs specific metabolic processes.

3.1 Loss of Neurotransmitter Homeostasis: Animal Models of Unpredictable Chronic Stress

To illustrate loss of neurotransmitter homeostasis due to high allostatic loads, let us consider studies on rodent models of unpredictable stress. Due to the intense increase in glutamate neurotransmission, astroglial uptake of glutamate and conversion to glutamine is overwhelmed, leading to regional increases in interstitial glutamate to pathological levels (Popoli et al. 2012). After a week, however, interstitial glutamate levels return to normal values in the most affected hippocampus and prefrontal cortex regions.

This adaptation is associated with a decrease in synaptic density. Although this adaptation may be part of a normal protective response, from the standpoint of a limited energy budget model, it may also be due to the initial increase in functional activity due to stress displacing the energy required for the GMR functions (see Chapter 3) needed for proper synaptic turnover. Consistent with this possibility, there is an allostatic shift after the adaptation to a mental state with behavioral traits considered analogous to human unipolar depression (Willner 2017). Evidence for a similar maladaptive response in humans are findings of reduced glucose oxidation and glutamate/glutamine cycling in human depression (Murrough et al. 2017) and post-traumatic stress disorders (Abdallah et al. 2018; Averill et al. 2022), in addition to MRI evidence of morphological changes (e.g., reduced hippocampal volume) associated with synaptic loss (O'Doherty et al. 2015).

3.2 Ketamine: Reversing Dysfunctional Allostatic Shifts

The ability of chronic stress to shift the brain energy budget to a state with lower functional activity and lower amounts of energy used for synaptic turnover implies that other interventions may be able to shift the energy budget back to a more functional state. An example of such a challenge is the use of ketamine to treat unipolar depression. Studies using the drug ketamine have found that infusion over approximately an hour in rodents adapted to chronic unpredictable stress led to a rapid increase of approximately 25–30% in the fluxes of ^{13}C from glucose that lasted for the duration of the infusion. After approximately one week there was a restoration of synaptic density and normalization of glucose metabolism and neurotransmitter cycling fluxes, consistent with an increase in vital processes in the energy budget (see Chapter 3; Mishra et al. 2018). In both healthy and depressed humans, a similar percentage increase in the glutamate/glutamine cycle and neuronal TCA cycle has been measured after ketamine administration (Abdallah et al. 2018).

The ability to cycle back and forth between dysfunctional and functional states can become difficult over time. In rats exposed to repeated stress, the administration of ketamine eventually failed to return the rats to the control condition. This also occurs in people exposed to repeated stress episodes; regression after successful therapy often takes longer to recover from and may indeed prove intractable. An approach that specifically targets the components of the energy budget may be more effective in inducing lasting healthy homeostatic mental states (Voineskos et al. 2020).

3.3 Disruptions in Glutamate and GABA Homeostasis and Neurotransmitter Cycling in Schizophrenia and Other Disorders

Proper synaptic function requires homeostasis of interstitial glutamate and GABA concentrations. In healthy brain, there is a stoichiometric relationship between the glutamate/glutamine cycle (which accounts for most neuronal glutamate release) and glucose metabolism (Sibson et al. 1998), and a similar relationship between GABAergic neuron glucose metabolism and the GABA/glutamine cycle (Duarte and Gruetter 2013; Patel et al. 2005). Thus, metabolism of glutamate and GABA is critical for supporting their roles as the key excitatory and inhibitory neurotransmitters. Furthermore, the high neurotransmitter cycle fluxes, and associated glucose metabolism, effectively couple the cellular and neurotransmitter pools (see Dienel et al. 2023).

As expected from this tight relationship, magnetic resonance spectroscopy (MRS) studies of a variety of psychiatric and neurological diseases, as well as different behavioral states in healthy subjects, have found regional shifts in bulk tissue glutamate and GABA concentrations (Merritt et al. 2023). In disease states, these shifts may reflect a dysfunctional loss of metabolic homeostasis in the processes regulating GABA and glutamate levels needed for neuromodulation and vesicular neurotransmission and may contribute to disease progression.

In the context of schizophrenia, findings from a recent meta- and mega-analysis (Merritt et al. 2023) suggest that compared to controls, individuals with schizophrenia show greater variability in both glutamate and glutamine, as measured by 1H-MRS from voxels in the medial frontal cortex, dorsolateral prefrontal cortex, striatum, and thalamus, depending on age and symptomatic burden. At the individual level, meta-analysis suggests lower glutamate levels in the medial frontal cortex but higher levels in the basal ganglia and thalamus relative to controls. Higher hippocampal glutamate levels have been reported, especially early in the disease and in subjects at high risk of developing schizophrenia (Basavaraju et al. 2022; Guo et al. 2023). These results differed to some extent by sex (Merritt et al. 2023). Evidence also suggests that anterior cingulate levels of the combined resonance of glutamate and glutamine (Glx) are predictive of antipsychotic response (Egerton et al. 2023).

In the adult cingulate cortex and hippocampus, glutamate levels (as measured by 1H-MRS) are positively correlated with an *in vivo* PET measure of synaptic density ([11C]-UCB-J) (Onwordi et al. 2021). These data suggest that synaptic density is related to the MRS-based glutamatergic signal in the healthy brain. Importantly, this relationship is not observed in patients with schizophrenia (Onwordi et al. 2021).

3.4 Elevated Lactate and Decreased pH in Schizophrenia

MRS studies have consistently reported elevated lactate and decreased pH in schizophrenia (Dogan et al. 2018; Liu et al. 2024). The source of the elevated lactate and decreased pH could be a combination of allostatic load reducing the oxygen available to support vital signaling and GMR processes, a failure of lactate and associated proton clearance, or both. The importance of maintaining homeostatic brain lactate levels is highlighted by studies in which blood and, consequently, brain lactate levels were elevated by infusion of lactate. In rats, Ido et al. (2001) reported that injecting lactate potentiated somatosensory cerebral blood flow (CBF) in response to whisker stimulation, whereas CBF was attenuated when pyruvate was injected. Subsequently, Mintun et al. (2004) and Vlassenko et al. (2006) performed parallel experiments in healthy humans using 15O-H2O PET to measure CBF. Injection of lactate resulting in a peak blood concentration of approximately 10 mM did not affect resting CBF but significantly augmented the visual stimulus-induced increase in local CBF by 38–53% (Mintun et al. 2004). In contrast, injection of pyruvate attenuated the stimulus-induced CBF response by approximately 19% (Vlassenko et al. 2006). These results are consistent with a model for neurovascular coupling, where stimulus-induced changes in local CBF are partly related to maintaining tissue homeostasis, including that arising from non-oxidative glycolysis (DiNuzzo et al. 2024). These results suggest that changes in how the brain produces, responds to, and disposes of lactate might directly affect the metabolic and blood flow response to stimuli and tasks, and thereby brain state, function, and behavior. Further, consistent with the importance of lactate homeostasis, intravenous infusion of lactate can elicit panic attack symptoms, though this may be limited to predisposed individuals (van Gemert et al. 2022).

3.5 Shifts in Energy Budget Assignment: Allostatic Shifts in Bipolar Disorder

Although the drivers for state shift may be different, a similar sort of relative change in energy assignment could be taking place in bipolar disorder: the manic phase could be viewed as excessive functional activity. Mania represents a condition of heightened cerebral energy metabolism facilitated by hyperglycolysis and glutaminolysis (Ketter et al. 2001; Kishimoto et al. 1987; Öngür et al. 2008). When oxidative glucose metabolism becomes impaired in the brain, neurons can utilize glutamate as an alternative substrate to generate energy through oxidative phosphorylation. In this manner, it has been hypothesized that the upregulation of glycolysis and glutaminolysis causes the brain to enter a state of heightened metabolism and excitatory activity, which may underlie the subjective experience of mania (Campbell and Campbell 2024). This would result in a deficit in dispensable maintenance activity which, over time, would cause functional activity to fail and lead to a period of depression during which potential recovery could be initiated through system repair, shifting the energy budget from functional activity to undertake the by now necessary maintenance activity.

3.6 Metabolic Alterations Due to Chronic Stress: Dysfunctional Allostatic Shifts in States

As described above, chronic stress with associated increased levels of cortisol can result in perturbations in brain energy metabolism (van der Kooij 2020), with relationships between plasma cortisol levels, blood flow, and glucose metabolism reported in various mental health conditions (Åhs et al. 2006; Bouhuys et al. 2006; Drevets et al. 2002). Chronic stress alters expression of glucose transporters (GLUT1) in an age- and sex-dependent manner (Kelly et al. 2014). Particularly strong effects occur in the vulnerable adolescence period.

Cortisol, the main stress-responsive glucocorticoid hormone, is intimately involved in regulating energy metabolism in the body and brain (Piroli et al. 2007). Some of the central effects of glucocorticoids on energy metabolism may be mediated by their effect on insulin. For example, repeated treatments with corticosterone (for seven days) resulted in impaired hippocampal signaling mediated by the insulin receptor, a decreased level of glucose transporter type 4 (GLUT4), and a decreased translocation of GLUT4 to the plasma membrane in response to glucose administration (Piroli et al. 2007). Within an hour of acute cortisol administration in humans, a decreased glucose uptake was found in the hippocampus but not in other brain regions (De Leon et al. 1997). In the longer term (4–8 hours), glucocorticoids produce a consistent inhibition of neuronal and astrocytic glucose uptake (Homer et al. 1990; Virgin Jr et al. 1991) and increase the expression of pyruvate dehydrogenase kinase (Juszczak and Stankiewicz 2018)—a key enzyme that inhibits the conversion of pyruvate to acetyl-CoA—thereby blocking glucose oxidation but increasing the pyruvate-to-lactate conversion. This likely shifts metabolism toward aerobic glycolysis. Notably, acute neuropsychiatric symptoms, including but not limited to psychosis, are also a well-established, typically reversible side effect of high dose glucocorticoid administration (Dubovsky et al. 2012).

In contrast, chronic exposure to cortisol and other glucocorticoids has consistently produced increased brain glucose uptake and content, either through the peripheral effects of glucocorticoids triggering rise in blood glucose or increasing the efficiency of glucose transport across the blood-brain barrier (Jaszczyk and Juszczak 2021). Similarly, glucocorticoids increase blood level of a number of bioenergetically relevant metabolites, such as 3-hydroxybutyrate (ketone bodies), mannose, pyruvate, and lactate, that can be used by the brain to produce energy (Jaszczyk and Juszczak 2021). Further links between stress, cortisol, and brain bioenergetics are highlighted by the fact that mitochondria mediate the stress response partially by sensing the levels of glucocorticoids (Manoli et al. 2007). Glucocorticoids, through their genomic effects, induce profound changes in mitochondrial physiology. They bind to potential mitochondrial glucocorticoid-response elements localized within the genes that encode the catalytic subunits of cytochrome c, the terminal oxidase of the mitochondrial electron transport chain (Demonacos et al. 1996). They produce adaptive changes, including increased mitochondrial biogenesis and ATP production, in the short term but maladaptive, deleterious effects (e.g., decreased mitochondrial biogenesis and ATP production, as well as increased production of reactive oxygen species, disturbances in OXPHOS, fusion/fission dysregulation, and increased apoptosis) from long-term stress or chronic glucocorticoid administration (reviewed by Głombik et al. 2021). These findings support the allostatic concept of stress and link chronic stress to bioenergetic dysregulations and psychiatric disorders through the mitochondria (Daniels et al. 2020; Eisner et al. 2018; Picard et al. 2018).

3.7 Estrogen: Protection against Stress

Estrogen exerts a protective effect on the brain against the impact of stress of different kinds (Albert and Newhouse 2019). These effects, including increasing hippocampal dendritic spine density, dendritic arborization, and adult neurogenesis, are opposite to that of cortisol, but we have no evidence showing that estrogen has a direct antagonistic/inhibitory interaction with cortisol. However, while investigating the effects of estrogen replacement therapy on cortisol response to a physical stressor (cold pressor test) in postmenopausal women to see whether estrogen attenuates stress effects on working memory, Herrera et al. (2017) found that women assigned to estradiol exhibited blunted cortisol responses to the cold pressor test compared with placebo and lesser negative effects of stress on working memory. This suggests that estrogen may protect certain types of cognition in the presence of stress.

Estrogens regulate important pathways implicated in schizophrenia, such as dopamine activity, mitochondrial function, and the stress response system (Brand et al. 2021). Estrogen deficiency is common in both women and men with schizophrenia and is associated with increased psychotic symptoms. Women with schizophrenia experience worsening of psychotic symptoms during low estrogen phases of the menstrual cycle, suggesting that estrogen exerts a protective effect against the development and severity of schizophrenia (Gogos et al. 2015; Mu et al. 2024; Riecher-Rössler and Kulkarni 2011). Premenopausal women require lower doses of antipsychotics compared to men, as estrogens raise the availability and efficacy of these medications. Conversely, postmenopausal women may require higher doses due to declining estrogen levels (Brand et al. 2021; Mu et al. 2024).

4 Cardiovascular Impairment in Chronic High Allostatic Load May Impact Brain Function

In addition to permeability of the blood-brain barrier, proper energy supply and clearance of by-products from the brain depend on the ability of the cardiovascular system to deliver adequate blood flow and the pulmonary system to provide sufficient oxygenation and balance blood pH, pCO_2, and pO_2 (DiNuzzo et al., 2024). Furthermore, synaptic and metabolic homeostasis depends on the liver and kidney working in coordination to maintain constant blood levels of ions such as K^+ and Na^+, ammonia, and critical amino acids. Consequently, chronic high allostatic loads that lead to chronic diseases, which impact systemic blood homeostasis and vascular delivery, may also lead to impaired brain metabolic and synaptic homeostasis, resulting in allostatic shifts to dysfunctional mental states. We provide several examples below.

4.1 Cardiovascular Disease and Schizophrenia

In the context of brain energy budget being tightly linked to blood flow and supply, it is worth noting that cardiovascular disease is prevalent and a major cause of early mortality in schizophrenia. There is evidence of cardiac fibrosis and/or inflammation using cardiac MRI in medicated patients with schizophrenia compared with matched healthy controls, but separating illness from medication-related effects remains challenging (Pillinger et al. 2019; Andreou et al. 2020). Importantly, insulin resistance is seen both before the advent of antipsychotic medications and at onset of psychosis (see Chapter 11). Supporting an illness effect, high polygenic risk scores for schizophrenia are associated with cardiac structural changes that are linked to worse cardiac outcomes (Pillinger et al. 2023). Further work is required to determine whether these associations are specific to schizophrenia or are also present in other psychiatric conditions.

4.2 Loss of Blood CO_2 and Lactate Homeostasis

Donald F. Klein, one of the pioneers of neuropsychopharmacology, proposed the false suffocation alarm theory to explain the underlying mechanism of panic attacks (Klein 1993). According to this theory, panic attacks result from the misfiring of an evolved suffocation alarm system triggered by signals of impending suffocation, such as increased carbon dioxide (CO_2) levels or respiratory obstruction. Klein suggested that individuals with panic disorder have a hypersensitive suffocation alarm system that is easily triggered by minor respiratory changes, leading to the sudden onset of panic symptoms (Klein 1994). This theory is supported by the observation that panic disorder patients are more sensitive to the panic-inducing effects of CO_2 inhalation compared to healthy individuals (Gorman et al. 1994). Furthermore, the false suffocation alarm theory provides a framework for understanding the effectiveness of certain treatments for panic disorder, such as breathing retraining and exposure to CO_2 (Klein 1993; Meuret and Ritz 2010).

Brain and blood lactate levels are also abnormal in other mental health disorders such as schizophrenia (Liu et al. 2024; Pruett and Meador-Woodruff 2020). In an Exercise II study (Keller-Varady et al. 2016), blood lactate levels were measured in patients with multi-episode schizophrenia and control subjects from the start of an exercise intervention and then continued over three months of training. The study showed that lactate levels increased over time in healthy controls, but patients with schizophrenia never reached the same level even if they seemed to profit from the intervention. This finding suggests a defect in the regulation of lactate metabolism and clearance, which as described above may result in dysfunctional shifts in cellular redox and pH levels. These alterations may contribute to allostatic mental state shifts due to the brain not being able to support metabolically the neuronal signaling required to support healthy homeostatic states.

4.3 Loss of Homeostasis of Blood Ammonia and Ion Levels

Hyperammonemia, which itself has multiple causes, generates increased glutamine production, increased anaplerosis, and increased glutamate release into the extracellular fluid (Cooper and Jeitner 2016). People who suffer from hyperammonemia commonly present with psychotic symptoms (Llansola et al. 2013) and encephalopathy, particularly in the context of liver failure. Similarly, hypokalemia impacts proper maintenance of K^+ balance during depolarization. Repolarization leads to psychotic episodes that can be mistaken for schizophrenia (Lam et al. 2009).

5 Mitochondrial Health in Energy Budget Assignment and Local Metabolic Homeostasis

Oxidative metabolism is the main source of ATP, particularly for neurons. Mitochondrial health is therefore important for restoring brain homeostasis as well as for providing energy to power the dispensable brain activity (allostasis and GMR) that maintains brain functionality. In addition, through apoptosis and other mechanisms, mitochondria play a key role in diverse processes, including synaptic biosynthesis and turnover beyond their metabolic functions. Here, we discuss the unique aspects of mitochondria in leading to functional and dysfunctional allostatic transfers.

5.1 Mitochondrial Diversity in the Brain

Just as there are different types of cells populating the brain and other organs, there are different types of mitochondria (for a review of mitochondrial phenotypes, or mitotypes, see Monzel et al. 2023). Across the brain (mouse and human), mitochondria specialize in their molecular composition, morphology, functions, and behaviors. Mitochondrial diversity is driven both by regional differences—some regions have more mitochondria of a certain type, regardless of the cell type—and by cell type, where some cell types contain vastly different mitochondrial phenotypes (Mosharov et al. 2025). In the mouse brain, mitochondrial energy transformation capacity in a specific corticostriatal network (group of brain regions), but not in other brain regions, correlates strongly with anxiety-related behaviors and social interaction (Rosenberg et al. 2023). Manipulating nucleus accumbens mitochondrial respiratory capacity in rats also modulates social behaviors and anxiety (Hollis et al. 2015). Thus, both the functional diversity of mitochondria and the regional differences/recalibrations in mitochondria could contribute to sustaining or altering brain functions, and perhaps ultimately to abnormal behaviors and symptoms.

The most consistent findings in schizophrenia relative to controls are reductions in expression of mitochondrial complex I subunits in postmortem brain tissue (Cohen's $d > 0.8$), reductions in complex I activity in postmortem brains ($d > 0.7$), and reductions in neural glucose utilization ($d = 0.3$ to 0.6) (Whitehurst and Howes 2022). Overall these findings are consistent with complex I dysfunction in schizophrenia and other neuropsychiatric and neurodegenerative disorders (Holper et al. 2019). They also highlight the need for well-designed *in vivo* studies to determine the link between oxidative phosphorylation dysfunction and symptoms in patients and to identify the variables associated with medication treatment or adverse effects (Bar-Yosef et al. 2020; Panizzutti et al. 2024; Turkheimer et al. 2019).

Postmortem brain samples of individuals with schizophrenia show dysregulated expression levels of genes encoding enzyme complexes comprising the mitochondrial electron transport chain, including ATP synthase. Shroitman et al. (2023) conducted a systematic meta-analysis of the expression of 16 ATP synthase-encoding genes in postmortem brain samples of individuals with schizophrenia versus healthy controls in three regions: Brodmann area 10 (BA10), BA22/superior temporal gyrus, and the cerebellum. Eight independent datasets were integrated (overall 294 brain samples: 145 of individuals with schizophrenia and 149 age-matched controls). The meta-analysis was applied to all individuals, with separate analysis of men and women. A significant downregulation of two ATP synthase-encoding genes was detected in schizophrenia, ATP5A1 and ATP5H, and a trend toward downregulation of five further ATP synthase genes. The downregulation tendency was shown for both men and women with schizophrenia (Shroitman et al. 2023).

5.2 Dopaminergic Transmission and Energy Budget Interactions

The importance of energy metabolism and supply in maintaining healthy brain states and in the etiology of psychiatric disease was long neglected due to the view, based on early success in treating the active symptoms of schizophrenia, that the primary cause of the disease is altered dopamine transmission (Meltzer and Stahl 1976). Studies conducted at a receptor, genetic, and metabolic level revealed the importance of glutamatergic function (Javitt 1987) in the etiology of the disease and led to a broader view of the potential causes of and treatments for schizophrenia and other psychiatric disorders (McCutcheon et al. 2020). Here, we go further and explore the importance of direct interactions between dopamine signaling and the brain energy budget at both a general assignment level and for specific processes.

Research has indicated that dopamine plays a significant role in regulating mitochondrial movement, thereby affecting the distribution of energy resources within neurons, a process crucial for maintaining neuronal health and function (Chen et al. 2008). Additionally, dopamine D2 receptors have been specifically associated with modulating glucose metabolism (Chen et al. 2024), further linking dopamine activity to broader metabolic processes (Palacios and Wiederhold 1985). The relationship is bidirectional: dopaminergic neurons in the substantia nigra are highly dependent on intact mitochondrial activity, as demonstrated in animal and human studies of toxin-induced and genetic forms of Parkinson disease (Morris et al. 2024; Ni and Ernst 2022; Schapira 2008). These interactions underscore the complex relationship between dopamine signaling, energy metabolism, and the pathophysiology of severe psychiatric conditions, setting the stage for understanding similar mechanisms in disorders such as anorexia nervosa.

Anorexia nervosa is an eating disorder characterized by ongoing self-starvation. Evidence from human and animal studies has yet to clarify whether putative dopamine abnormalities precede the disorder as a risk factor or emerge as a result of starvation (Barry and Klawans 1976; Branch et al. 2013; Cowdrey et al. 2011; Gelegen et al. 2008; Kaye et al. 1999). The disorder can be hypothesized to progress through two stages, each distinguished by specific changes in dopamine levels that underlie each phase (Beeler and Burghardt 2022). Initially, caloric restriction combined with physical exercise leads to increased midbrain dopamine, which enhances behavioral flexibility and reinforces weight-loss behaviors (Bergh and Södersten 1996; O'Hara et al. 2015). This increase is likely driven by stress-induced activation of the HPA axis, heightened insulin sensitivity, decreased leptin levels, changes in ghrelin, and increased orexin (Chen et al. 2019; Cone et al. 2015; Douma and de Kloet 2020;

Fernandes et al. 2015). As the disorder advances, continuous self-starvation reduces dopamine levels, promoting behavioral rigidity and solidifying entrenched behaviors as the brain adapts (Ambwani et al. 2020). In this hypodopaminergic state, dopamine receptors may upregulate, increasing their expression and sensitivity (Kostrzewa et al. 2008). Understanding these stages as mediated by distinct neuroadaptations suggests that pharmacological treatments could vary by stage.

This proposed sequence of the disorder—an initial increase followed by a decrease in dopamine—may represent a common pathway for various behavioral disorders characterized by shifts between reinforcement learning and behavioral rigidity. If validated, this hypothesis could significantly impact the clinical approach to treating and researching anorexia nervosa and other related disorders such as addiction and obesity, indicating that treatment strategies might need to be stage specific.

6 Four Pillars of Mental Health for a Homeostasis-Limited Energy Budget

Sleep, physical activity, diet, and social interactions (Gorelick et al. 2017; Mintzer et al. 2019) support brain energy metabolism. Sleep, a necessary requirement in most animals, underpins brain function (Palagini et al. 2022; Scott et al. 2021). Disturbed sleep shows bioenergetic deficits (Miller et al. 2017; Rae et al. 2009; Vakulin et al. 2021), whereas exercise and an enriched environment improve blood flow and promote synaptic formation through release of factors such as brain-derived neurotrophic factor (Gómez-Pinilla et al. 2002) and vascular endothelial growth factor (Cao et al. 2004). Social interaction is particularly important at vulnerable periods of development, such as early life (Heidbreder et al. 2000) and adolescence (Orben et al. 2020), with loneliness in the general population impacting the brain (Cacioppo et al. 2014) and contributing significantly to mental disorders (Beutel et al. 2017).

Since 1975, the global prevalence of obesity has more than tripled in males and more than doubled in females. Consistent with the global trend, the prevalence of obesity has increased rapidly in China in recent decades (Phelps et al. 2024), doubling between 2004 (3.1%) and 2018 (8.1%) (Pan et al. 2021; Wang et al. 2021). Obesity is associated with higher risks of hypertension, type 2 diabetes, coronary heart disease, stroke, and certain cancers (Chen et al. 2018; Mi et al. 2020; Momin et al. 2020; Pang et al. 2018). Based on the estimated 2018 prevalence, obesity affects approximately 85 million people in China. The growing societal burden of overweight and obese people may be driven by economic developments, sociocultural norms, and policies that have influenced individual-level risk factors through urbanization, urban planning, and the built environment, as well as food systems and environments (Du et al. 2014). In China, substantial changes in dietary patterns have occurred, marked by increased consumption of animal-source foods, refined grains, and highly processed, high-sugar, and high-fat foods. Meanwhile, physical activity levels in all major domains have decreased due to rising sedentary behaviors. These dietary and physical inactivity effects intersect with other individual-level risk factors, such as genetic susceptibility, psychosocial factors, obesogens, as well as *in utero* and early-life exposures.

Unfavorable environmental conditions during pregnancy have been shown to promote the onset of mental disorders in the offspring via epigenetic mechanisms (Babenko et al. 2015; Brannigan et al. 2020; Kleinhaus et al. 2013; Linnér and Almgren 2020). Undernutrition in pregnant women is one such condition that can affect intrauterine development and has been shown to have long-lasting detrimental consequences for the mental health of the offspring later in life (Roseboom 2019). The effect of prenatal undernutrition on mental health can be analyzed through retrospective analysis of events, such as the Dutch famine during World War II and the Chinese famine (1959–1961), where undernutrition occurred historically in specific populations (Bleker et al. 2021; Li and Sunder 2021). Meta-analytic results have demonstrated an increased risk of psychotic, affective, and personality disorders in adults exposed to famine during prenatal development (Dana et al. 2019).

The psychosocial environment (e.g., low socioeconomic status or repeated exposure to traumatic events) contributes substantially in important ways to the prevalence and onset of psychopathology and associated macroscale brain change disorders in youth (Du et al. 2014). Understanding how such individual experiences shape brain metabolism could help us develop personalized treatments (e.g., through environmental enrichment). For example, it is possible to determine individual behavioral differences in mice that have been exposed to a large, enriched environment for three months. Specifically, the activity of the mice followed a continuum from continuous high activity to habituation. Using structural MRI, a striking correlation was observed between regional brain volume and the activity of the mice. Furthermore, using network-based statistics, distinct subnetworks of murine structural covariance, which underlie these differences in behavioral activity, were identified. Together, these results reveal that differentiated behavioral trajectories of mice in an enriched environment are associated with differences in brain volume and covariance, which could offer a useful system to investigate individualized responses to enrichment from a brain metabolism perspective (Bogado Lopes et al. 2023). In humans, an MRI analysis of 5,306 participants across 15 countries, designed to tease apart the most influential drivers of brain aging, identified structural socioeconomic inequality, pollution, and health disparities as the most influential predictors of brain aging (e.g., atrophy) (Moguilner et al. 2024).

7 Where Do We Go from Here?

Measurement of energy metabolism and blood flow in neuropsychiatric disorders, particularly in a developmental context, is key to helping us understand the different bioenergetic states and the process(es) that lead to particular state shifts. It would be informative to undertake these measurements in conditions that are not "steady-state," under conditions where stressors are applied or tasks are performed. These measurements in different states could be performed in healthy people—where the stressor is known and can be controlled using, for example, environmental stressors (sleep deprivation, drug administration, hypercarbia) to study controlled allostatic shifts—and compared to shifts in people who are in altered state conditions. During the stages of aging and adolescence, brain energy metabolism comes under particular stress. Adolescence is a particularly critical period for stressors to introduce allostatic alterations that can have long-term mental health consequences if not identified and rectified (Alloy et al. 2006; Dahl 2004; Dayan et al. 2010; O'Connor and Cryan 2014). Conducting studies during these stages would allow us to directly test our hypothesis that vulnerability is associated with a reduced margin between maximum energy supply and enhanced needs.

We need to better understand how state shifts arise at the neurochemical level, how to rectify them when they occur, and when repeated failed or exaggerated attempts to shift states can lead to them becoming intractable. A better understanding of basic brain energy metabolism (Rae et al. 2024) that underlies brain state shifts at rest and under stress in human cohorts will be particularly relevant. Moreover, developing methods to measure the brain's total metabolic capacity within individuals might help to identify those who are more or less vulnerable to metabolic stress and, thereby, mental and cognitive illnesses. By leveraging ongoing developments in our ability to measure various features of brain energy metabolism (e.g., oxygen and glucose use, ATP, lactate, brain pH, NADH/NAD+, glutamate/GABA/glutamine cycling) via advances in neuroimaging (e.g., MRI/MRS, PET, and optical techniques) and *ex vivo*/*in situ* measurements (e.g., mitochondrial assays and multi-omic fluid analysis), we can now address these questions in both humans and animal models.

Acknowledgments The authors gratefully acknowledge the contributions of Peter Falkai and Lily Mujica-Parodi.

References

Abdallah CG, De Feyter HM, Averill LA, et al. (2018) The Effects of Ketamine on Prefrontal Glutamate Neurotransmission in Healthy and Depressed Subjects. Neuropsychopharmacology 43 (10):2154–2160. https://doi.org/10.1038/s41386-018-0136-3

Åhs F, Furmark T, Michelgård Å, et al. (2006) Hypothalamic Blood Flow Correlates Positively with Stress-Induced Cortisol Levels in Subjects with Social Anxiety Disorder. Psychosom Med 68 (6):859–862. https://doi.org/10.1097/01.psy.0000242120.91030.d8

Albert KM, and Newhouse PA (2019) Estrogen, Stress, and Depression: Cognitive and Biological Interactions. Annu Rev Clin Psychol 15:399–423. https://doi.org/10.1146/annurev-clinpsy-050718-095557

Alloy LB, Abramson LY, Walshaw PD, Keyser J, and Gerstein RK (2006) A Cognitive Vulnerability–Stress Perspective on Bipolar Spectrum Disorders in a Normative Adolescent Brain, Cognitive, and Emotional Development Context. Dev Psychopathol 18 (4):1055–1103. https://doi.org/10.1017/S0954579406060524

Ambwani S, Cardi V, Albano G, et al. (2020) A Multicenter Audit of Outpatient Care for Adult Anorexia Nervosa: Symptom Trajectory, Service Use, and Evidence in Support of "Early Stage" versus "Severe and Enduring" Classification. Int J Eat Disord 53 (8):1337–1348. https://doi.org/10.1002/eat.23246

Andreou D, Saetre P, Fors BM, et al. (2020) Cardiac Left Ventricular Ejection Fraction in Men and Women with Schizophrenia on Long-Term Antipsychotic Treatment. Schizophr Res 218:226–232. https://doi.org/10.1016/j.schres.2019.12.042

Averill LA, Jiang L, Purohit P, et al. (2022) Prefrontal Glutamate Neurotransmission in PTSD: A Novel Approach to Estimate Synaptic Strength *in Vivo* in Humans. Chronic Stress 6:24705470221092734. https://doi.org/10.1177/24705470221092734

Babenko O, Kovalchuk I, and Metz GA (2015) Stress-Induced Perinatal and Transgenerational Epigenetic Programming of Brain Development and Mental Health. Neurosci Biobehav Rev 48:70–91. https://doi.org/10.1016/j.neubiorev.2014.11.013

Bar-Yosef T, Hussein W, Yitzhaki O, et al. (2020) Mitochondrial Function Parameters as a Tool for Tailored Drug Treatment of an Individual with Psychosis: A Proof of Concept Study. Sci Rep 10 (1):12258. https://doi.org/10.1038/s41598-020-69207-4

Barry VC, and Klawans H (1976) On the Role of Dopamine in the Pathophysiology of Anorexia Nervosa. J Neural Transm 38:107–122. https://doi.org/10.1007/BF01262969

Basavaraju R, Guo J, Small SA, et al. (2022) Hippocampal Glutamate and Positive Symptom Severity in Clinical High Risk for Psychosis. JAMA Psychiatry 79 (2):178–179. http://dx.doi.org/10.1001/jamapsychiatry.2021.3710

Beeler JA, and Burghardt NS (2022) The Rise and Fall of Dopamine: A Two-Stage Model of the Development and Entrenchment of Anorexia Nervosa. Front Psychiatry 12:799548. https://doi.org/10.3389/fpsyt.2021.799548

Bergh C, and Södersten P (1996) Anorexia Nervosa, Self–Starvation and the Reward of Stress. Nat Med 2 (1):21–22. https://doi.org/10.1038/nm0196-21

Beutel ME, Klein EM, Brähler E, et al. (2017) Loneliness in the General Population: Prevalence, Determinants and Relations to Mental Health. BMC Psychiatry 17 (1):97. https://doi.org/10.1186/s12888-017-1262-x

Bleker LS, de Rooij SR, Painter RC, Ravelli AC, and Roseboom TJ (2021) Cohort Profile: The Dutch Famine Birth Cohort (DFBC): A Prospective Birth Cohort Study in the Netherlands. BMJ Open 11 (3):e042078. https://doi.org/10.1136/bmjopen-2020-042078

Bobba-Alves N, Juster R-P, and Picard M (2022) The Energetic Cost of Allostasis and Allostatic Load. Psychoneuroendocrinology 146:105951. https://doi.org/10.1016/j.psyneuen.2022.105951

Bogado Lopes J, Senko AN, Bahnsen K, et al. (2023) Individual Behavioral Trajectories Shape Whole-Brain Connectivity in Mice. eLife 12:e80379. https://doi.org/10.7554/eLife.80379

Bouhuys AL, Bos EH, Geerts E, van Os TW, and Ormel J (2006) The Association between Levels of Cortisol Secretion and Fear Perception in Patients with Remitted Depression Predicts Recurrence. J Nerv Ment Dis 194 (7):478–484. http://dx.doi.org/10.1097/01.nmd.0000228502.52864.ce

Branch SY, Goertz RB, Sharpe AL, et al. (2013) Food Restriction Increases Glutamate Receptor-Mediated Burst Firing of Dopamine Neurons. J Neurosci 33 (34):13861–13872. https://doi.org/10.1523/JNEUROSCI.5099-12.2013

Brand BA, de Boer JN, and Sommer IE (2021) Estrogens in Schizophrenia: Progress, Current Challenges and Opportunities. Curr Opin Psychiatry 34 (3):228–237. http://dx.doi.org/10.1097/YCO.0000000000000699

Brannigan R, Tanskanen A, Huttunen MO, et al. (2020) The Role of Prenatal Stress as a Pathway to Personality Disorder: Longitudinal Birth Cohort Study. The British Journal of Psychiatry 216 (2):85–89. http://dx.doi.org/10.1192/bjp.2020.50

Cacioppo S, Capitanio JP, and Cacioppo JT (2014) Toward a Neurology of Loneliness. Psychol Bull 140 (6):1464–1504. https://psycnet.apa.org/doi/10.1037/a0037618

Campbell IH, and Campbell H (2024) The Metabolic Overdrive Hypothesis: Hyperglycolysis and Glutaminolysis in Bipolar Mania. Mol Psychiatr 29 (5):1521–1527. https://doi.org/10.1038/s41380-024-02431-w

Cao L, Jiao X, Zuzga DS, et al. (2004) VEGF Links Hippocampal Activity with Neurogenesis, Learning and Memory. Nat Genet 36 (8):827–835. https://doi.org/10.1038/ng1395

Chen B, Zhang Q, Zhong X, et al. (2024) Dopamine Modification of Glycolytic Enzymes Impairs Glycolysis: Possible Implications for Parkinson's Disease. Cell Commun Signal 22 (1):75. https://doi.org/10.1186/s12964-024-01478-0

Chen S, Owens GC, and Edelman DB (2008) Dopamine Inhibits Mitochondrial Motility in Hippocampal Neurons. PLoS One 3 (7):e2804. https://doi.org/10.1371/journal.pone.0002804

Chen W, Li J, Liu J, Wang D, and Hou L (2019) Aerobic Exercise Improves Food Reward Systems in Obese Rats via Insulin Signaling Regulation of Dopamine Levels in the Nucleus Accumbens. ACS Chem Neurosci 10 (6):2801–2808. https://doi.org/10.1021/acschemneuro.9b00022

Chen Z, Iona A, Parish S, et al. (2018) Adiposity and Risk of Ischaemic and Haemorrhagic Stroke in 0·5 Million Chinese Men and Women: A Prospective Cohort Study. Lancet Glob Health 6 (6):e630–e640. https://doi.org/10.1016/S2214-109X(18)30216-X

Cone JJ, Roitman JD, and Roitman MF (2015) Ghrelin Regulates Phasic Dopamine and Nucleus Accumbens Signaling Evoked by Food-Predictive Stimuli. J Neurochem 133 (6):844–856. https://doi.org/10.1111/jnc.13080

Cooper AJ, and Jeitner TM (2016) Central Role of Glutamate Metabolism in the Maintenance of Nitrogen Homeostasis in Normal and Hyperammonemic Brain. Biomolecules 6 (2):16. https://doi.org/10.3390/biom6020016

Cowdrey FA, Park RJ, Harmer CJ, and McCabe C (2011) Increased Neural Processing of Rewarding and Aversive Food Stimuli in Recovered Anorexia Nervosa. Biol Psychiatry 70 (8):736–743. https://doi.org/10.1016/j.biopsych.2011.05.028

Dahl RE (2004) Adolescent Brain Development: A Period of Vulnerabilities and Opportunities. Keynote Address. Ann N Y Acad Sci 1021 (1):1–22. https://doi.org/10.1196/annals.1308.001

Dana K, Finik J, Koenig S, et al. (2019) Prenatal Exposure to Famine and Risk for Development of Psychopathology in Adulthood: A Meta-Analysis. J Psychiatry Psychiatr Disord 3 (5):227–240. http://dx.doi.org/10.26502/jppd.2572-519X0077

Daniels TE, Olsen EM, and Tyrka AR (2020) Stress and Psychiatric Disorders: The Role of Mitochondria. Annu Rev Clin Psychol 16:165–186. https://doi.org/10.1146/annurev-clinpsy-082719-104030

Dayan J, Bernard A, Olliac B, Mailhes A-S, and Kermarrec S (2010) Adolescent Brain Development, Risk-Taking and Vulnerability to Addiction. J Physiol 104 (5):279–286. https://doi.org/10.1016/j.jphysparis.2010.08.007

De Leon M, McRae T, Rusinek H, et al. (1997) Cortisol Reduces Hippocampal Glucose Metabolism in Normal Elderly, but Not in Alzheimer's Disease. The Journal of Clinical Endocrinology & Metabolism 82 (10):3251–3259. https://doi.org/10.1210/jcem.82.10.4305

Demonacos CV, Karayanni N, Hatzoglou E, et al. (1996) Mitochondrial Genes as Sites of Primary Action of Steroid Hormones. Steroids 61 (4):226–232. https://doi.org/10.1016/0039-128X(96)00019-0

Dienel GA, Schousboe A, McKenna MC, and Rothman DL (2023) A Tribute to Leif Hertz: The Historical Context of His Pioneering Studies of the Roles of Astrocytes in Brain Energy Metabolism, Neurotransmission, Cognitive Functions, and Pharmacology Identifies Important, Unresolved Topics for Future Studies. J Neurochem 168 (5):461–495. https://doi.org/10.1111/jnc.15812

DiNuzzo M, Dienel GA, Behar KL, et al. (2024) Neurovascular Coupling Is Optimized to Compensate for the Increase in Proton Production from Nonoxidative Glycolysis and Glycogenolysis during Brain Activation and Maintain Homeostasis of pH, pCO_2, and pO_2. J Neurochem 168 (5):632–662. https://doi.org/10.1111/jnc.15839

Dogan AE, Yuksel C, Du F, Chouinard VA, and Öngür D (2018) Brain Lactate and pH in Schizophrenia and Bipolar Disorder: A Systematic Review of Findings from Magnetic Resonance Studies. Neuropsychopharmacology 43 (8):1681–1690. https://doi.org/10.1038/s41386-018-0041-9

Douma EH, and de Kloet ER (2020) Stress-Induced Plasticity and Functioning of Ventral Tegmental Dopamine Neurons. Neurosci Biobehav Rev 108:48–77. https://doi.org/10.1016/j.neubiorev.2019.10.015

Drevets WC, Price JL, Bardgett ME, et al. (2002) Glucose Metabolism in the Amygdala in Depression: Relationship to Diagnostic Subtype and Plasma Cortisol Levels. Pharmacol Biochem Behav 71 (3):431–447. https://doi.org/10.1016/S0091-3057(01)00687-6

Du S, Wang H, Zhang B, Zhai F, and Popkin BM (2014) China in the Period of Transition from Scarcity and Extensive Undernutrition to Emerging Nutrition-Related Non-Communicable Diseases, 1949–1992. Obes Rev 15 (S1):8–15. https://doi.org/10.1111/obr.12122

Duarte JM, and Gruetter R (2013) Glutamatergic and GABA Ergic Energy Metabolism Measured in the Rat Brain by ^{13}C NMR Spectroscopy at 14.1 T. J Neurochem 126 (5):579–590. https://doi.org/10.1111/jnc.12333

Dubovsky AN, Arvikar S, Stern TA, and Axelrod L (2012) The Neuropsychiatric Complications of Glucocorticoid Use: Steroid Psychosis Revisited. Psychosomatics 53 (2):103–115. https://doi.org/10.1016/j.psym.2011.12.007

Egerton A, Griffiths K, Casetta C, et al. (2023) Anterior Cingulate Glutamate Metabolites as a Predictor of Antipsychotic Response in First Episode Psychosis: Data from the Strata Collaboration. Neuropsychopharmacology 48 (3):567–575. https://doi.org/10.1038/s41386-022-01508-w

Eisner V, Picard M, and Hajnóczky G (2018) Mitochondrial Dynamics in Adaptive and Maladaptive Cellular Stress Responses. Nat Cell Biol 20 (7):755–765. https://doi.org/10.1038/s41556-018-0133-0

Fernandes MFA, Matthys D, Hryhorczuk C, et al. (2015) Leptin Suppresses the Rewarding Effects of Running via STAT3 Signaling in Dopamine Neurons. Cell Metab 22 (4):741–749. https://doi.org/10.1016/j.cmet.2015.08.003

Gelegen C, Van Den Heuvel J, Collier D, et al. (2008) Dopaminergic and Brain-Derived Neurotrophic Factor Signalling in Inbred Mice Exposed to a Restricted Feeding Schedule. Genes Brain Behav 7 (5):552–559. https://doi.org/10.1111/j.1601-183X.2008.00394.x

Głombik K, Detka J, and Budziszewska B (2021) Hormonal Regulation of Oxidative Phosphorylation in the Brain in Health and Disease. Cells 10 (11):2937. https://doi.org/10.3390/cells10112937

Gogos A, Sbisa AM, Sun J, et al. (2015) A Role for Estrogen in Schizophrenia: Clinical and Preclinical Findings. Int J Endocrinol 2015:615356. https://doi.org/10.1155/2015/615356

Gómez-Pinilla F, Ying Z, Roy RR, Molteni R, and Edgerton VR (2002) Voluntary Exercise Induces a BDNF-Mediated Mechanism That Promotes Neuroplasticity. J Neurophysiol 88 (5):2187–2195. https://doi.org/10.1152/jn.00152.2002

Gorelick PB, Furie KL, Iadecola C, et al. (2017) Defining Optimal Brain Health in Adults: A Presidential Advisory from the American Heart Association/American Stroke Association. Stroke 48 (10):e284–e303. https://doi.org/10.1161/STR.0000000000000148

Gorman JM, Papp LA, Coplan JD, et al. (1994) Anxiogenic Effects of CO_2 and Hyperventilation in Patients with Panic Disorder. Am J Psychiatry 151 (4):547–553. https://doi.org/10.1176/ajp.151.4.547

Gu S, Cieslak M, Baird B, et al. (2018) The Energy Landscape of Neurophysiological Activity Implicit in Brain Network Structure. Sci Rep 8 (1):2507. https://doi.org/10.1038/s41598-018-20123-8

Guo J, Rothman DL, and Small SA (2023) Why Hippocampal Glutamate Levels Are Elevated in Schizophrenia. JAMA Psychiatry 80 (3):274–275. http://dx.doi.org/10.1001/jamapsychiatry.2022.3849

Heidbreder C, Weiss I, Domeney A, et al. (2000) Behavioral, Neurochemical and Endocrinological Characterization of the Early Social Isolation Syndrome. Neuroscience 100 (4):749–768. https://doi.org/10.1016/S0306-4522(00)00336-5

Herrera AY, Hodis HN, Mack WJ, and Mather M (2017) Estradiol Therapy after Menopause Mitigates Effects of Stress on Cortisol and Working Memory. The Journal of Clinical Endocrinology & Metabolism 102 (12):4457–4466. https://doi.org/10.1210/jc.2017-00825

Hollis F, van der Kooij MA, Zanoletti O, et al. (2015) Mitochondrial Function in the Brain Links Anxiety with Social Subordination. PNAS 112 (50):15486–15491. https://doi.org/10.1073/pnas.1512653112
Holper L, Ben-Shachar D, and Mann JJ (2019) Multivariate Meta-Analyses of Mitochondrial Complex I and IV in Major Depressive Disorder, Bipolar Disorder, Schizophrenia, Alzheimer Disease, and Parkinson Disease. Neuropsychopharmacology 44 (5):837–849. https://doi.org/10.1038/s41386-018-0090-0
Homer HC, Packan DR, and Sapolsky RM (1990) Glucocorticoids Inhibit Glucose Transport in Cultured Hippocampal Neurons and Glia. Neuroendocrinology 52 (1):57–64. https://doi.org/10.1159/000125539
Ido Y, Chang K, Woolsey TA, and Williamson JR (2001) NADH: Sensor of Blood Flow Need in Brain, Muscle, and Other Tissues. FASEB J 15 (8):1419–1421. https://doi.org/10.1096/fj.00-0652fje
Jaszczyk A, and Juszczak GR (2021) Glucocorticoids, Metabolism and Brain Activity. Neurosci Biobehav Rev 126:113–145. https://doi.org/10.1016/j.neubiorev.2021.03.007
Javitt D (1987) Negative Schizophrenic Symptomatology and the PCP (Phencyclidine) Model of Schizophrenia. Hillside J Clin Psychiatry 9 (1):12–35
Juszczak GR, and Stankiewicz AM (2018) Glucocorticoids, Genes and Brain Function. Prog Neuropsychopharmacol Biol Psychiatry 82:136–168. https://doi.org/10.1016/j.pnpbp.2017.11.020
Kaye WH, Frank GK, and McConaha C (1999) Altered Dopamine Activity after Recovery from Restricting-Type Anorexia Nervosa. Neuropsychopharmacology 21 (4):503–506. https://doi.org/10.1016/S0893-133X(99)00053-6
Keller-Varady K, Hasan A, Schneider-Axmann T, et al. (2016) Endurance Training in Patients with Schizophrenia and Healthy Controls: Differences and Similarities. Eur Arch Psychiatry Clin Neurosci 266 (5):461–473. https://doi.org/10.1007/s00406-015-0651-8
Kelly SD, Harrell CS, and Neigh GN (2014) Chronic Stress Modulates Regional Cerebral Glucose Transporter Expression in an Age-Specific and Sexually-Dimorphic Manner. Physiol Behav 126:39–49. https://doi.org/10.1016/j.physbeh.2013.12.002
Ketter TA, Kimbrell TA, George MS, et al. (2001) Effects of Mood and Subtype on Cerebral Glucose Metabolism in Treatment-Resistant Bipolar Disorder. Biol Psychiatry 49 (2):97–109. https://doi.org/10.1016/S0006-3223(00)00975-6
Kishimoto H, Takazu O, Ohno S, et al. (1987) 11C-Glucose Metabolism in Manic and Depressed Patients. Psychiatry Res 22 (1):81–88. https://doi.org/10.1016/0165-1781(87)90053-9
Klein DF (1993) False Suffocation Alarms, Spontaneous Panics, and Related Conditions: an Integrative Hypothesis. Arch Gen Psychiatry 50 (4):306–317. https://doi.org/10.1001/archpsyc.1993.01820160076009
——— (1994) Testing the Suffocation False Alarm Theory of Panic Disorder. Anxiety 1 (1):1–7. https://doi.org/10.1002/anxi.3070010103
Kleinhaus K, Harlap S, Perrin M, et al. (2013) Prenatal Stress and Affective Disorders in a Population Birth Cohort. Bipolar Disord 15 (1):92–99. https://doi.org/10.1111/bdi.12015
Kostrzewa RM, Kostrzewa JP, Brown RW, Nowak P, and Brus R (2008) Dopamine Receptor Supersensitivity: Development, Mechanisms, Presentation, and Clinical Applicability. Neurotox Res 14 (2–3):121–128. https://doi.org/10.1007/BF03033804
Lam MH-B, Chau SW-H, and Wing Y-K (2009) High Prevalence of Hypokalemia in Acute Psychiatric Inpatients. Gen Hosp Psychiatry 31 (3):262–265. https://doi.org/10.1016/j.genhosppsych.2009.02.007
Li Y, and Sunder N (2021) What Doesn't Kill Her, Will Make Her Depressed. Econ Hum Biol 43:101064. https://doi.org/10.1016/j.ehb.2021.101064
Linnér A, and Almgren M (2020) Epigenetic Programming—the Important First 1000 Days. Acta Paediatr 109 (3):443–452. https://doi.org/10.1111/apa.15050

Liu S, Zhang L, Fan X, et al. (2024) Lactate Levels in the Brain and Blood of Schizophrenia Patients: A Systematic Review and Meta-Analysis. Schizophr Res 264:29–38. https://doi.org/10.1016/j.schres.2023.11.013

Llansola M, Montoliu C, Cauli O, et al. (2013) Chronic Hyperammonemia, Glutamatergic Neurotransmission and Neurological Alterations. Metab Brain Dis 28 (2):151–154. https://doi.org/10.1007/s11011-012-9337-3

Manoli I, Alesci S, Blackman MR, et al. (2007) Mitochondria as Key Components of the Stress Response. Trends Endocrinol Metab 18 (5):190–198. https://doi.org/10.1016/j.tem.2007.04.004

McCutcheon RA, Krystal JH, and Howes OD (2020) Dopamine and Glutamate in Schizophrenia: Biology, Symptoms and Treatment. World Psychiatry 19 (1):15–33. https://doi.org/10.1002/wps.20693

Meltzer HY, and Stahl SM (1976) The Dopamine Hypothesis of Schizophrenia: A Review. Schizophr Bull 2 (1):19–76. https://doi.org/10.1093/schbul/2.1.19

Merritt K, McCutcheon RA, Aleman A, et al. (2023) Variability and Magnitude of Brain Glutamate Levels in Schizophrenia: A Meta and Mega-Analysis. Mol Psychiatr 28 (5):2039–2048. https://doi.org/10.1038/s41380-023-01991-7

Meuret AE, and Ritz T (2010) Hyperventilation in Panic Disorder and Asthma: Empirical Evidence and Clinical Strategies. Int J Psychophysiol 78 (1):68–79. https://doi.org/10.1016/j.ijpsycho.2010.05.006

Mi B, Wu C, Gao X, et al. (2020) Long-Term BMI Change Trajectories in Chinese Adults and Its Association with the Hazard of Type 2 Diabetes: Evidence from a 20-Year China Health and Nutrition Survey. BMJ Open Diabetes Res Care 8 (1):e000879. https://doi.org/10.1136/bmjdrc-2019-000879

Miller CB, Rae CD, Green MA, et al. (2017) An Objective Short-Sleep Insomnia Disorder Subtype Is Associated with Reduced Brain Metabolite Concentrations *in vivo*: A Preliminary Magnetic Resonance Spectroscopy Assessment. Sleep 40 (11):zsx148. https://doi.org/10.1093/sleep/zsx148

Mintun MA, Vlassenko AG, Rundle MM, and Raichle ME (2004) Increased Lactate/Pyruvate Ratio Augments Blood Flow in Physiologically Activated Human Brain. PNAS 101 (2):659–664. https://doi.org/10.1073/pnas.0307457100

Mintzer J, Donovan KA, Kindy AZ, et al. (2019) Lifestyle Choices and Brain Health. Front Med 6:204. https://doi.org/10.3389/fmed.2019.00204

Mishra PK, Kumar A, Behar KL, and Patel AB (2018) Subanesthetic Ketamine Reverses Neuronal and Astroglial Metabolic Activity Deficits in a Social Defeat Model of Depression. J Neurochem 146 (6):722–734. https://doi.org/10.1111/jnc.14544

Moguilner S, Baez S, Hernandez H, et al. (2024) Brain Clocks Capture Diversity and Disparities in Aging and Dementia across Geographically Diverse Populations. Nat Med 30 (12):3646–3657. https://doi.org/10.1038/s41591-024-03209-x

Momin M, Fan F, Li J, et al. (2020) Joint Effects of Body Mass Index and Waist Circumference on the Incidence of Hypertension in a Community-Based Chinese Population. Obes Facts 13 (2):245–255. https://doi.org/10.1159/000506689

Monzel AS, Enríquez JA, and Picard M (2023) Multifaceted Mitochondria: Moving Mitochondrial Science Beyond Function and Dysfunction. Nat Metab 5 (4):546–562. https://doi.org/10.1038/s42255-023-00783-1

Morris HR, Spillantini MG, Sue CM, and Williams-Gray CH (2024) The Pathogenesis of Parkinson's Disease. Lancet 403 (10423):293–304. https://doi.org/10.1016/S0140-6736(23)01478-2

Mosharov E, Rosenberg A, Monzel A, et al. (2025) A Human Brain Map of Mitochondrial Respiratory Capacity and Diversity. Nature. https://doi.org/10.1038/s41586-025-08740-6

Mu E, Gurvich C, and Kulkarni J (2024) Estrogen and Psychosis—a Review and Future Directions. Arch Womens Ment Health 27:877–885. https://doi.org/10.1007/s00737-023-01409-x

Murrough JW, Abdallah CG, and Mathew SJ (2017) Targeting Glutamate Signalling in Depression: Progress and Prospects. Nat Rev Drug Discov 16 (7):472–486. https://doi.org/10.1038/nrd.2017.16

Nakama N, Usui N, Doi M, and Shimada S (2023) Early Life Stress Impairs Brain and Mental Development during Childhood Increasing the Risk of Developing Psychiatric Disorders. Prog Neuropsychopharmacol Biol Psychiatry 126:110783. https://doi.org/10.1016/j.pnpbp.2023.110783

Ni A, and Ernst C (2022) Evidence That Substantia Nigra Pars Compacta Dopaminergic Neurons Are Selectively Vulnerable to Oxidative Stress Because They Are Highly Metabolically Active. Front Cell Neurosci 16:826193. https://doi.org/10.3389/fncel.2022.826193

O'Doherty DC, Chitty KM, Saddiqui S, Bennett MR, and Lagopoulos J (2015) A Systematic Review and Meta-Analysis of Magnetic Resonance Imaging Measurement of Structural Volumes in Posttraumatic Stress Disorder. Psychiatry Res Neuroimaging 232 (1):1–33. https://doi.org/10.1016/j.pscychresns.2015.01.002

O'Connor RM, and Cryan JF (2014) Adolescent Brain Vulnerability and Psychopathology through the Generations: Role of Diet and Dopamine. Biol Psychiatry 75 (1):4–6. https://doi.org/10.1016/j.biopsych.2013.10.022

O'Hara CB, Campbell IC, and Schmidt U (2015) A Reward-Centred Model of Anorexia Nervosa: A Focussed Narrative Review of the Neurological and Psychophysiological Literature. Neurosci Biobehav Rev 52:131–152. https://doi.org/10.1016/j.neubiorev.2015.02.012

Öngür D, Jensen JE, Prescot AP, et al. (2008) Abnormal Glutamatergic Neurotransmission and Neuronal-Glial Interactions in Acute Mania. Biol Psychiatry 64 (8):718–726. https://doi.org/10.1016/j.biopsych.2008.05.014

Onwordi EC, Whitehurst T, Mansur A, et al. (2021) The Relationship between Synaptic Density Marker SV2A, Glutamate and N-Acetyl Aspartate Levels in Healthy Volunteers and Schizophrenia: A Multimodal PET and Magnetic Resonance Spectroscopy Brain Imaging Study. Transl Psychiatry 11 (1):393. https://doi.org/10.1038/s41398-021-01515-3

Orben A, Tomova L, and Blakemore S-J (2020) The Effects of Social Deprivation on Adolescent Development and Mental Health. Lancet Child Adolesc Health 4 (8):634–640. https://doi.org/10.1016/S2352-4642(20)30186-3

Palacios J, and Wiederhold K-H (1985) Dopamine D2 Receptor Agents, but Not Dopamine D1, Modify Brain Glucose Metabolism. Brain Res 327 (1–2):390–394. https://doi.org/10.1016/0006-8993(85)91543-4

Palagini L, Hertenstein E, Riemann D, and Nissen C (2022) Sleep, Insomnia and Mental Health. J Sleep Res 31 (4):e13628. https://doi.org/10.1111/jsr.13628

Pan X-F, Wang L, and Pan A (2021) Epidemiology and Determinants of Obesity in China. Lancet Diabetes Endocrinol 9 (6):373–392. https://doi.org/10.1016/S2213-8587(21)00045-0

Pang Y, Kartsonaki C, Guo Y, et al. (2018) Adiposity and Risks of Colorectal and Small Intestine Cancer in Chinese Adults: A Prospective Study of 0.5 Million People. Br J Cancer 119 (2):248–250. https://doi.org/10.1038/s41416-018-0124-8

Panizzutti B, Bortolasci CC, Spolding B, et al. (2024) Effects of Antipsychotic Drugs on Energy Metabolism. Eur Arch Psychiatry Clin Neurosci 274 (5):1125–1135. https://doi.org/10.1007/s00406-023-01727-2

Patel AB, de Graaf RA, Mason GF, et al. (2005) The Contribution of GABA to Glutamate/Glutamine Cycling and Energy Metabolism in the Rat Cortex *in Vivo*. PNAS 102 (15):5588–5593. https://doi.org/10.1073/pnas.0501703102

Phelps NH, Singleton RK, Zhou B, et al. (2024) Worldwide Trends in Underweight and Obesity from 1990 to 2022: A Pooled Analysis of 3663 Population-Representative Studies with 222 Million Children, Adolescents, and Adults. Lancet 403 (10431):1027–1050. https://doi.org/10.1016/S0140-6736(23)02750-2

Picard M, McEwen BS, Epel ES, and Sandi C (2018) An Energetic View of Stress: Focus on Mitochondria. Front Neuroendocrinol 49:72–85. https://doi.org/10.1016/j.yfrne.2018.01.001

Pillinger T, Osimo EF, de Marvao A, et al. (2019) Cardiac Structure and Function in Patients with Schizophrenia Taking Antipsychotic Drugs: An MRI Study. Transl Psychiatry 9 (1):163. https://doi.org/10.1038/s41398-019-0502-x

Pillinger T, Osimo EF, de Marvao A, et al. (2023) Effect of Polygenic Risk for Schizophrenia on Cardiac Structure and Function: A UK Biobank Observational Study. Lancet Psychiatry 10 (2):98–107. https://doi.org/10.1016/S2215-0366(22)00403-5

Piroli GG, Grillo CA, Reznikov LR, et al. (2007) Corticosterone Impairs Insulin-Stimulated Translocation of GLUT4 in the Rat Hippocampus. Neuroendocrinology 85 (2):71–80. https://doi.org/10.1159/000101694

Popoli M, Yan Z, McEwen BS, and Sanacora G (2012) The Stressed Synapse: The Impact of Stress and Glucocorticoids on Glutamate Transmission. Nat Rev Neurosci 13 (1):22–37. https://doi.org/10.1038/nrn3138

Pruett BS, and Meador-Woodruff JH (2020) Evidence for Altered Energy Metabolism, Increased Lactate, and Decreased pH in Schizophrenia Brain: A Focused Review and Meta-Analysis of Human Postmortem and Magnetic Resonance Spectroscopy Studies. Schizophr Res 223:29–42. https://doi.org/10.1016/j.schres.2020.09.003

Rae CD, Bartlett D, Yang Q, et al. (2009) Dynamic Changes in Brain Bioenergetics during Obstructive Sleep Apnea. Journal of Cerebral Blood Flow & Metabolism 29 (8):1421–1428. https://doi.org/10.1038/jcbfm.2009.57

Rae CD, Baur JA, Borges K, et al. (2024) Brain Energy Metabolism: A Roadmap for Future Research. J Neurochem 168 (5):910–954. https://doi.org/10.1111/jnc.16032

Riecher-Rössler A, and Kulkarni J (2011) Estrogens and Gonadal Function in Schizophrenia and Related Psychoses. In: Neill JC, Kulkarni J (eds) Biological Basis of Sex Differences in Psychopharmacology, vol 8. Current Topics in Behavioural Neuroscience. Springer, Berlin, pp 155–171. https://doi.org/10.1007/7854_2010_100

Roseboom TJ (2019) Epidemiological Evidence for the Developmental Origins of Health and Disease: Effects of Prenatal Undernutrition in Humans. J Endocrinol 242 (1):T135–T144. https://doi.org/10.1530/JOE-18-0683

Rosenberg AM, Saggar M, Monzel AS, et al. (2023) Brain Mitochondrial Diversity and Network Organization Predict Anxiety-Like Behavior in Male Mice. Nat Commun 14 (1):4726. https://doi.org/10.1038/s41467-023-39941-0

Schaefer JD, Cheng TW, and Dunn EC (2022) Sensitive Periods in Development and Risk for Psychiatric Disorders and Related Endpoints: A Systematic Review of Child Maltreatment Findings. Lancet Psychiatry 9 (12):978–991. https://doi.org/10.1016/S2215-0366(22)00362-5

Schapira AH (2008) Mitochondria in the Aetiology and Pathogenesis of Parkinson's Disease. Lancet Neurol 7 (1):97–109. https://doi.org/10.1016/S1474-4422(07)70327-7

Scott AJ, Webb TL, Martyn-St James M, Rowse G, and Weich S (2021) Improving Sleep Quality Leads to Better Mental Health: A Meta-Analysis of Randomised Controlled Trials. Sleep Med Rev 60:101556. https://doi.org/10.1016/j.smrv.2021.101556

Shroitman NK, Yitzhaky A, Shachar DB, Gurwitz D, and Hertzberg L (2023) Meta-Analysis of Brain Samples of Individuals with Schizophrenia Detects Down-Regulation of Multiple ATP Synthase Encoding Genes in Both Females and Males. J Psychiatr Res 158:350–359. https://doi.org/10.1016/j.jpsychires.2023.01.005

Sibson NR, Dhankhar A, Mason GF, et al. (1998) Stoichiometric Coupling of Brain Glucose Metabolism and Glutamatergic Neuronal Activity. PNAS 95 (1):316–321. https://doi.org/10.1073/pnas.95.1.316

Sohal VS, and Rubenstein JL (2019) Excitation-Inhibition Balance as a Framework for Investigating Mechanisms in Neuropsychiatric Disorders. Mol Psychiatr 24 (9):1248–1257. https://doi.org/10.1038/s41380-019-0426-0

Stern Y, Arenaza-Urquijo EM, Bartrés-Faz D, et al. (2020) Whitepaper: Defining and Investigating Cognitive Reserve, Brain Reserve, and Brain Maintenance. Alzheimers Dement 16 (9):1305–1311. https://doi.org/10.1016/j.jalz.2018.07.219
Turkheimer FE, Selvaggi P, Mehta MA, et al. (2019) Normalizing the Abnormal: Do Antipsychotic Drugs Push the Cortex into an Unsustainable Metabolic Envelope? Schizophr Bull 46 (3):484–495. https://doi.org/10.1093/schbul/sbz119
Vakulin A, Green MA, D'Rozario AL, et al. (2021) Brain Mitochondrial Dysfunction and Driving Simulator Performance in Untreated Obstructive Sleep Apnea. J Sleep Res 31 (2):e13482. https://doi.org/10.1111/jsr.13482
van der Kooij MA (2020) The Impact of Chronic Stress on Energy Metabolism. Mol Cell Neurosci 107:103525. https://doi.org/10.1016/j.mcn.2020.103525
van Gemert LA, de Galan BE, Wevers RA, Ter Heine R, and Willemsen MA (2022) Lactate Infusion as Therapeutical Intervention: A Scoping Review. Eur J Pediatr 181 (6):2227–2235. https://doi.org/10.1007/s00431-022-04446-3
Virgin Jr CE, Ha TPT, Packan DR, et al. (1991) Glucocorticoids Inhibit Glucose Transport and Glutamate Uptake in Hippocampal Astrocytes: Implications for Glucocorticoid Neurotoxicity. J Neurochem 57 (4):1422–1428. https://doi.org/10.1111/j.1471-4159.1991.tb08309.x
Vlassenko AG, Rundle MM, Raichle ME, and Mintun MA (2006) Regulation of Blood Flow in Activated Human Brain by Cytosolic NADH/NAD+ Ratio. PNAS 103 (6):1964–1969. https://doi.org/10.1073/pnas.0510632103
Voineskos D, Daskalakis ZJ, and Blumberger DM (2020) Management of Treatment-Resistant Depression: Challenges and Strategies. Neuropsychiatr Dis Treat 16:221–234. https://doi.org/10.2147/NDT.S198774
Wang L, Zhou B, Zhao Z, et al. (2021) Body-Mass Index and Obesity in Urban and Rural China: Findings from Consecutive Nationally Representative Surveys during 2004–18. Lancet 398 (10294):53–63. http://dx.doi.org/10.1016/S0140-6736(21)00798-4
Whitehurst T, and Howes O (2022) The Role of Mitochondria in the Pathophysiology of Schizophrenia: A Critical Review of the Evidence Focusing on Mitochondrial Complex One. Neurosci Biobehav Rev 132:449–464. https://doi.org/10.1016/j.neubiorev.2021.11.047
Willner P (2017) The Chronic Mild Stress (CMS) Model of Depression: History, Evaluation and Usage. Neurobiol Stress 6:78–93. https://doi.org/10.1016/j.ynstr.2016.08.002
Yang X, Heinemann M, Howard J, et al. (2021) Physical Bioenergetics: Energy Fluxes, Budgets, and Constraints in Cells. PNAS 118 (26):e2026786118. https://doi.org/10.1073/pnas.2026786118

Open Access This chapter is licensed under the terms of the Creative Commons Attribution-NonCommercial-NoDerivatives 4.0 International License (http://creativecommons.org/licenses/by-nc-nd/4.0/), which permits any noncommercial use, sharing, distribution and reproduction in any medium or format, as long as you give appropriate credit to the original author(s) and the source, provide a link to the Creative Commons license and indicate if you modified the licensed material. You do not have permission under this license to share adapted material derived from this chapter or parts of it.

The images or other third party material in this chapter are included in the chapter's Creative Commons license, unless indicated otherwise in a credit line to the material. If material is not included in the chapter's Creative Commons license and your intended use is not permitted by statutory regulation or exceeds the permitted use, you will need to obtain permission directly from the copyright holder.

Metabolic Abnormalities in Neuropsychiatric Disorders

5

What Is the Best Evidence for Bioenergetic Abnormalities in Severe Mental Illness?

Hunter Eby, Ali Sajid Imami, William G. Ryan V, Smita Sahay, John Vergis, Jennifer H. Nguyen, Taylen O. Arvay, Priyanka Pulvender, Nicholas Henkel, and Robert E. McCullumsmith

Abstract Recent work implicates abnormalities of metabolic pathways and processes in severe mental illness (SMI). SMI encompasses moderate to severe mood disorders, as well as schizophrenia spectrum psychotic disorders. Metabolic systems implicated in SMI include mitochondrial function, glucose utilization, and insulin signaling, among others, with evidence suggesting changes in these systems are influenced by genetic, epigenetic, and environmental factors. This chapter reviews seminal studies that explore these metabolic perturbations and disruptions in SMI, with a focus on schizophrenia, bipolar disorder, and major depressive disorder. Specific suggestions for future directions are proposed to advance our understanding of these often life-changing disorders.

Keywords Bioenergetic function, risk for severe mental illness, mitochondria, schizophrenia, bipolar disorder, major depressive disorder, insulin signaling, fates of glucose, psychotropic medications

1 Introduction

What is bioenergetic function? The term "bioenergetic" encompasses metabolism, mitochondria, energy, numerous metabolites, catabolism, anabolism, as well as ATP (adenosine triphosphate) synthesis and oxidative phosphorylation. Applied to

Robert E. McCullumsmith (✉)
Dept. of Neurosciences and Psychiatry, University of Toledo College of Medicine, Toledo, OH, 43614; Neurosciences Institute, ProMedica, Toledo, OH 43614, USA
Email: robert.mccullumsmith@utoledo.edu

Affiliations for the coauthors are available in the List of Contributors.

© Frankfurt Institute for Advanced Studies (FIAS) 2025

D. Öngür, J. M. Ford (eds.), *Metabolic Neuropsychiatry*, Strüngmann Forum Reports,
https://doi.org/10.1007/978-3-032-05630-6_5

the human brain, bioenergetic function permits the integration of these myriads of concepts, unifying diverse neurophysiological processes such as synaptic transmission and long-term potentiation that are highly interconnected. Brain disorders are divided into psychiatric and neurological buckets, with psychiatric disorders considered disorders of the "mind," or as having altered capacity for mentation, and thus called mental illnesses (David and Nicholson 2015). Severe mental illness (SMI) generally includes schizophrenia, schizoaffective disorder, bipolar affective disorder (BAD, typically type 1), and major depressive disorder (MDD, moderate to severe) diagnoses. While the risk for SMI overlaps to varying degrees between these diagnostic entities, psychiatric diagnoses are descriptive, syndromic, and not based on genetic or pathophysiological markers often found in neurological disorders. It follows that there is likely a significant overlap between SMI in terms of causal factors and biological mechanisms.

Regarding SMI, much of the evidence we present is focused on schizophrenia, with some examples from BAD and MDD. This is not to favor one diagnosis over another; rather, there are advantages to studying schizophrenia that have fostered several seminal findings, including the relative (compared to MDD, for example) homogeneity of schizophrenia, as well as historically salient hypotheses (such as the dopamine hypothesis) that drove work for many decades.

2 An Overarching Hypothesis

We posit that SMI is caused by a combination of genetic, epigenetic, and environmental risk factors that combine to drive diverse disease phenotypes that may share common pathophysiological processes (Figure 5.1). In schizophrenia, much of the genetic risk converges on excitatory synapse-associated genes (Trubetskoy et al. 2022). Perturbations of insulin signaling and glucose metabolism appear to follow this genetic risk, perhaps secondary to aberrant metabolic coupling between astrocytes and neurons driven by synaptic changes during brain development (Dear et al. 2024). In contrast, BAD is more closely associated with mitochondrial dysfunction (Ceylan et al. 2023; Gimenez-Palomo et al. 2024; Kuang et al. 2018). The pathophysiology of schizophrenia and severe mood disorders may converge on abnormalities of glycolytic pathways and glucose utilization, albeit arriving at this intermediate phenotype as a likely consequence of different primary causes associated with each diagnosis.

In this chapter, we present the "best" evidence available for bioenergetic dysfunction in SMI, reported in seminal papers (Sahay et al. 2024). Of course, what constitutes the "best" evidence is subjective and open to debate. We have divided this evidence into sections that roughly parallel our overarching hypothesis: genetic risk (Section 3), mitochondria (Section 4), evidence for a bioenergetic intermediate phenotype (Section 5), insulin signaling (Section 6), fates of glucose (Section 7), bioenergetic substrates (lactate, brain pH, and ATP; Section 8), lipids and fatty acids (Section 9), and psychotropic medications (Section 10). To evaluate the quality and novelty of these studies, we used a critical appraisal tool framework, adapted from the Centre for Evidence-Based Medicine (OCEBM Levels of Evidence Working

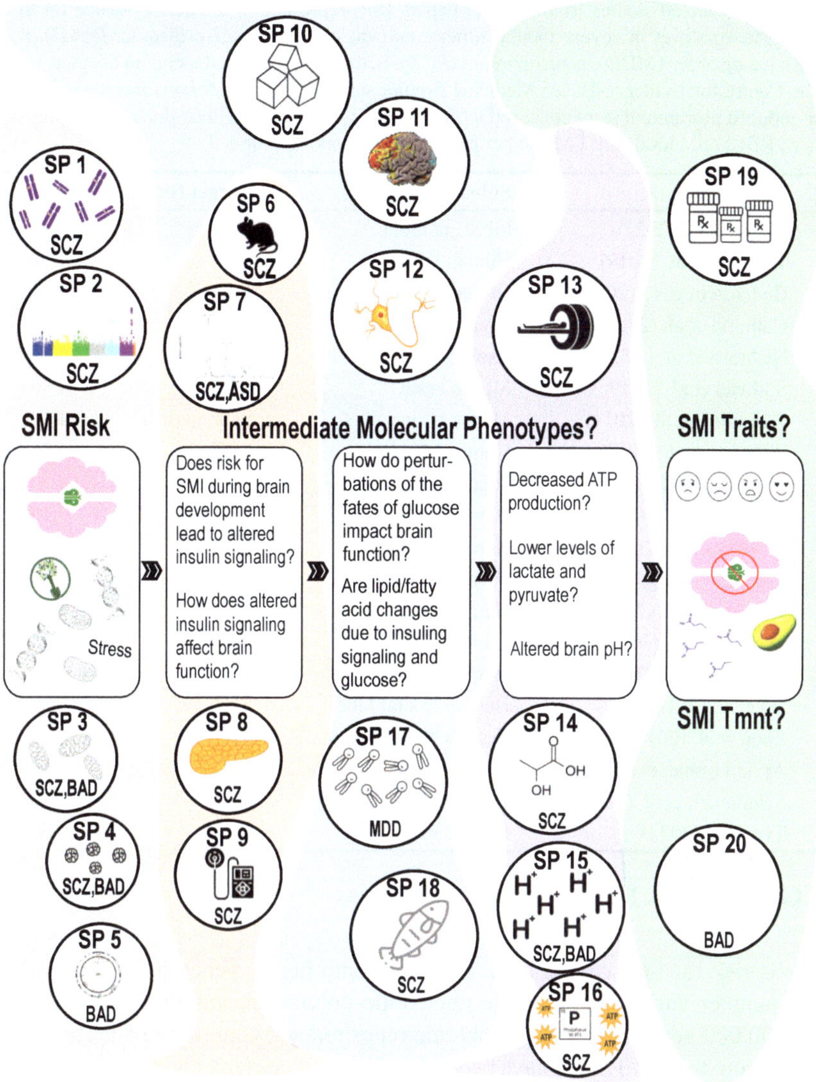

Figure 5.1 Risk for severe mental illness (SMI) includes genetic, epigenetic, and environmental factors. Under such risk, brain development and maturation may lead to an intermediate bioenergetic phenotype, characterized by abnormalities of mitochondria, insulin signaling, fates of glucose, as well as lipids and fatty acids. Compelling evidence for alterations in bioenergetic molecules, including lactate, ATP as well as decreased brain pH, supports this hypothesis. Psychotropic medications have complex effects on bioenergetic function, perhaps worsening some elements and improving others. Pharmacological and diet-based interventions have the potential to modify disease phenotypes, improving patient outcomes and quality of life for the afflicted. *Abbreviations* SP refers to the seminal papers presented in this chapter, numbered in succession; SCZ: schizophrenia; BAD: bipolar affective disorder; MDD: major depressive disorder; Tmnt: treatment.

Group 2011), to standardize evaluations and critical assessments of the contribution of research within its field (Sackett et al. 1996). Results are presented in Table 5.1. We end by synthesizing findings from these papers and make specific suggestions for how they might be leveraged to improve patient outcomes.

Table 5.1 Selected studies from seminal papers (SP) representing the best evidence for bioenergetic abnormalities in severe mental illness: autism, bipolar affective disorder (BAD), major depressive disorder (MDD), schizophrenia (SCZ). Scores were reached using an adapted version of the Centre for Evidence-Based Medicine 5-point scale (5 is best). *Abbreviations* hiPSC: human-induced pluripotent stem cells; mtDNA: mitochondrial DNA; dlPFC: dorsolateral prefrontal cortex; RBC: red blood cell; PMBC: peripheral blood mononuclear cell.

SP	Reference	Substrate Studied	Target Illness	Score
1	Chong et al. (2022)	hiPSC neurons	SCZ	4.56
2	Tomasik et al. (2019)	Human DNA	SCZ	4.22
3	Bodenstein et al. (2019)	Human brain mtDNA	SCZ	4.22
4	Kathuria et al. (2023)	Organoids	SCZ	4.33
5	Nishioka et al. (2023)	Exome	BAD	4.44
6	Sullivan et al. (2019c)	Mouse brain	SCZ	4.80
7	Wesseling et al. (2014)	Mouse brain	SCZ, autism	4.40
8	Pillinger et al. (2017)	Human blood	SCZ	4.56
9	Chouinard et al. (2019)	Human blood	SCZ	4.30
10	Lee et al. (2024)	Human transcripts	SCZ	4.70
11	Townsend et al. (2023)	Human brain	SCZ	4.40
12	Sullivan et al. (2019a)	Human dlPFC	SCZ	4.70
13	Rowland et al. (2016)	Human brain	SCZ	4.50
14	Sullivan et al. (2019b)	Human dlPFC	SCZ	4.40
15	Hagihara et al. (2018)	Human brain	SCZ, BAD	4.40
16	Du et al. (2014)	Human frontal lobe	SCZ	5.0
17	Zhou et al. (2019)	Human blood plasma	MDD	4.11
18	Arvindakshan et al. (2003)	Human RBC	SCZ	4.30
19	Schulmann et al. (2023)	Human PFC	SCZ	4.30
20	Tye et al. (2022)	Human PBMC	BAD	4.60

3 Genetic Risk for Severe Mental Illness

Genetic risk for SMI can be broadly divided into highly penetrant rare mutations, copy number variants, and single nucleotide polymorphisms (SNPs). With more than 100,000 genomes for schizophrenia reposited and counting (Trubetskoy et al. 2022), more than 200 SNPs have been identified that confer a higher risk for this illness. Each SNP confers an increased risk to the carrier, with multiple SNPs accounting for the majority of genetic risk in most persons with schizophrenia (Trubetskoy et al. 2022). As such, genetic risk is often an amalgamation of coincident risks across biological pathways and disease states (Richards et al. 2022). In general, the heritability of SMI is highly variable, and it is a challenge to connect specific genomic variants with disease phenotypes (Szoke et al. 2024; Mitteroecker and Merola 2024). An example of genetic risk for SMI is worth considering. In a Scottish family with a rare highly penetrant mutation (DISC1), there are 18 carriers with SMI, seven of which have a diagnosis of schizophrenia (Blackwood et al. 2001; Hennah et al. 2009; Hodgkinson et al. 2004). Eleven carriers of the DISC1 mutation in this family do not have schizophrenia but instead have been diagnosed with a mood disorder (BAD or MDD). Importantly, SNPs within the DISC locus may confer heterogeneity to the clinical phenotype (Hennah et al. 2009). This example argues in favor of considering that mechanisms and pathophysiological processes "found" in one diagnostic entity should be considered across the SMI phenotypic spectrum. Evidence shows that the DISC translocation impacts bioenergetic function via direct effects on mitochondrial function (Norkett et al. 2017, 2020; Park et al. 2017). Below we present two additional recent studies that link genetic risk for SMI with bioenergetic perturbations.

3.1 Seminal Paper 1: Chong et al. (2022)

Haploinsufficiency results when one copy of a gene is deleted or has a loss-of-function mutation and the single wild-type gene is not sufficient for normal biological function. SETD1A is a histone-lysine methyl transferase involved in regulating mitosis and cell proliferation. In the study by Chong et al. (2022), performed in human-induced pluripotent stem cells (hiPSCs) derived from BJ fibroblasts and H7 human embryonic stem cells, cells were genetically edited using RNAs targeting exon 7 of SETD1A. These edited hiPSCs were then differentiated into cortical spheroids and neurally induced. Subsequent experiments involved biochemical and metabolic approaches to study neuronal differentiation and functional properties. Altered neurite outgrowth and diminished spontaneous neuronal action potentials were found in hiPSCs with haploinsufficiency of the SETD1A gene. Neural progenitor cell migration was also altered. Functional metabolic changes were observed in these cells including reduced glycolytic enzyme expression and lactate release. Metabolic flux assays showed lower respiratory capacity and basal glycolysis. Supplementation with pyruvate improved these observed deficits and thus identified bioenergetic changes as being critically important to the downstream effects of SETD1A haploinsufficiency. Chong et al. (2022) highlight an underappreciated mechanism where genomic variants with little apparent known connection to bioenergetic function may have an unexpected and significant impact.

3.2 Seminal Paper 2: Tomasik et al. (2019)

Does genetic risk for SMI confer risk for diabetes? Investigating the relationship between higher polygenic risk score (PRS) for schizophrenia and insulin resistance, recent work by Tomasik et al. (2019) supports this hypothesis. The homeostatic model assessment 2 (HOMA2, an updated version of the original homeostatic model assessment; Levy et al. 1998), is a tool that models the glucose-insulin feedback system to predict C-peptide, insulin, and glucose levels in the fasting steady state. This in turn allows for the deduction of insulin resistance and/or β-cell function following measurement of fasting glucose and insulin levels (Levy et al. 1998). Further, Tomasik et al. (2019) found that subjects with schizophrenia exhibited elevated baseline levels of HOMA2-IR, HOMA2 β-cell function, and fasting insulin levels while insulin sensitivity and fasting glucose levels did not significantly differ. Adjusted for covariates, HOMA2-IR remained significantly elevated in schizophrenia subjects and positively correlated with schizophrenia PRS.

Other well-characterized examples also support this connection. For instance, genome-wide association studies identified T cell factor 4 (TCF4, also known as TCF7L2)—a major effector of the canonical WNT pathway via dimerization with β-catenin—as a susceptibility gene for schizophrenia and diabetes (Dastani et al. 2012; Florez 2007; Grant et al. 2006; Schizophrenia Working Group of the Psychiatric Genomics Consortium 2014).

4 Mitochondria in SMI

Mitochondria are abundant within neuronal dendrites and synaptic terminals in the brain. Abnormalities of mitochondria may lead to impaired oxidative phosphorylation as well as generation of reactive oxygen species, contributing to oxidative stress within neurons and glia. Dysregulation of energy production and high levels of oxidative stress are commonly reported in SMI (Shao et al. 2008). Disparate findings support mitochondrial dysfunction in schizophrenia and MDD. In BAD, mitochondrial dysfunction is considered a likely contributor to its etiopathogenesis (Kato 2005; Lam et al. 2023).

4.1 Seminal Paper 3: Bodenstein et al. (2019)

While the precise etiopathologies of SMI remains unknown, accumulating evidence suggests that mitochondrial dysfunction has a central role in BAD (Andreazza and Nierenberg 2018). A positive correlation was found in BAD between the electron transport chain complex I subunit NDUFS7 protein levels and mtDNA content across multiple brain regions, including the hippocampus, cerebellum, and prefrontal cortex (Bodenstein et al. 2019). In their work, Bodenstein et al. noted an elevation in mtDNA content in the hippocampus of patients with BAD, while a reduction in mtDNA oxidation was observed in the hippocampus of individuals with BAD and schizophrenia. This reinforces the role of mitochondrial dysfunction in SMI (Ben-Shachar and Karry 2008).

4.2 Seminal Paper 4: Kathuria et al. (2023)

By comparing transcriptomic profiles of cerebral organoids from subjects with SMI, Kathuria et al. (2023) found that an upregulation of genes was involved in mitochondrial and oxidative phosphorylation in schizophrenia, but not BAD. Interestingly, functional assessments in their study show deficits in basal oxygen consumption rate and ATP production, implicating mitochondrial dysfunction in schizophrenia, but not BAD, derived organoids. However, cortical neurons from BAD subjects exhibited fewer mitochondria-endoplasmic reticulum contact sites compared to those from schizophrenia subjects, highlighting disease-specific differences in mitochondrial dynamics. These findings highlight the involvement of mitochondrial dysfunction in schizophrenia and BAD pathogenesis, although with distinct molecular signatures and functional consequences.

4.3 Seminal Paper 5: Nishioka et al. (2023)

Investigating the role of mosaic variants in BAD, Nishioka et al. (2023) identified significant enrichment of mitochondrial heteroplasmic variants, particularly in mitochondrial tRNA genes, through extensive deep exome sequencing of BAD subjects. In particular, recurrent m.3243A>G variants, known for their causal role in mitochondrial diseases, were detected in two unrelated BAD probands, albeit at lower allele fractions than typically observed in mitochondrial diseases. Despite being limited to peripheral tissues, these findings emphasize the potential contribution of deleterious mosaic variants in the nuclear genome, predominantly those associated with mitochondrial dysfunction in the pathogenesis of BAD. This study reveals a novel avenue for understanding the genetic architecture of BAD, emphasizing the relevance of mitochondrial dysfunction in its etiology.

5 Intermediate Bioenergetic Phenotypes in SMI

Taken together, the heritable risk factors for SMI appear to converge *indirectly* on bioenergetic dysfunction via genetic mechanisms and *directly* via mitochondrial dysfunction (when present). While examples of specific genes directly involved in bioenergetic function conferring risk for SMI exist, accumulating evidence reveals an intermediate bioenergetic disease phenotype secondary to genetic risk that does not map to metabolic genes. In this section, we discuss evidence for such an intermediate phenotype at the molecular level.

Recent animal model work suggests that growing a brain with "broken" synapses leads to bioenergetic defects that are likely secondary to improper bioenergetic coupling of astrocytes and neurons (Henkel et al. 2022). This concept starts with the observations that (a) blockade of NMDA receptors with the open channel blocker phencyclidine can precipitate a schizophreniform psychosis that is indistinguishable from schizophrenia, including positive, negative, and cognitive symptoms (Castellani et al. 1982; Cohen et al. 1962); (b) anti-NMDA receptor antibodies may also simulate schizophrenia, ultimately (if severe) leading to an autoimmune encephalitis (Dalmau et al. 2008; Iizuka et al. 2008; Vincent and Bien 2008); and (c) NMDA-subtype glutamate receptor hypomorphs (aka Grin1KD mice) have smaller body mass and impaired glycolytic function (Mohn et al. 1999; Sullivan et al. 2019c). Finally, much of the genetic risk for schizophrenia (and to a lesser extent BAD) converges on genes that are involved with excitatory neurotransmission (Fromer et al. 2014; Lin et al. 2016), with NMDA receptors as the major excitatory neurotransmitter receptor in the central nervous system (CNS). The work described below suggests bioenergetic dysfunction as an intermediate, or secondary, pathological phenotype that is a consequence of aberrant excitatory neurotransmission during brain development and/or maturation (Henkel et al. 2022).

5.1 Seminal Paper 6: Sullivan et al. (2019c)

In mice, a congenital Grin1 loss-of-function allele exhibits behavioral deficits in assays that are endophenotypes of SMI, such as deficits in executive function and working memory, sensorimotor gating, and social interactions (Sullivan et al. 2019c; Mohn et al. 1999; Mielnik et al. 2021). Interestingly, Sullivan et al. (2019c) found that these mice have decreased mRNA expression of glucose transporters in the frontal cortex and deficits in glycolytic pathways in synaptic proteomes isolated from the frontal cortex. Specific pathways implicated in Grin1 knockdown (Grin1KD) mice included gluconeogenesis, glucose metabolism, pyruvate and citric acid metabolism, metabolism of carbohydrates, and glycogen storage disease (Sullivan et al. 2019c). Restoring the endogenous Grin1 locus with a genetic rescue approach restored behavioral deficits and corrected electrophysiological measures, including hippocampal LTP (Mielnik et al. 2021).

5.2 Seminal Paper 7: Wesseling et al. (2014)

In their detailed assessment of the molecular changes in the Grin1KD mouse, Wesseling et al. (2014) used liquid chromatography–mass spectrometry (LCMS) to assess changes in protein levels in the frontal cortex and hippocampus. They found that mitochondrial pyruvate kinase expression was one of the top differentially expressed proteins. In this proteomic dataset from Grin1KD mice, the top hippocampal pathways that were perturbed included glycolysis, gluconeogenesis, and tricarboxylic acid (TCA) cycle, while fatty acid metabolism was found to be perturbed in the frontal cortex. In addition, serum analysis found decreases in insulin-like growth factor 1. These data connect an animal model of SMI with a genetic lesion causing "broken" excitatory synapses with altered bioenergetic function in adult animals, leading to the question: Which bioenergetic downstream substrates are perturbed in SMI?

6 Insulin Signaling in SMI

Insulin signaling is perhaps the best-studied, particularly systemically, biological mechanism in biomedical research. Although less studied in the CNS, insulin receptors and their complex associated signaling pathways regulate a myriad of biological pathways essential for neuroplasticity and brain function (Derakhshan and Toth 2013; Zeng et al. 2016). For example, insulin regulates molecular correlates of learning and memory in the hippocampus, an effect mediated by protein kinase C and other factors (Zhao et al. 2019).

Historical data reveals a complex relationship between insulin resistance and schizophrenia dating back almost a century, predating the use of antipsychotics (Agarwal et al. 2020). Studies as far back as the 1920s highlight the observation of metabolic disturbances in association with schizophrenia (Barrett and Serre 1924; Kasanin 1926; Kooy 1919; Raphael and Parsons 1921). One early study showed that patients with schizophrenic psychosis required higher doses of insulin to induce hypoglycemia during insulin coma therapy (Appel and Farr 1929). There is also evidence that insulin in the CNS may be synthesized locally (Havrankova et al. 1978). Insulin signaling in the CNS is mediated via activation of receptor tyrosine kinases that trigger a cascade of molecular changes, including protein kinases and transcription factors that regulate bioenergetic gene expression, immune function pathways, and synaptic composition (Fernandez and Torres-Alemán 2012; Lemche et al. 2024).

6.1 Seminal Paper 8: Pillinger et al. (2017)

A meta-analysis examining glucose homeostasis in first-episode schizophrenia reported elevated fasting glucose and insulin levels, higher insulin resistance, and lower glucose tolerance before initiation of antipsychotic medication (Pillinger et al. 2017). When matched for diet and exercise status, fasting plasma glucose levels remained elevated in antipsychotic naïve schizophrenia subjects compared to controls. These findings have been bolstered by other studies in schizophrenia that have noted increases in fasting glucose, insulin, cortisol, IL-6, as well as abnormalities in cholesterol levels compared to matched controls.

6.2 Seminal Paper 9: Chouinard et al. (2019)

Another study examined the differences in insulin sensitivity between subjects with first-episode psychosis (FEP), their unaffected siblings, and unrelated healthy controls (Chouinard et al. 2019). The unaffected siblings were antipsychotic naïve. Insulin sensitivity was measured using the Oral Minimal Model Method. Choudinard et al. found that while insulin sensitivity did not differ significantly between FEP subjects and their siblings, the healthy controls exhibited the highest insulin sensitivity, compared to FEP subjects and their unaffected siblings. This study suggests that bioenergetic disturbances are present before administration of psychotropic medications and supports the hypothesis that genetic risk for bioenergetic dysfunction and SMI overlap.

7 Fates of Glucose in SMI

Glucose maintains global excitatory neural network function in the CNS (Tourigny et al. 2019). To meet high-energy demands, neurons and astrocytes metabolize glucose, the brain's primary energy source, through glycolysis to produce ATP and supply substrates for oxidative phosphorylation (Mergenthaler et al. 2013). To meet the demands of synaptic transmission, glycolytic enzymes localize to the synaptic terminal to maintain ATP levels for vesicle loading and acidification (Ikemoto et al. 2003; Ishida et al. 2009; Jang et al. 2016; Knull 1980). Astrocytes and neurons take up glucose through glucose transporters (GLUTs) (Jurcovicova 2014). Glucose is phosphorylated by hexokinase, the first committed step to glycolysis. Interestingly, subcellular localization of hexokinase is altered in schizophrenia (Shan et al. 2014). Hexokinase normally localizes to the outer mitochondrial membrane to couple cytosolic glycolysis to mitochondrial oxidative phosphorylation (Abu-Hamad et al. 2008; Wilson 2003). In schizophrenia, the cytosolic to mitochondrial hexokinase ratio was increased in the dorsolateral prefrontal cortex (dlPFC), suggesting uncoupling of bioenergetic processes that produce ATP (Regenold et al. 2012; Shan et al. 2014).

Glucose can enter the pentose-phosphate pathway (PPP), be converted to lactate, or be converted to pyruvate to enter the TCA cycle. The activity of the PPP rate-limiting enzyme glucose-6-phosphate dehydrogenase was not changed in the somatosensory association cortex (Puthumana and Regenold 2019) but was negatively correlated with the expression of hexokinase-1 in a mitochondrial fraction (Regenold et al. 2012). It is through the TCA cycle that intermediate metabolites are produced to fuel ATP production via oxidative phosphorylation. Several studies have reported an association between oxidative stress and the pathophysiology of schizophrenia (Chowdari et al. 2011; Flatow et al. 2013; Schulz et al. 2000), perhaps as a reflection of mitochondrial dysfunction (Prabakaran et al. 2004). A major function of the PPP shunt is to produce NADPH generated from the oxidation of glucose-6-phosphate to reduce glutathione. The enzyme 6-phosphogluconolactonase, which produces NAPDH (the cofactor used to reduce oxidized glutathione), was increased in the dlPFC in schizophrenia (Pennington et al. 2008). Consistent with these findings, a proteomic analysis of 11 schizophrenia subjects found increased abundance of transketolase and glutathione-S-transferase, as well as increased NADPH levels (Martins-de-Souza et al. 2010).

7.1 Seminal Paper 10: Lee et al. (2024)

To examine the relationship between glucose dysregulation and psychosis spectrum disorders, Lee et al. (2024) preformed a meta-analysis in antipsychotic naïve FEP patients. Using transcriptomic data from FEP patients, they then compared this information to data from individuals with early dysglycemia and identified 221 common gene expression signatures. Their findings suggest an intrinsic link between psychosis spectrum disorders and dysglycemic states, independent of the bioenergetic effects of antipsychotic drugs (Lee et al. 2024).

Other work showed an increase in cerebrospinal fluid glucose in individuals with schizophrenia compared to nonpsychotic controls (Warren et al. 2024). Pathway enrichment analysis found connections to endoplasmic reticulum stress and abnormal brain bioenergetics. Additional studies have explored potential pharmacological treatments as candidates for addressing metabolic dysfunction in psychosis spectrum disorder. Metformin was identified as a promising candidate for suggesting that early intervention may be beneficial for patients (Battini et al. 2023).

7.2 Seminal Paper 11: Townsend et al. (2023)

This seminal paper analyzed 36 datasets comprising 1,335 subjects to investigate metabolic dysfunction in glucose metabolism associated with schizophrenia. Using ^{18}FDG-PET (F-18-deoxyglucose positron emission tomography) to measure glucose uptake rates, regional hypometabolism in the frontal cortex in schizophrenia was found to be more pronounced in chronic schizophrenia patients and those on medication. The meta-analysis did not find consistent metabolic alterations in other brain regions except the basal ganglia. These data support the hypothesis that impairment of glucose uptake and utilization may underlie cortical dysfunction in SMI (Townsend et al. 2023).

7.3 Seminal Paper 12: Sullivan et al. (2019a)

Additional work on the fate of glucose was conducted by Sullivan et al. (2019a), who examined metabolic disturbances at a cellular level in schizophrenia. Utilizing laser-capture microdissection and quantitative polymerase chain reaction, significant reductions in transcripts for key glycolytic enzymes, including hexokinase and phosphofructokinase, were identified in dlPFC pyramidal neurons but not astrocytes. They also reported decreased neuronal mRNA expression of the glucose transporters GLUT1 and GLUT3, which suggests that impaired glucose uptake is a neuron-specific mechanism in schizophrenia. Region-level decreases in hexokinase and phosphofructokinase activity were also found in the frontal cortex, suggesting impaired glycolysis. Similar changes were not observed in rodents treated for nine months with haloperidol (Sullivan et al. 2019a). Taken together, findings by Sullivan et al. suggest neuron-specific deficits in the dlPFC in glucose utilization and metabolism that are independent of medication status.

8 Lactate, Brain pH, and ATP in SMI

Glutamate released at the synapse prompts astrocytes to ramp up their energy production and leads to an increase in the production of bioenergetic metabolites, such as lactate. Lactate may then be transferred from astrocytes to neurons via monocarboxylate transporters. Within neurons, lactate serves as a fuel source to generate energy through the TCA cycle. This process, known as the astrocyte-neuron lactate shuttle, facilitates the support of neuronal oxidative phosphorylation. Remarkably, this mechanism enables neurons to sustain energy production through lactate even under high-energy demand conditions, where neurons struggle to boost glycolysis effectively in aerobic conditions.

Lactate plays a crucial role in cognition; during learning tasks in rats, there is a noticeable rise in extracellular lactate levels in the hippocampus. Disrupting the production of lactate or its transfer into neurons leads to memory impairment or amnesia (Suzuki et al. 2011). These data suggest that the astrocyte-neuron lactate shuttle mechanism represents a critical element of brain function that may be disrupted in SMI. As lactate is a weak acid, it is possible that changes in lactate levels may be associated with alterations in brain pH. It is tempting to predict that having a compensatory biological state with increased lactate may reveal an underlying impairment in ATP production. However, debate is ongoing as to whether alterations in lactate levels are indicative of a pathological process linked to the illness or merely a result of medication effects or other postmortem factors influencing lactate levels (Henkel et al. 2022).

8.1 Seminal Paper 13: Rowland et al. (2016)

Using 7T magnetic resonance spectroscopy (MRS), Rowland et al. (2016) measured brain lactate levels *in vivo* in 29 controls and 27 subjects with schizophrenia. Higher lactate levels were found in the anterior cingulate cortex in schizophrenia and were associated with lower scores on the Measurement and Treatment Research to Improve Cognition in Schizophrenia (MATRICS) Consensus Cognitive Battery, which assesses cognitive function, as well as on the University of California, San Diego Performance-Based Skills Assessment, a measure of real-world functional capacity—indicating poorer cognitive and functional performance. This study was the first to show increased lactate levels in living humans with SMI (Rowland et al. 2016).

8.2 Seminal Paper 14: Sullivan et al. (2019b)

Lactate has also been measured in postmortem brain and animal models of SMI. In their study, Sullivan et al. (2019b) found increased dlPFC lactate levels in schizophrenia (N = 16) and in IPSC-derived cortical neurons from a subject with schizophrenia who had the DISC mutation. In the postmortem cohort, lactate levels did not correlate with postmortem interval, pH levels, or age. In contrast, in the mutant GFAP-DISC1 mice (expressing humanized DISC only in astrocytes), lactate levels were reduced in the frontal cortex. Changes in lactate were not detected in Grin1KD mice, CAMK-DISC1 mice (expressing humanized DISC only in frontal cortical neurons), or in rats treated with haloperidol-decanoate for nine months. Taken together, their data suggest that changes in lactate may be dependent on having common perturbations of astrocytes and neurons in the same disease substrate to manifest an environment where excess lactate is produced.

Other work is consistent with the findings described above *in vivo* and in postmortem brain. Lactate and pyruvate (another monocarboxylate and lactate metabolite) were increased in the striatum in schizophrenia (Dean et al. 2016), whereas pyruvate levels were increased in serum and in urine collected from schizophrenia subjects (Yang et al. 2013). Elevated lactate in sporadic SMI and SMI with a rare highly penetrant mutation suggest a common intermediate bioenergetic phenotype driving bioenergetic changes (Sullivan et al. 2019b).

8.3 Seminal Paper 15: Hagihara et al. (2018)

Since lactate is a weak acid, it has been proposed that there may be changes in brain pH in SMI. In a meta-analysis, Hagihara et al. (2018) examined brain tissue pH data from several SMI studies (9 schizophrenia and 5 BAD datasets) and found decreased brain pH in schizophrenia and BAD, compared to healthy controls. Similar to the lactate study discussed in Section 8.2, age and postmortem interval did not impact the findings. Brain pH, lactate, and pyruvate levels were also assessed in several animal models of SMI. Interestingly, decreases in brain pH and increases in lactate were found, consistent with the findings in human subjects. Notably, assessment of rodent models avoids factors such as agonal status, medication treatment, and other variables that may impact the brain in human subjects. These data support the conclusion that lower brain pH and elevated lactate levels are not artifacts but integral to SMI, thus opening avenues for novel therapeutic approaches targeting bioenergetic dysfunction.

8.4 Seminal Paper 16: Du et al. (2014)

While direct biochemical measurements of ATP are quite difficult in the laboratory and highly unfeasible in postmortem tissues, cortical bioenergetic function was innovatively assessed by Du et al. (2014) using ^{31}P MRS. In this study, 26 subjects with schizophrenia or schizoaffective disorder and 26 age-matched gender-matched healthy control subjects underwent ^{31}P-MT-MRS and brain anatomic imaging, utilizing a novel half-helmet head coil with dual-tuned frequency channels. Primary outcomes included measuring the forward rate constant (kf) of creatine kinase enzyme activity, intracellular pH, and several steady-state metabolite ratios of high-energy phosphate-containing compounds. Significant decreases in creatine kinase kf and intracellular pH were found in subjects from the SMI cohort. There was also a significant decrease in the phosphodiester to ATP ratio, suggesting possible accelerated phospholipid metabolism. Taken together, these findings in living patients with SMI suggest a profound impairment in bioenergetic function involving a creatine kinase, an enzyme involved in distributing high-energy phosphate bonds around the cell via synthesis of creatine phosphate.

9 Lipids and Fatty Acids in SMI

Assessment of fatty acids and lipids in SMI has revealed complex and often contradictory findings that may influence disease pathology and patient outcomes (Perica and Delaš 2011). As critical components of cell membranes and signaling pathways, fatty acids and lipids have central roles in a myriad of biological functions (De Carvalho and Caramujo 2018; Yaqoob and Calder 2007). Importantly, proteins are often sequestered via anchoring mechanisms that rely on specialized lipid moieties (Jiang et al. 2018), whereas fatty acids are the building blocks for phospholipids and a critical source of energy via fatty acid oxidation (Bebernitz and Schuster 2002). Research in this field has focused on perturbations in fatty acid metabolism and metabolites in terms of pathological mechanisms, biomarker development, and targets for therapeutic intervention (Bharti et al. 2021; Hu et al. 2023; Liu et al. 2016). Reduced levels of essential polyunsaturated fatty acids are associated with psychopathology in chronically medicated and antipsychotic naïve schizophrenia subjects (Berger et al. 2019; Hu et al. 2023). Studies conducted with omega 6 and/or omega 3 supplementation had varying degrees of effectiveness. Some studies showed a significant decline in Positive and Negative Syndrome Scale (PANSS) scores in patients treated with omega 3 compared to patients treated with placebo (Peet et al. 2001; Mahadik and Evans 1997). However, other randomized controlled trials, found no significant differences in patients' PANSS scores when supplemented with omega 3 (Fenton et al. 2001).

9.1 Seminal Paper 17: Zhou et al. (2019)

In a recent investigation of adolescents with MDD, Zhou et al. (2019) conducted metabolic profiling of blood samples using mass spectrometry on subjects between the age of 6–18 years. Pathway analyses were performed on subjects with MDD who were medication-naïve (N = 52) or who were being treated with psychotropic medications (N = 32). Decreased levels of capric acid, cis-9-palmitoleic acid, dodecanoic acid, oleic acid, and palmitic acid were found in children and adolescents with MDD who were drug-naïve compared with healthy controls. Pathway analysis found altered fatty acid biosynthesis and several amino acid degradation pathways. This study suggests that unique fatty acid metabolic disturbances are associated with MDD in young populations compared to adults (Zhou et al. 2019).

9.2 Seminal Paper 18: Arvindakshan et al. (2003)

Recent work has focused on refining fatty acid supplementation strategies for individuals with schizophrenia. Employing a novel study design, Arvindakshan et al. (2003) administered specific combinations of eicosapentaenoic acid and docosapentaenoic acid (180:120 ratio) twice daily along with antioxidants, such as vitamins E and C, to subjects with schizophrenia (N = 28) and healthy controls (N = 45). During this four-month study, subjects had monthly follow-up visits to assess the impact and efficacy of the supplementation regimen. Fatty acid profiles in red blood cell membranes were significantly improved in schizophrenia, without an increase in plasma lipid peroxides. Clinically, there was a notable reduction in psychopathology, as measured by the Brief Psychiatric Rating Scale (BPRS) and PANSS, along with improvements in quality-of-life scales. This reduction persisted for four months after supplementation was discontinued. The findings of Arvindakshan et al. underscore the potential of targeted nutritional supplementation as an adjunct therapy for SMI.

10 Bioenergetic Function and Medications for SMI

Over the past decade, second-generation (aka "atypical") antipsychotic medications have been widely used to treat SMI. Notably, use of these medications has been formally extended to MDD and BAD, drastically increasing the number of patients being treated with these medications. The history of the rollout of atypical medications offers a cautionary tale. In an early study of olanzapine, Beasley et al. (1996) reported little to no weight gain following six weeks of treatment with olanzapine. Using a tortuous method to report adverse effects, they include a "zero" in the treatment-emergent adverse effects table for weight gain (see Table 8 in Beasley et al. 1996). About five years later, a second study (with some of the same authors) reported an average weight gain of 6.3 kg after about 39 weeks of treatment, suggesting that the atypical antipsychotic medications increased appetite and weight (Kinon et al. 2001). Interestingly, weight gain was not correlated with olanzapine dosage (Kinon et al. 2001). However, the relief at having a much lower rate of extrapyramidal side effects and tardive dyskinesia was transformative. Since then, a substantial body of preclinical and clinical literature indicates that antipsychotic medications—particularly the atypical class—affect metabolism, energy sensing, and inflammation throughout the CNS (Hahn et al. 2011; Henderson 2007; Henderson et al. 2005, 2006; Kowalchuk et al. 2019a, b; Leucht et al. 2012; Ren et al. 2019). Evaluation of the impact of psychotropic medications on bioenergetic processes presents significant challenges, due primarily to the difficulty of isolating the effects of psychotropic medications from those attributable directly to disease.

10.1 Seminal Paper 19: Schulmann et al. (2023)

Are the transcriptomic changes found in postmortem brain from subjects with SMI the result of treatment with medications, or are they part of the underlying disease process? To address this question, Schulmann et al. (2023) analyzed transcriptional profiles from medicated and unmedicated subjects (Schulmann et al. 2023), using reposited RNAseq datasets from the National Institutes of Health Human Brain Core Collection: unmedicated subjects with schizophrenia (N = 23), subjects with schizophrenia treated with antipsychotic medication (N = 65), and untreated healthy controls (N = 113). A transcriptional profile for unmedicated schizophrenia subjects was generated and compared with medicated subjects (Schulmann et al. 2023). In addition, datasets from the Common Mind Consortium were also assessed, including rhesus macaque monkeys treated with vehicle, low-dose haloperidol, high-dose haloperidol, and clozapine for extended periods to simulate the effect of long-term antipsychotic therapy (Hoffman et al. 2019). The study found that exposure to antipsychotics had a distinct effect on biological pathways, including synaptic function, inflammation, and glucose metabolism. The assessment of antipsychotic treatment in rhesus macaque monkeys utilized weighted gene co-expression network analysis to identify gene expression modules. This method revealed both convergent and divergent modules when comparing antipsychotic treatment to gene expression patterns in schizophrenia. Among the divergent modules, glucose homeostasis stood out as having the most significant divergence. This suggests that SMI pathophysiology and antipsychotic medications generate opposing effects on this central bioenergetic function in brain cells.

10.2 Seminal Paper 20: Tye et al. (2022)

Here, insulin-stimulated mTOR/GSK3 signaling in peripheral immune cells was examined in 34 BAD subjects (Tye et al. 2022). Subjects were stratified using the Alda score, a widely used and well-validated clinical index that measures overall and lithium-specific treatment response in BAD studies (Scott et al. 2020). Baseline and insulin-induced protein levels were measured by western blot in buffy coats and peripheral blood mononuclear cells. Increases in phosphorylated mTOR and GSK3β were found only in lithium-responsive patients, suggesting a relationship between insulin-signaling dynamics and lithium efficacy. These findings indicate that immune-metabolic responses may be useful as predictive markers for personalized lithium treatment strategies in BAD and support the hypothesis that insulin-signaling perturbations are a common intermediate molecular phenotype in SMI.

11 Conclusions

Genetic risk for SMI converges on common bioenergetic processes: insulin signaling, glucose uptake and metabolism, lipid and fatty acid metabolism, as well as changes in the levels of bioenergetic molecules including monocarboxylates and ATP. Such changes may be a consequence of mitochondrial dysfunction, acquired or inherited. Notably, SMI may include genetic risk factors and mitochondrial dysfunction that act causally in concert, perhaps leading to more severe disease phenotypes, such as those seen in patients with the DISC translocation. Much of the work presented here focuses on cross-sectional, brain region-specific measures of bioenergetic function, which leaves several important questions open for future resolution:

Are there cell subtype-specific changes in bioenergetic function in SMI? Few studies have this resolution. The advent of iPSCs and mini-brain technology provides an important avenue to assess cell-level changes, with important caveats: iPSCs or mini-brains do not contain all the cell types and networks found *in vivo*. Findings described above, in particular for changes in lactate levels, suggest that some features of bioenergetic dysfunction may only be apparent *in vivo*. Further, iPSCs and cultured mini-brains are quite young (weeks to months old at best), whereas pyramidal neurons *in vivo*, for example, may be decades old (or older!) at the onset of SMI.

Are there micro- and/or macro-circuit changes in bioenergetic function in SMI? Again, most studies do not have the resolution to determine whether there are cell-level changes in bioenergetic function in human SMI substrates. Imaging studies are quite informative, but typical voxels contain hundreds if not thousands of cells, making assessment of cell-to-cell interactions and their perturbations difficult. Notable work in animals has shown how bioenergetic function and neuroplasticity are intertwined. An area that may be better explored involves bioenergetic function and presynaptic vesicle release mechanisms. Pulido and Ryan (2021) found that synaptic vesicle pools are a major source of presynaptic basal energy consumptions. The basal metabolic processes occur due to H^+ efflux from the vesicle lumen, creating compensatory V-ATPase activity, resulting in a constant energetic burden. Small changes in the capacity to synthesize ATP or other bioenergetic intermediates could impact presynaptic release mechanisms. Further, small changes in intracellular pH could also change the dynamics of vesicular packaging of neurotransmitters such as glutamate, which depend on a H^+ gradient (Egashira et al. 2016; Eriksen et al. 2020).

Do bioenergetic changes impact brain oscillations? Oscillatory activity is disrupted in SMI, in particular in schizophrenia. It is unclear, however, if metabolic interventions alone might impact oscillations, which are an indirect measure of healthy and coordinated brain activity.

Are there developmental windows or milestones that may be impacted by bioenergetic dysfunction? MDD and BAD are often episodic, with onset often well after the end of typical brain development and maturation. Further, persons with schizophrenia typically do not manifest the classic signs and symptoms until well into adulthood. Could it be that bioenergetic changes at different points during or post brain development impact the disease phenotypes? Timing of successful metabolic interventions might differ based on disease phenotypes, making development of assessment tools and biomarkers for bioenergetic traits a critical next step for the field.

Are there metabolic interventions for bioenergetic dysfunction? There are several promising areas related to interventions, including the repurposing of FDA approved drugs. For example, metformin is often prescribed along with antipsychotics to help manage glucose abnormalities (Cernea et al. 2020; Hasnain et al. 2011), and there are studies of the PPAR agonist pioglitazone in subjects with schizophrenia that show improvement in negative symptom scales (Iranpour et al. 2016). Other approaches involve dietary modification. Persons afflicted with type 2 diabetes may become healthy with changes in eating habits and dietary intake (Astrup 2001; Steyn et al. 2004). More precise targeting via changes in diet includes the ketogenic diet, which switches the bioenergetic supply of energy from glucose to ketones (Sinha and Kossoff 2005). Such a diet has well-proven efficacy for pediatric seizures (from six randomized controlled trials), suggesting that the ketogenic diet alters brain plasticity (Desli et al. 2022; IJff et al. 2016; Lambrechts et al. 2017; Pizzo et al. 2022). If plasticity can be changed via an intervention for seizures, such an approach may be deployed for other brain diseases that have neuroplastic elements, which includes all SMI diagnoses. A recent open clinical trial of the ketogenic diet showed efficacy for SMI (Longhitano et al. 2024; Sethi et al. 2024); this study supports a growing literature of case reports and case series suggesting the potential promise of ketogenic interventions.

We do wish to point out, however, some important caveats: While there is no question regarding the efficacy of antipsychotics for mood and psychosis, metabolic changes are hard to disentangle in patients with SMI who take antipsychotics. Work described above suggests opposing contributions of antipsychotics and genetic risk for glucose metabolism. In addition, FDG-PET studies have found that antipsychotics (including haloperidol, olanzapine, and sertindole) increase the metabolic rate (as measured by glucose uptake) in the frontal cortices (Buchsbaum et al. 1992, 2007, 2009; Turkheimer et al. 2019). This may occur without secondary normalization of "oxidative capacity" (i.e., the ability for mitochondria to oxidize substrates through oxidative phosphorylation) in the brain (Turkheimer et al. 2019). That is, glucose can be taken up more effectively in the presence of antipsychotics, but the neuron or astrocyte may be unequipped to manage the glucose-load downstream through either glycolysis or oxidative phosphorylation (Turkheimer et al. 2019). Such findings may help reconcile the divergent effects found in transcriptional profiles associated with SMI.

Finally, while it is somewhat easier to link bioenergetic defects with cognitive measures, connection to psychosis, thought disorder, mood, and mood cycling is more difficult. This is in part due to the difficulty of recapitulating complex human behaviors, such as suicidal ideation, low mood, and anhedonia in model systems. Put simply, it is hard to ask a rodent if they hear voices. Nevertheless, there is growing evidence for circadian (and possibly ultradian) rhythm abnormalities in SMI and, in particular, BAD. Loss of coordination of bioenergetic processes with gene expression and other biological processes may be a significant driver of disease phenotypes, especially mood. For example, a recent review describes a "metabolic overdrive hypothesis" for BAD (Campbell and Campbell 2024). This hypothesis suggests that manic states are a specific type of bioenergetic dysfunction involving glutaminolysis and compensatory hyperglycolysis. Glutaminolysis involves the breakdown of the amino acid glutamine to generate ATP and lactate, followed by a compensatory hyperglycolytic state (Mansur et al. 2020). Other brain disorders, including traumatic brain injury and epilepsy, are associated with manic states with similar physiological signs. Shifting away from glucose as a fuel source to ketones might provide a metabolic intervention for the postulated overdrive hypothesis. Hope for helping the afflicted is supported by the observation that genetic rescue of adult Grin1KD animals partially rescued neuroplastic measures of brain function, suggesting that bioenergetic changes may be reversible even in adults following closure of developmental windows in the brain (Venkatesan et al. 2023). It follows that development of dietary, pharmacological, and even genetic interventions may lead to better patient outcomes for these often devastating disorders.

Acknowledgments Hunter Eby, Ali Sajid Imami, and Robert McCullumsmith contributed equally to this work.

References

Abu-Hamad S, Zaid H, Israelson A, Nahon E, and Shoshan-Barmatz V (2008) Hexokinase-I Protection against Apoptotic Cell Death Is Mediated via Interaction with the Voltage-Dependent Anion Channel-1: Mapping the Site of Binding. J Biol Chem 283 (19):13482–13490. https://doi.org/10.1074/jbc.M708216200

Agarwal SM, Caravaggio F, Costa-Dookhan KA, et al. (2020) Brain Insulin Action in Schizophrenia: Something Borrowed and Something New. Neuropharmacology 163:107633. https://doi.org/10.1016/j.neuropharm.2019.05.010

Andreazza AC, and Nierenberg AA (2018) Mitochondrial Dysfunction: At the Core of Psychiatric Disorders? Biol Psychiatry 83 (9):718–719. https://doi.org/10.1016/j.biopsych.2018.03.004

Appel KE, and Farr CB (1929) The Blood Sugar Reaction to Insulin in Psychoses. Arch Neurol Psychiatry 21 (1):145–148. https://doi.org/10.1001/archneurpsyc.1929.02210190148009

Arvindakshan M, Ghate M, Ranjekar PK, Evans DR, and Mahadik SP (2003) Supplementation with a Combination of Ω-3 Fatty Acids and Antioxidants (Vitamins E and C) Improves the Outcome of Schizophrenia. Schizophr Res 62 (3):195–204. https://doi.org/10.1016/S0920-9964(02)00284-0

Astrup A (2001) Healthy Lifestyles in Europe: Prevention of Obesity and Type II Diabetes by Diet and Physical Activity. Public Health Nutr 4 (2B):499–515. http://dx.doi.org/10.1079/PHN2001136

Barrett TM, and Serre P (1924) Blood Analysis and Sugar Tolerance Tests in Mental Disease. J Nerv Ment Dis 59 (6):561–570

Battini V, Cirnigliaro G, Leuzzi R, et al. (2023) The Potential Effect of Metformin on Cognitive and Other Symptom Dimensions in Patients with Schizophrenia and Antipsychotic-Induced Weight Gain: A Systematic Review, Meta-Analysis, and Meta-Regression. Front Psychiatry 14. https://doi.org/10.3389/fpsyt.2023.1215807

Beasley CM, Jr., Tollefson G, Tran P, et al. (1996) Olanzapine versus Placebo and Haloperidol: Acute Phase Results of the North American Double-Blind Olanzapine Trial. Neuropsychopharmacology 14 (2):111–123. https://doi.org/10.1016/0893-133X(95)00069-P

Bebernitz GR, and Schuster HF (2002) The Impact of Fatty Acid Oxidation on Energy Utilization: Targets and Therapy. Curr Pharm Des 8 (14):1199–1227. http://dx.doi.org/10.2174/1381612023394692

Ben-Shachar D, and Karry R (2008) Neuroanatomical Pattern of Mitochondrial Complex I Pathology Varies between Schizophrenia, Bipolar Disorder and Major Depression. PLoS One 3 (11):e3676. https://doi.org/10.1371/journal.pone.0003676

Berger M, Nelson B, Markulev C, et al. (2019) Relationship between Polyunsaturated Fatty Acids and Psychopathology in the Neurapro Clinical Trial. Front Psychiatry 10:393. https://doi.org/10.3389/fpsyt.2019.00393

Bharti V, Bhardwaj A, Hood K, et al. (2021) A Systematic Review and Meta-Analysis of Lipid Metabolomic Signatures of Major Depressive Disorder. J Psychiatr Res 139:197–205. https://doi.org/10.1016/j.jpsychires.2021.05.036

Blackwood DH, Fordyce A, Walker MT, et al. (2001) Schizophrenia and Affective Disorders: Cosegregation with a Translocation at Chromosome 1q42 That Directly Disrupts Brain-Expressed Genes: Clinical and P300 Findings in a Family. Am J Hum Genet 69 (2):428–433. https://doi.org/10.1086/321969

Bodenstein DF, Kim HK, Brown NC, et al. (2019) Mitochondrial DNA Content and Oxidation in Bipolar Disorder and Its Role across Brain Regions. NPJ Schizophr 5:21. https://doi.org/10.1038/s41537-019-0089-5

Buchsbaum MS, Haznedar MM, Aronowitz J, et al. (2007) FDG-PET in Never-Previously Medicated Psychotic Adolescents Treated with Olanzapine or Haloperidol. Schizophr Res 94 (1–3):293–305. https://doi.org/10.1016/j.schres.2007.04.027

Buchsbaum MS, Haznedar MM, Newmark RE, et al. (2009) FDG-PET and MRI Imaging of the Effects of Sertindole and Haloperidol in the Prefrontal Lobe in Schizophrenia. Schizophr Res 114 (1–3):161–171. https://doi.org/10.1016/j.schres.2009.07.015

Buchsbaum MS, Potkin SG, Siegel BV, Jr., et al. (1992) Striatal Metabolic Rate and Clinical Response to Neuroleptics in Schizophrenia. Arch Gen Psychiatry 49 (12):966–974. https://psycnet.apa.org/doi/10.1001/archpsyc.1992.01820120054008

Campbell IH, and Campbell H (2024) The Metabolic Overdrive Hypothesis: Hyperglycolysis and Glutaminolysis in Bipolar Mania. Mol Psychiatr 29 (5):1521–1527. https://doi.org/10.1038/s41380-024-02431-w

Castellani S, Giannini AJ, Boeringa JA, and Adams PM (1982) Phencyclidine Intoxication: Assessment of Possible Antidotes. J Toxicol Clin Toxicol 19 (3):313–319. https://doi.org/10.3109/15563658209025737

Cernea S, Dima L, Correll CU, and Manu P (2020) Pharmacological Management of Glucose Dysregulation in Patients Treated with Second-Generation Antipsychotics. Drugs 80 (17):1763–1781. https://doi.org/10.1007/s40265-020-01393-x

Ceylan D, Karacicek B, Tufekci KU, et al. (2023) Mitochondrial DNA Oxidation, Methylation, and Copy Number Alterations in Major and Bipolar Depression. Front Psychiatry 14:1304660. https://doi.org/10.3389/fpsyt.2023.1304660

Chong ZS, Khong ZJ, Tay SH, and Ng SY (2022) Metabolic Contributions to Neuronal Deficits Caused by Genomic Disruption of Schizophrenia Risk Gene SETD1A. Schizophrenia 8 (1):115. https://doi.org/10.1038/s41537-022-00326-9

Chouinard V-A, Henderson DC, Dalla Man C, et al. (2019) Impaired Insulin Signaling in Unaffected Siblings and Patients with First-Episode Psychosis. Mol Psychiatr 24 (10):1513–1522. https://doi.org/10.1038/s41380-018-0045-1

Chowdari KV, Bamne MN, and Nimgaonkar VL (2011) Genetic Association Studies of Antioxidant Pathway Genes and Schizophrenia. Antioxid Redox Signal 15 (7):2037–2045. https://doi.org/10.1089/ars.2010.3508

Cohen BD, Rosenbaum G, Luby ED, and Gottlieb JS (1962) Comparison of Phencyclidine Hydrochloride (Sernyl) with Other Drugs. Simulation of Schizophrenic Performance with Phencyclidine Hydrochloride (Sernyl), Lysergic Acid Diethylamide (LSD-25), and Amobarbital (Amytal) Sodium; II. Symbolic and Sequential Thinking. Arch Gen Psychiatry 6 (5):395–401. https://doi.org/10.1001/archpsyc.1962.01710230063007

Dalmau J, Gleichman AJ, Hughes EG, et al. (2008) Anti-NMDA-Receptor Encephalitis: Case Series and Analysis of the Effects of Antibodies. Lancet Neurol 7 (12):1091–1098. https://doi.org/10.1016/S1474-4422(08)70224-2

Dastani Z, Hivert MF, Timpson N, et al. (2012) Novel Loci for Adiponectin Levels and Their Influence on Type 2 Diabetes and Metabolic Traits: A Multi-Ethnic Meta-Analysis of 45,891 Individuals. PLoS Genet 8 (3):e1002607. https://doi.org/10.1371/journal.pgen.1002607

David AS, and Nicholson T (2015) Are Neurological and Psychiatric Disorders Different? Br J Psychiatry 207 (5):373–374. https://doi.org/10.1192/bjp.bp.114.158550

De Carvalho CC, and Caramujo MJ (2018) The Various Roles of Fatty Acids. Molecules 23 (10):2583. https://doi.org/10.3390/molecules23102583

Dean B, Thomas N, Scarr E, and Udawela M (2016) Evidence for Impaired Glucose Metabolism in the Striatum, Obtained Postmortem, from Some Subjects with Schizophrenia. Transl Psychiatry 6 (11):e949. https://doi.org/10.1038/tp.2016.226

Dear R, Wagstyl K, Seidlitz J, et al. (2024) Cortical Gene Expression Architecture Links Healthy Neurodevelopment to the Imaging, Transcriptomics and Genetics of Autism and Schizophrenia. Nat Neurosci 27 (6):1075–1086. https://doi.org/10.1038/s41593-024-01624-4

Derakhshan F, and Toth C (2013) Insulin and the Brain. Curr Diabetes Rev 9 (2):102–116. http://dx.doi.org/10.2174/15733998113090200002

Desli E, Spilioti M, Evangeliou A, et al. (2022) The Efficacy and Safety of Ketogenic Diets in Drug-Resistant Epilepsy in Children and Adolescents: A Systematic Review of Randomized Controlled Trials. Curr Nutr Rep 11 (2):102–116. https://doi.org/10.1007/s13668-022-00405-4

Du F, Cooper AJ, Thida T, et al. (2014) *In Vivo* Evidence for Cerebral Bioenergetic Abnormalities in Schizophrenia Measured Using ^{31}P Magnetization Transfer Spectroscopy. JAMA Psychiatry 71 (1):19–27. https://doi.org/10.1001/jamapsychiatry.2013.2287

Egashira Y, Takase M, Watanabe S, et al. (2016) Unique pH Dynamics in GABAergic Synaptic Vesicles Illuminates the Mechanism and Kinetics of GABA Loading. PNAS 113 (38):10702–10707. https://doi.org/10.1073/pnas.1604527113

Eriksen J, Li F, and Edwards RH (2020) The Mechanism and Regulation of Vesicular Glutamate Transport: Coordination with the Synaptic Vesicle Cycle. Biochim Biophys Acta Biomembr 1862 (12):183259. https://doi.org/10.1016/j.bbamem.2020.183259

Fenton WS, Dickerson F, Boronow J, Hibbeln JR, and Knable M (2001) A Placebo-Controlled Trial of Omega-3 Fatty Acid (Ethyl Eicosapentaenoic Acid) Supplementation for Residual Symptoms and Cognitive Impairment in Schizophrenia. Am J Psychiatry 158 (12):2071–2074. https://doi.org/10.1176/appi.ajp.158.12.2071

Fernandez AM, and Torres-Alemán I (2012) The Many Faces of Insulin-Like Peptide Signalling in the Brain. Nat Rev Neurosci 13 (4):225–239. https://doi.org/10.1038/nrn3209

Flatow J, Buckley P, and Miller BJ (2013) Meta-Analysis of Oxidative Stress in Schizophrenia. Biol Psychiatry 74 (6):400–409. https://doi.org/10.1016/j.biopsych.2013.03.018

Florez JC (2007) The New Type 2 Diabetes Gene *TCF7L2*. Curr Opin Clin Nutr Metab Care 10 (4):391–396. https://doi.org/10.1097/mco.0b013e3281e2c9be

Fromer M, Pocklington AJ, Kavanagh DH, et al. (2014) De Novo Mutations in Schizophrenia Implicate Synaptic Networks. Nature 506 (7487):179–184. https://doi.org/10.1038/nature12929

Gimenez-Palomo A, Guitart-Mampel M, Roque G, et al. (2024) Aerobic Capacity and Mitochondrial Function in Bipolar Disorder: A Longitudinal Study during Acute Phases and after Clinical Remission. Front Psychiatry 15:1386286. https://doi.org/10.3389/fpsyt.2024.1386286

Grant SF, Thorleifsson G, Reynisdottir I, et al. (2006) Variant of Transcription Factor 7-Like 2 (*TCF7L2*) Gene Confers Risk of Type 2 Diabetes. Nat Genet 38 (3):320–323. https://doi.org/10.1038/ng1732

Hagihara H, Catts VS, Katayama Y, et al. (2018) Decreased Brain pH as a Shared Endophenotype of Psychiatric Disorders. Neuropsychopharmacology 43 (3):459–468. https://doi.org/10.1038/npp.2017.167

Hahn M, Chintoh A, Giacca A, et al. (2011) Atypical Antipsychotics and Effects of Muscarinic, Serotonergic, Dopaminergic and Histaminergic Receptor Binding on Insulin Secretion *in Vivo*: an Animal Model. Schizophr Res 131 (1–3):90–95. https://doi.org/10.1016/j.schres.2011.06.004

Hasnain M, Fredrickson SK, and Vieweg WV (2011) Metformin for Obesity and Glucose Dysregulation in Patients with Schizophrenia Receiving Antipsychotic Drugs. J Psychopharm 25 (6):715–721. https://doi.org/10.1177/0269881110389214

Havrankova J, Schmechel D, Roth J, and Brownstein M (1978) Identification of Insulin in Rat Brain. PNAS 75 (11):5737–5741. https://doi.org/10.1073/pnas.75.11.5737

Henderson DC (2007) Weight Gain with Atypical Antipsychotics: Evidence and Insights. J Clin Psychiatry 68 (Suppl 12):18–26

Henderson DC, Cagliero E, Copeland PM, et al. (2005) Glucose Metabolism in Patients with Schizophrenia Treated with Atypical Antipsychotic Agents: A Frequently Sampled Intravenous Glucose Tolerance Test and Minimal Model Analysis. Arch Gen Psychiatry 62 (1):19–28. https://doi.org/10.1001/archpsyc.62.1.19

Henderson DC, Copeland PM, Borba CP, et al. (2006) Glucose Metabolism in Patients with Schizophrenia Treated with Olanzapine or Quetiapine: A Frequently Sampled Intravenous Glucose Tolerance Test and Minimal Model Analysis. J Clin Psychiatry 67 (5):789–797. https://doi.org/10.4088/jcp.v67n0513

Henkel ND, Wu X, O'Donovan SM, et al. (2022) Schizophrenia: A Disorder of Broken Brain Bioenergetics. Mol Psychiatr 27 (5):2393–2404. https://doi.org/10.1038/s41380-022-01494-x

Hennah W, Thomson P, McQuillin A, et al. (2009) DISC1 Association, Heterogeneity and Interplay in Schizophrenia and Bipolar Disorder. Mol Psychiatr 14 (9):865–873. https://doi.org/10.1038/mp.2008.22

Hodgkinson CA, Goldman D, Jaeger J, et al. (2004) Disrupted in Schizophrenia 1 (DISC1): Association with Schizophrenia, Schizoaffective Disorder, and Bipolar Disorder. Am J Hum Genet 75 (5):862–872. https://doi.org/10.1086/425586

Hoffman GE, Bendl J, Voloudakis G, et al. (2019) CommonMind Consortium Provides Transcriptomic and Epigenomic Data for Schizophrenia and Bipolar Disorder. Sci Data 6 (1):180. https://doi.org/10.1038/s41597-019-0183-6

Hu J, Ji Y, Lang X, and Zhang X-Y (2023) Prevalence and Clinical Correlates of Abnormal Lipid Metabolism in First-Episode and Drug-Naïve Patients with Major Depressive Disorder: A Large-Scale Cross-Sectional Study. J Psychiatr Res 163:55–62. https://doi.org/10.1016/j.jpsychires.2023.05.016

Iizuka T, Sakai F, Ide T, et al. (2008) Anti-NMDA Receptor Encephalitis in Japan: Long-Term Outcome without Tumor Removal. Neurology 70 (7):504–511. https://doi.org/10.1212/01.wnl.0000278388.90370.c3

IJff D, Postulart D, Lambrechts D, et al. (2016) Cognitive and Behavioral Impact of the Ketogenic Diet in Children and Adolescents with Refractory Epilepsy: A Randomized Controlled Trial. Epilepsy Behav 60:153–157. https://doi.org/10.1016/j.yebeh.2016.04.033

Ikemoto A, Bole DG, and Ueda T (2003) Glycolysis and Glutamate Accumulation into Synaptic Vesicles. Role of Glyceraldehyde Phosphate Dehydrogenase and 3-Phosphoglycerate Kinase. J Biol Chem 278 (8):5929–5940. https://doi.org/10.1074/jbc.M211617200

Iranpour N, Zandifar A, Farokhnia M, et al. (2016) The Effects of Pioglitazone Adjuvant Therapy on Negative Symptoms of Patients with Chronic Schizophrenia: A Double-Blind and Placebo-Controlled Trial. Hum Psychopharmacol 31 (2):103–112. https://doi.org/10.1002/hup.2517

Ishida A, Noda Y, and Ueda T (2009) Synaptic Vesicle-Bound Pyruvate Kinase Can Support Vesicular Glutamate Uptake. Neurochem Res 34 (5):807–818. https://doi.org/10.1007/s11064-008-9833-3

Jang S, Nelson JC, Bend EG, et al. (2016) Glycolytic Enzymes Localize to Synapses under Energy Stress to Support Synaptic Function. Neuron 90 (2):278–291. https://doi.org/10.1016/j.neuron.2016.03.011

Jiang H, Zhang X, Chen X, et al. (2018) Protein Lipidation: Occurrence, Mechanisms, Biological Functions, and Enabling Technologies. Chem Rev 118 (3):919–988. https://doi.org/10.1021/acs.chemrev.6b00750

Jurcovicova J (2014) Glucose Transport in Brain: Effect of Inflammation. Endocr Regul 48 (1):35–48. https://doi.org/10.4149/endo_2014_01_35

Kasanin J (1926) The Blood Sugar Curve in Mental Disease: II. The Schizophrenic (Dementia Praecox) Groups. Arch Neurol Psychiatry 16 (4):414–419. https://doi.org/10.1001/ARCHNEURPSYC.1926.02200280022002

Kathuria A, Lopez-Lengowski K, McPhie D, Cohen BM, and Karmacharya R (2023) Disease-Specific Differences in Gene Expression, Mitochondrial Function and Mitochondria-Endoplasmic Reticulum Interactions in iPSC-Derived Cerebral Organoids and Cortical Neurons in Schizophrenia and Bipolar Disorder. Discov Ment Health 3 (1):8. https://doi.org/10.1007/s44192-023-00031-8

Kato T (2005) Mitochondrial Dysfunction in Bipolar Disorder: From ^{31}P-Magnetic Resonance Spectroscopic Findings to Their Molecular Mechanisms. Int Rev Neurobiol 63:21–40. https://doi.org/10.1016/S0074-7742(05)63002-4

Kinon BJ, Basson BR, Gilmore JA, and Tollefson GD (2001) Long-Term Olanzapine Treatment: Weight Change and Weight-Related Health Factors in Schizophrenia. J Clin Psychiatry 62 (2):92–100. http://dx.doi.org/10.4088/JCP.v62n0204

Knull HR (1980) Compartmentation of Glycolytic-Enzymes in Nerve-Endings as Determined by Glutaraldehyde Fixation. J Biol Chem 255 (13):6439–6444. https://doi.org/10.1016/S0021-9258(18)43758-1

Kooy FH (1919) Hyperglyaemia in Mental Disorders. Brain 42 (3):214–290.

Kowalchuk C, Castellani LN, Chintoh A, et al. (2019a) Antipsychotics and Glucose Metabolism: How Brain and Body Collide. Am J Physiol Endocrinol Metab 316 (1):E1–E15. https://doi.org/10.1152/ajpendo.00164.2018

Kowalchuk C, Kanagasundaram P, Belsham DD, and Hahn MK (2019b) Antipsychotics Differentially Regulate Insulin, Energy Sensing, and Inflammation Pathways in Hypothalamic Rat Neurons. Psychoneuroendocrinology 104:42–48. https://doi.org/10.1016/j.psyneuen.2019.01.029

Kuang H, Duong A, Jeong H, Zachos K, and Andreazza AC (2018) Lactate in Bipolar Disorder: A Systematic Review and Meta-Analysis. Psychiatry Clin Neurosci 72 (8):546–555. https://doi.org/10.1111/pcn.12671

Lam XJ, Xu B, Yeo PL, Cheah PS, and Ling KH (2023) Mitochondria Dysfunction and Bipolar Disorder: From Pathology to Therapy. IBRO Neurosci Rep 14:407–418. https://doi.org/10.1016/j.ibneur.2023.04.002

Lambrechts DA, de Kinderen RJ, Vles JS, et al. (2017) A Randomized Controlled Trial of the Ketogenic Diet in Refractory Childhood Epilepsy. Acta Neurol Scand 135 (2):231–239. https://doi.org/10.1111/ane.12592

Lee J, Xue X, Au E, et al. (2024) Glucose Dysregulation in Antipsychotic-Naive First-Episode Psychosis: In Silico Exploration of Gene Expression Signatures. Transl Psychiatry 14 (1):19. https://doi.org/10.1038/s41398-023-02716-8

Lemche E, Killick R, Mitchell J, et al. (2024) Molecular Mechanisms Linking Type 2 Diabetes Mellitus and Late-Onset Alzheimer's Disease: A Systematic Review and Qualitative Meta-Analysis. Neurobiol Dis 196:106485. https://doi.org/10.1016/j.nbd.2024.106485

Leucht S, Tardy M, Komossa K, et al. (2012) Maintenance Treatment with Antipsychotic Drugs for Schizophrenia. Cochrane Database Syst Rev 16 (5):CD008016. https://doi.org/10.1002/14651858.CD008016.pub2

Levy JC, Matthews DR, and Hermans MP (1998) Correct Homeostasis Model Assessment (HOMA) Evaluation Uses the Computer Program. Diabetes Care 21 (12):2191–2192. https://doi.org/10.2337/diacare.21.12.2191

Lin JR, Cai Y, Zhang Q, et al. (2016) Integrated Post-GWAS Analysis Sheds New Light on the Disease Mechanisms of Schizophrenia. Genetics 204 (4):1587–1600. https://doi.org/10.1534/genetics.116.187195

Liu X, Li J, Zheng P, et al. (2016) Plasma Lipidomics Reveals Potential Lipid Markers of Major Depressive Disorder. Anal Bioanal Chem 408 (23):6497–6507. https://doi.org/10.1007/s00216-016-9768-5

Longhitano C, Finlay S, Peachey I, et al. (2024) The Effects of Ketogenic Metabolic Therapy on Mental Health and Metabolic Outcomes in Schizophrenia and Bipolar Disorder: A Randomized Controlled Clinical Trial Protocol. Front Nutr 11:1444483. https://doi.org/10.3389/fnut.2024.1444483

Mahadik S, and Evans D (1997) Essential Fatty Acids in the Treatment of Schizophrenia. Med Actual 33 (1):5–17

Mansur RB, Lee Y, McIntyre RS, and Brietzke E (2020) What Is Bipolar Disorder? A Disease Model of Dysregulated Energy Expenditure. Neurosci Biobehav Rev 113:529–545. https://doi.org/10.1016/j.neubiorev.2020.04.006

Martins-de-Souza D, Maccarrone G, Wobrock T, et al. (2010) Proteome Analysis of the Thalamus and Cerebrospinal Fluid Reveals Glycolysis Dysfunction and Potential Biomarkers Candidates for Schizophrenia. J Psychiatr Res 44 (16):1176–1189. https://doi.org/10.1016/j.jpsychires.2010.04.014

Mergenthaler P, Lindauer U, Dienel GA, and Meisel A (2013) Sugar for the Brain: The Role of Glucose in Physiological and Pathological Brain Function. Trends Neurosci 36 (10):587–597. https://doi.org/10.1016/j.tins.2013.07.001

Mielnik CA, Binko MA, Chen Y, et al. (2021) Consequences of NMDA Receptor Deficiency Can Be Rescued in the Adult Brain. Mol Psychiatr 26 (7):2929–2942. https://doi.org/10.1038/s41380-020-00859-4

Mitteroecker P, and Merola GP (2024) The Cliff Edge Model of the Evolution of Schizophrenia: Mathematical, Epidemiological, and Genetic Evidence. Neurosci Biobehav Rev 160:105636. https://doi.org/10.1016/j.neubiorev.2024.105636

Mohn AR, Gainetdinov RR, Caron MG, and Koller BH (1999) Mice with Reduced NMDA Receptor Expression Display Behaviors Related to Schizophrenia. Cell 98 (4):427–436. https://doi.org/10.1016/S0092-8674(00)81972-8

Nishioka M, Takayama J, Sakai N, et al. (2023) Deep Exome Sequencing Identifies Enrichment of Deleterious Mosaic Variants in Neurodevelopmental Disorder Genes and Mitochondrial tRNA Regions in Bipolar Disorder. Mol Psychiatr 28 (10):4294–4306. https://doi.org/10.1038/s41380-023-02096-x

Norkett R, Lesept F, and Kittler JT (2020) DISC1 Regulates Mitochondrial Trafficking in a Miro1-GTP-Dependent Manner. Front Cell Dev Biol 8:449. https://doi.org/10.3389/fcell.2020.00449

Norkett R, Modi S, and Kittler JT (2017) Mitochondrial Roles of the Psychiatric Disease Risk Factor DISC1. Schizophr Res 187:47–54. https://doi.org/10.1016/j.schres.2016.12.025

OCEBM Levels of Evidence Working Group (2011) The Oxford 2011 Levels of Evidence. http://www.cebm.net/index.aspx?o=5653

Park SJ, Lee SB, Suh Y, et al. (2017) DISC1 Modulates Neuronal Stress Responses by Gate-Keeping ER-Mitochondria Ca^{2+} Transfer through the MAM. Cell Rep 21 (10):2748–2759. https://doi.org/10.1016/j.celrep.2017.11.043

Peet M, Brind J, Ramchand C, Shah S, and Vankar G (2001) Two Double-Blind Placebo-Controlled Pilot Studies of Eicosapentaenoic Acid in the Treatment of Schizophrenia. Schizophr Res 49 (3):243–251. https://doi.org/10.1016/S0920-9964(00)00083-9

Pennington K, Beasley CL, Dicker P, et al. (2008) Prominent Synaptic and Metabolic Abnormalities Revealed by Proteomic Analysis of the Dorsolateral Prefrontal Cortex in Schizophrenia and Bipolar Disorder. Mol Psychiatr 13 (12):1102–1117. https://doi.org/10.1038/sj.mp.4002098

Perica MM, and Delaš I (2011) Essential Fatty Acids and Psychiatric Disorders. Nutr Clin Pract 26 (4):409–425. https://doi.org/10.1177/0884533611411306

Pillinger T, Beck K, Gobjila C, et al. (2017) Impaired Glucose Homeostasis in First-Episode Schizophrenia: A Systematic Review and Meta-Analysis. JAMA Psychiatry 74 (3):261–269. https://doi.org/10.1001/jamapsychiatry.2016.3803

Pizzo F, Collotta AD, Di Nora A, et al. (2022) Ketogenic Diet in Pediatric Seizures: A Randomized Controlled Trial Review and Meta-Analysis. Expert Rev Neurother 22 (2):169–177. https://doi.org/10.1080/14737175.2022.2030220

Prabakaran S, Swatton JE, Ryan MM, et al. (2004) Mitochondrial Dysfunction in Schizophrenia: Evidence for Compromised Brain Metabolism and Oxidative Stress. Mol Psychiatr 9 (7):684–697. https://doi.org/10.1038/sj.mp.4001511

Pulido C, and Ryan TA (2021) Synaptic Vesicle Pools Are a Major Hidden Resting Metabolic Burden of Nerve Terminals. Sci Adv 7 (49):eabi9027. https://doi.org/10.1126/sciadv.abi9027

Puthumana JS, and Regenold WT (2019) Glucose-6-Phosphate Dehydrogenase Activity in Bipolar Disorder and Schizophrenia: Relationship to Mitochondrial Impairment. J Psychiatr Res 112:99–103. https://doi.org/10.1016/j.jpsychires.2019.03.004

Raphael T, and Parsons JP (1921) Blood Sugar Studies in Dementia Praecox and Manic-Depressive Insanity. Arch Neurol Psychiatry 5 (6):687–709. https://doi.org/10.1001/archneurpsyc.1921.02180300052005

Regenold WT, Pratt M, Nekkalapu S, et al. (2012) Mitochondrial Detachment of Hexokinase 1 in Mood and Psychotic Disorders: Implications for Brain Energy Metabolism and Neurotrophic Signaling. J Psychiatr Res 46 (1):95–104. https://doi.org/10.1016/j.jpsychires.2011.09.018

Ren L, Zhou X, Huang X, Wang C, and Li Y (2019) The IRS/PI3K/Akt Signaling Pathway Mediates Olanzapine-Induced Hepatic Insulin Resistance in Male Rats. Life Sci 217:229–236. https://doi.org/10.1016/j.lfs.2018.12.015

Richards AL, Cardno A, Harold G, et al. (2022) Genetic Liabilities Differentiating Bipolar Disorder, Schizophrenia, and Major Depressive Disorder, and Phenotypic Heterogeneity in Bipolar Disorder. JAMA Psychiatry 79 (10):1032–1039. https://doi.org/10.1001/jamapsychiatry.2022.2594

Rowland LM, Pradhan S, Korenic S, et al. (2016) Elevated Brain Lactate in Schizophrenia: A 7T Magnetic Resonance Spectroscopy Study. Transl Psychiatry 6 (11):e967. https://doi.org/10.1038/tp.2016.239

Sackett DL, Rosenberg WM, Gray JA, Haynes RB, and Richardson WS (1996) Evidence Based Medicine: What It Is and What It Isn't. BMJ 312 (7023):71–72. https://doi.org/10.1136/bmj.312.7023.71

Sahay S, Pulvender P, Rami Reddy MVSR, McCullumsmith RE, and O'Donovan SM (2024) Metabolic Insights into Neuropsychiatric Illnesses and Ketogenic Therapies: A Transcriptomic View. Int J Mol Sci 25 (15):8266. https://doi.org/10.3390/ijms25158266

Schizophrenia Working Group of the Psychiatric Genomics Consortium (2014) Biological Insights from 108 Schizophrenia-Associated Genetic Loci. Nature 511 (7510):421–427. https://doi.org/10.1038/nature13595

Schulmann A, Marenco S, Vawter MP, et al. (2023) Antipsychotic Drug Use Complicates Assessment of Gene Expression Changes Associated with Schizophrenia. Transl Psychiatry 13 (1):93. https://doi.org/10.1038/s41398-023-02392-8

Schulz JB, Lindenau J, Seyfried J, and Dichgans J (2000) Glutathione, Oxidative Stress and Neurodegeneration. Eur J Biochem 267 (16):4904–4911. https://doi.org/10.1046/j.1432-1327.2000.01595.x

Scott J, Etain B, Manchia M, et al. (2020) An Examination of the Quality and Performance of the Alda Scale for Classifying Lithium Response Phenotypes. Bipolar Disord 22 (3):255–265. https://doi.org/10.1111/bdi.12829

Sethi S, Wakeham D, Ketter T, et al. (2024) Ketogenic Diet Intervention on Metabolic and Psychiatric Health in Bipolar and Schizophrenia: A Pilot Trial. Psychiatry Res 335:115866. https://doi.org/10.1016/j.psychres.2024.115866

Shan D, Mount D, Moore S, et al. (2014) Abnormal Partitioning of Hexokinase 1 Suggests Disruption of a Glutamate Transport Protein Complex in Schizophrenia. Schizophr Res 154 (1–3):1–13. https://doi.org/10.1016/j.schres.2014.01.028

Shao L, Martin MV, Watson SJ, et al. (2008) Mitochondrial Involvement in Psychiatric Disorders. Ann Med 40 (4):281–295. https://doi.org/10.1080/07853890801923753

Sinha SR, and Kossoff EH (2005) The Ketogenic Diet. Neurologist 11 (3):161–170. https://doi.org/10.1097/01.nrl.0000160818.58821.d2

Steyn NP, Mann J, Bennett PH, et al. (2004) Diet, Nutrition and the Prevention of Type 2 Diabetes. Public Health Nutr 7 (1A):147–165. https://doi.org/10.1079/phn2003586

Sullivan CR, Koene RH, Hasselfeld K, et al. (2019a) Neuron-Specific Deficits of Bioenergetic Processes in the Dorsolateral Prefrontal Cortex in Schizophrenia. Mol Psychiatr 24 (9):1319–1328. https://doi.org/10.1038/s41380-018-0035-3

Sullivan CR, Mielnik CA, Funk A, et al. (2019b) Measurement of Lactate Levels in Postmortem Brain, iPSCs, and Animal Models of Schizophrenia. Sci Rep 9 (1):5087. https://doi.org/10.1038/s41598-019-41572-9

Sullivan CR, Mielnik CA, O'Donovan SM, et al. (2019c) Connectivity Analyses of Bioenergetic Changes in Schizophrenia: Identification of Novel Treatments. Mol Neurobiol 56 (6):4492–4517. https://doi.org/10.1007/s12035-018-1390-4

Suzuki A, Stern SA, Bozdagi O, et al. (2011) Astrocyte-Neuron Lactate Transport Is Required for Long-Term Memory Formation. Cell 144 (5):810–823. https://doi.org/10.1016/j.cell.2011.02.018

Szoke A, Pignon B, Godin O, et al. (2024) Multimorbidity and the Etiology of Schizophrenia. Curr Psychiatry Rep 26 (5):253–263. https://doi.org/10.1007/s11920-024-01500-9

Tomasik J, Lago SG, Vázquez-Bourgon J, et al. (2019) Association of Insulin Resistance with Schizophrenia Polygenic Risk Score and Response to Antipsychotic Treatment. JAMA Psychiatry 76 (8):864–867. https://doi.org/10.1001/jamapsychiatry.2019.0304

Tourigny DS, Karim MKA, Echeveste R, Kotter MRN, and O'Neill JS (2019) Energetic Substrate Availability Regulates Synchronous Activity in an Excitatory Neural Network. PLoS One 14 (8):e0220937. https://doi.org/10.1371/journal.pone.0220937

Townsend L, Pillinger T, Selvaggi P, et al. (2023) Brain Glucose Metabolism in Schizophrenia: A Systematic Review and Meta-Analysis of ^{18}FDG-PET Studies in Schizophrenia. Psychol Med 53 (11):4880–4897. https://doi.org/10.1017/s003329172200174x

Trubetskoy V, Pardinas AF, Qi T, et al. (2022) Mapping Genomic Loci Implicates Genes and Synaptic Biology in Schizophrenia. Nature 604 (7906):502–508. https://doi.org/10.1038/s41586-022-04434-5

Turkheimer FE, Selvaggi P, Mehta MA, et al. (2019) Normalizing the Abnormal: Do Antipsychotic Drugs Push the Cortex into an Unsustainable Metabolic Envelope? Schizophr Bull 46 (3):484–495. https://doi.org/10.1093/schbul/sbz119

Tye SJ, Borreggine K, Price JB, et al. (2022) Dynamic Insulin-Stimulated mTOR/GSK3 Signaling in Peripheral Immune Cells: Preliminary Evidence for an Association with Lithium Response in Bipolar Disorder. Bipolar Disord 24 (1):39–47. https://doi.org/10.1111/bdi.13081

Venkatesan S, Binko MA, Mielnik CA, Ramsey AJ, and Lambe EK (2023) Deficits in Integrative NMDA Receptors Caused by *Grin1* Disruption Can Be Rescued in Adulthood. Neuropsychopharmacology 48 (12):1742–1751. https://doi.org/10.1038/s41386-023-01619-y

Vincent A, and Bien CG (2008) Anti-NMDA-Receptor Encephalitis: A Cause of Psychiatric, Seizure, and Movement Disorders in Young Adults. Lancet Neurol 7 (12):1074–1075. https://doi.org/10.1016/s1474-4422(08)70225-4

Warren N, O'Gorman C, Horgan I, et al. (2024) Inflammatory Cerebrospinal Fluid Markers in Schizophrenia Spectrum Disorders: A Systematic Review and Meta-Analysis of 69 Studies with 5710 Participants. Schizophr Res 266:24–31. https://doi.org/10.1016/j.schres.2024.02.001

Wesseling H, Guest PC, Lee CM, et al. (2014) Integrative Proteomic Analysis of the NMDA NR1 Knockdown Mouse Model Reveals Effects on Central and Peripheral Pathways Associated with Schizophrenia and Autism Spectrum Disorders. Mol Autism 5:38. https://doi.org/10.1186/2040-2392-5-38

Wilson JE (2003) Isozymes of Mammalian Hexokinase: Structure, Subcellular Localization and Metabolic Function. J Exp Biol 206 (Pt 12):2049–2057. https://doi.org/10.1242/jeb.00241

Yang J, Chen T, Sun L, et al. (2013) Potential Metabolite Markers of Schizophrenia. Mol Psychiatr 18 (1):67–78. https://doi.org/10.1038/mp.2011.131

Yaqoob P, and Calder PC (2007) Fatty Acids and Immune Function: New Insights into Mechanisms. Br J Nutr 98 (S1):S41–S45. https://doi.org/10.1017/s0007114507832995

Zeng Y, Zhang L, and Hu Z (2016) Cerebral Insulin, Insulin Signaling Pathway, and Brain Angiogenesis. Neurol Sci 37 (1):9–16. https://doi.org/10.1007/s10072-015-2386-8

Zhao F, Siu JJ, Huang W, Askwith C, and Cao L (2019) Insulin Modulates Excitatory Synaptic Transmission and Synaptic Plasticity in the Mouse Hippocampus. Neuroscience 411:237–254. https://doi.org/10.1016/j.neuroscience.2019.05.033

Zhou X, Liu L, Lan X, et al. (2019) Polyunsaturated Fatty Acids Metabolism, Purine Metabolism and Inosine as Potential Independent Diagnostic Biomarkers for Major Depressive Disorder in Children and Adolescents. Mol Psychiatr 24 (10):1478–1488. https://doi.org/10.1038/s41380-018-0047-z

Open Access This chapter is licensed under the terms of the Creative Commons Attribution-NonCommercial-NoDerivatives 4.0 International License (http://creativecommons.org/licenses/by-nc-nd/4.0/), which permits any noncommercial use, sharing, distribution and reproduction in any medium or format, as long as you give appropriate credit to the original author(s) and the source, provide a link to the Creative Commons license and indicate if you modified the licensed material. You do not have permission under this license to share adapted material derived from this chapter or parts of it.

The images or other third party material in this chapter are included in the chapter's Creative Commons license, unless indicated otherwise in a credit line to the material. If material is not included in the chapter's Creative Commons license and your intended use is not permitted by statutory regulation or exceeds the permitted use, you will need to obtain permission directly from the copyright holder.

6

Role of Metabolism in the Emergence of Neuropsychiatric Disorders

Corey Weistuch, Anthony G. Chesebro, Botond B. Antal, and Lilianne R. Mujica-Parodi

Abstract This chapter explores the critical role of metabolic dysregulation in the emergence of neuropsychiatric disorders, emphasizing the complex interplay between molecular, cellular, and brain-wide processes. Key metabolic control circuits are examined, including mTOR-AMPK signaling, glucose-insulin regulation, and mitochondrial function, as well as their impact on neural dynamics. Also highlighted are how emergent phenomena (e.g., criticality, metastability, and traveling waves) arise from interactions between microscopic components, providing insights into brain function and dysfunction. Discussion focuses on how combining mathematical modeling with measurements of these emergent properties can reveal mechanistic information about underlying metabolic processes. Finally, a

Lilianne R. Mujica-Parodi (✉)
Dept. of Biomedical Engineering and Laufer Center for Physical and Quantitative Biology, Renaissance School of Medicine, State University of New York at Stony Brook, Stony Brook, NY 11794; Martinos Center for Biomedical Imaging, Massachusetts General Hospital, Boston, MA 02114, USA
Email: mujica@lcneuro.org

Anthony G. Chesebro
Dept. of Biomedical Engineering and Laufer Center for Physical and Quantitative Biology, Renaissance School of Medicine, State University of New York at Stony Brook, Stony Brook, NY 11794; Martinos Center for Biomedical Imaging, Massachusetts General Hospital, Boston, MA 02114, USA

Botond B. Antal
Dept. of Biomedical Engineering and Laufer Center for Physical and Quantitative Biology, Renaissance School of Medicine, State University of New York at Stony Brook, Stony Brook, NY 11794; Martinos Center for Biomedical Imaging, Massachusetts General Hospital, Boston, MA 02114, USA

Corey Weistuch
Dept. of Medical Physics, Memorial Sloan Kettering Cancer Center, New York, NY 10065, USA

© Frankfurt Institute for Advanced Studies (FIAS) 2025
D. Öngür, J. M. Ford (eds.), *Metabolic Neuropsychiatry*, Strüngmann Forum Reports,
https://doi.org/10.1007/978-3-032-05630-6_6

systematic pipeline is proposed to disentangle the driving mechanisms behind these emergent processes, aiming to use data for the development of more precise and personalized interventions for neuropsychiatric disorders.

Key Words Criticality, computational psychiatry, multiscale modeling, emergent phenomena, aging neurometabolism

1 Introduction

Growing evidence suggests that metabolic dysregulation may play a critical role in the development and progression of numerous neuropsychiatric disorders. Because we know that "metabolic regulation" involves not one, but many (mutually interacting) homeostatic control circuits, there may be many distinct points of potential failure. Dissociating these individual points of failure is likely to be critical to understanding how a specific type of disrupted metabolism at the mechanistic scale can lead to very different clinical presentations at the emergent scale (e.g., as seen with epilepsy vs. dementia vs. cancer vs. psychiatric disorders). Here we review key circuits implicated in metabolic control and outline what is currently known about how metabolism impacts the emergent dynamics of the key neural circuits implicated in psychiatric disorders (see Section 7 for a glossary of key associated terms). We then provide an overview of emergence within biological systems with a focus on metabolism and properties of neuron populations, followed by a discussion of multiscale approaches that allow one to probe the impact of different types of metabolic dysregulation at the mechanistic scale on emergent brain dynamics at the neuroimaging scale. Finally, we consider the problem of dissociating individual points of failure, in distinguishing driving mechanisms versus downstream effects. In doing so, we discuss how hypotheses, experimental designs, analytic strategies, and diagnostics/therapeutics might differ from standard statistical approaches employed in the biomedical sciences when evaluating a complex system involving regulation across multiple interconnected feedback loops, both neural and physiological.

2 Neurometabolism: Mechanistic Control Circuits and Their Interactions

2.1 Homeostatic Regulation

Homeostasis is the ability of an organism to maintain a stable internal environment despite changes in external conditions. It involves regulating various physiological excitatory and inhibitory processes through control circuits interconnected both within and across scales: these feedback mechanisms keep key parameters (e.g., core body temperature, pH levels, fluid balance) within optimal ranges. Dynamic equilibrium is thus essential for the proper functioning and survival of living organisms, as it allows adaption to environmental fluctuations while preserving internal stability. Disease states reflect breakdowns in homeostasis and, by implication, they also involve breakdowns in the mechanisms that maintain homeostatic regulation.

Thus, breakdowns in homeostatic regulation of metabolism in the brain are hypothesized to contribute directly to many disorders, such as epilepsy, dementia, brain cancer, and psychiatric disorders. Yet the fact that all four sets of disorders have markedly different emergent signs, symptoms, and trajectories underscores how the same physiological control circuit (e.g., metabolism) can be dysregulated in more than one way, with distinct etiologies, resulting in divergent clinical features. Identifying the driving mechanism is a key challenge but also the ultimate aim in establishing the most efficacious clinical targets for intervention.

2.2 Key Metabolic Control Circuits

Given the critical dependency of every cell on stable access to energy, and the brain's disproportionate energetic demands, homeostatic regulation of neurometabolism involves a complex set of compensatory pathways (Figure 6.1), each playing complementary roles. Conceptually, homeostatic control circuits consist of three basic components: a *sensor* that signals *excitatory* and *inhibitory* components, which in turn feed back to the sensor. This is analogous to a house thermostat's role in continuously activating either heating or cooling systems to regulate temperature.

2.2.1 Cellular Energy Sensing via mTOR-AMPK Regulation

The mammalian target of rapamycin (mTOR) and AMP-activated protein kinase (AMPK) are two critical cellular energy sensors that play antagonistic roles in regulating cellular metabolism, growth, and survival. mTOR forms two distinct multiprotein complexes, mTORC1 and mTORC2, with mTORC1 being particularly sensitive to nutrient availability and energy status. When cellular energy levels are high, as indicated by an elevated ATP/AMP ratio, mTORC1 is activated, promoting anabolic processes such as protein synthesis and cell growth. Conversely, when cellular energy levels are low, as indicated by an increased AMP/ATP ratio, AMPK is activated through phosphorylation. Once activated, AMPK initiates a cascade of signaling events that promote catabolic processes to generate ATP while inhibiting anabolic processes that consume ATP.

Because the mTOR-AMPK axis is a central energy sensor, it is a driving mechanism that affects every other metabolic control circuit downstream, including regulating mitochondrial function and glucose-insulin-glucagon control circuits (Herzig and Shaw 2018). mTOR is activated when nutrients and energy are abundant, promoting anabolic processes and inhibiting autophagy, including mitophagy (Kim et al. 2011). Chronic mTOR activation can lead to the accumulation of dysfunctional mitochondria, insulin resistance, and impaired glucose homeostasis (Herzig and Shaw 2018). In contrast, AMPK is activated during cellular energy depletion and promotes catabolic processes, mitochondrial biogenesis, and autophagy (Hardie et al. 2012). AMPK activation enhances insulin sensitivity, glucose uptake, and regulates the secretion of insulin and glucagon (Herzig and Shaw 2018). The interplay between mTOR and AMPK is crucial for maintaining cellular energy homeostasis and mitochondrial quality control, with mTOR activation suppressing AMPK activity and vice versa (Kim

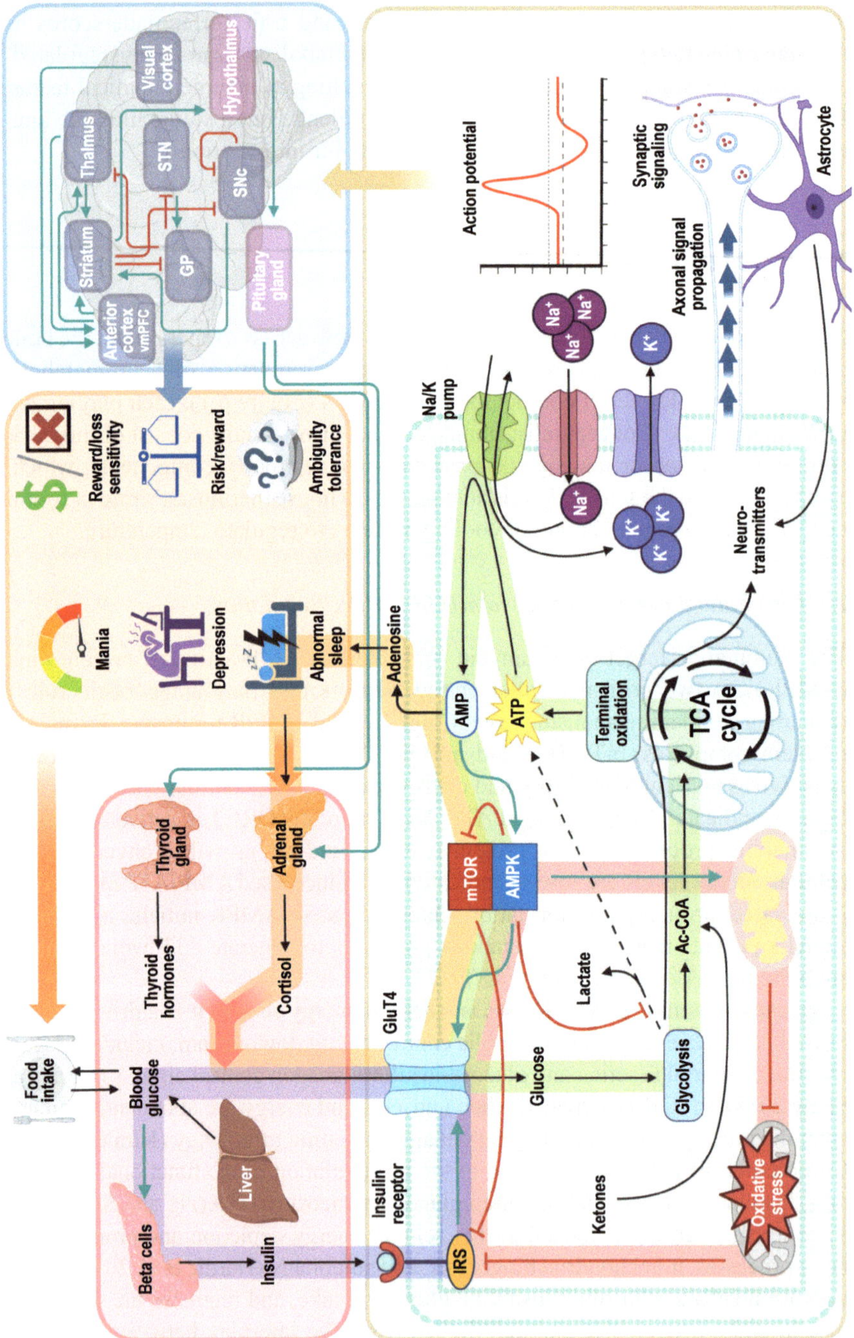
Visual cortex
Hypothalmus
Thalmus
STN
SNc
Striatum
GP
Pituitary gland
Anterior cortex vmPFC
Reward/loss sensitivity
Risk/reward
Ambiguity tolerance
Mania
Depression
Abnormal sleep
Adenosine
Thyroid gland
Adrenal gland
Thyroid hormones
Cortisol
Food intake
Blood glucose
Liver
Beta cells
Insulin
Insulin receptor
IRS
GluT4
Glucose
Glycolysis
Lactate
Ac-CoA
Ketones
Oxidative stress
mTOR
AMPK
AMP
ATP
Terminal oxidation
TCA cycle
Na/K pump
Neuro-transmitters
Action potential
Axonal signal propagation
Synaptic signaling
Astrocyte

Figure 6.1 Schematic of five key metabolic control circuits and their mechanistic impact on brain and behavior. These include mTOR/AMPK, which acts as an energy sensor, comparing ATP to the cells' energetic set-point and regulating all other metabolic control circuits accordingly; glucose-insulin regulation (purple circuit), gating entry of glucose into the cell via GLUT4; glycolytic regulation (green circuit) and mitochondrial regulation (red circuit), which together take up available fuel into the TCA cycle, producing ATP; production of metabolites, including glutamate and GABA neurotransmitters, which maintain the excitatory-inhibitory balance in the brain, and potassium ion gradients, which affect axon conduction velocity, and therefore regulation of neural control circuits, including the corticostriatal, prefrontal-limbic, and ventral stream circuits implicated in psychiatric cognitive and emotional signs and symptoms, with effects on behavior, including sleep pressure, which is regulated by adenosine (orange circuit). Sleep and stress (depression and mania), in turn, modulate cortisol, and thus neuronal glucose levels/insulin resistance (Sapolsky 1986). Note that, due to feedback between control circuits, dysregulation of any one component will lead to aberrant values for all other components; thus, dissociation of driving mechanisms from downstream effects requires experimental and analytic approaches that go beyond statistical comparison of a single measured parameter across populations. Created in BioRender. Mujica-Parodi, L. (2025) https://BioRender.com/cl6gs1g.

and Guan 2019). Mitochondrial function is closely linked to glucose-insulin-glucagon control circuits, and mitochondrial dysfunction can lead to insulin resistance and metabolic disorders (Szendroedi et al. 2011). The mTOR-AMPK axis regulates these processes by modulating cellular energy metabolism, insulin sensitivity, and glucose homeostasis (Hardie et al. 2012). This axis is therefore a key therapeutic target for treating mitochondrial dysfunction and metabolic diseases (Szendroedi et al. 2011).

2.2.2 Maintaining Cellular Access to Glucose via Insulin-Glucagon Regulation

Glucose homeostasis is maintained through a complex interplay of hormones, primarily insulin and glucagon, and the regulation of membrane glucose transporters, including glucose transporter 4 (GLUT4) translocation (Saltiel and Kahn 2001). Following a meal, elevated blood glucose levels stimulate the pancreatic β-cells to secrete insulin into circulation (Rorsman and Braun 2013). Insulin binds to its receptor on target cells, initiating a signaling cascade that leads to the activation of phosphatidylinositol 3-kinase (PI3K) and AKT/protein kinase B (Taniguchi et al. 2006). This signaling pathway triggers the translocation of GLUT4 from intracellular storage vesicles to the plasma membrane. The increased presence of GLUT4 on the cell surface facilitates glucose uptake from the bloodstream into the neurons, where it is either utilized for energy production or stored as glycogen in astrocytes (a type of glial cell) through glycogenesis. Conversely, during periods of fasting or hypoglycemia, the pancreatic α-cells secrete glucagon, which binds to its receptor on hepatocytes (Gromada et al. 2007). This interaction activates adenylate cyclase, leading to increased production of cyclic AMP (cAMP) and activation of protein kinase A (PKA) (Jiang and Zhang 2003). PKA phosphorylates key enzymes involved in glycogenolysis and gluconeogenesis, promoting the breakdown of glycogen and the synthesis of glucose from non-carbohydrate precursors, respectively (Pilkis and Granner 1992). The newly synthesized glucose is then released into the bloodstream to maintain euglycemia (Gerich et al. 2001). The intricate balance between insulin and glucagon, along with the precise regulation of transporter translocations to the cell membrane, ensures tight control over neuron glucose levels, maintaining cellular energy supply while preventing the deleterious effects of hyper- or hypoglycemia.

Beyond local storage of excess glucose as glycogen, astrocytes play a crucial role in the metabolic support of neurons by providing them with lactate, an essential energy substrate, through the astrocyte-neuron lactate shuttle (ANLS) system (Figure 6.2). Astrocytes also regulate the extracellular concentration of potassium ions (K^+) and glutamate, ensuring proper neuronal excitability and preventing excitotoxicity. Furthermore, astrocytes and other glial cells express transporters and enzymes essential for neurotransmitter recycling and synthesis, such as glutamine synthetase, which converts glutamate to glutamine for neuronal reuptake and subsequent conversion back to glutamate.

2.2.3 Converting Cellular Glucose and Ketones to ATP via Mitochondrial Density-Activity Regulation

Mitochondrial density and activity are regulated by complex control circuits involving various signaling pathways and transcription factors. The peroxisome proliferator-activated receptor gamma coactivator 1-alpha (PGC-1α) is a key regulator of mitochondrial biogenesis and function (Puigserver and Spiegelman 2003). PGC-1α is activated by cellular energy demands and stress signals, such as exercise, cold exposure, and caloric restriction (Handschin and Spiegelman 2006). Upon activation, PGC-1α interacts with transcription factors, including nuclear respiratory factors (NRF-1 and NRF-2) and estrogen-related receptors (ERRs), to promote the expression of genes involved in mitochondrial DNA replication, transcription, and translation (Scarpulla 2011; Wu et al. 1999). Additionally, the AMPK and sirtuin 1 (SIRT1) pathways sense cellular energy status and modulate PGC-1α activity through phosphorylation and deacetylation, respectively (Cantó and Auwerx 2009; Rodgers et al. 2005). These control circuits work in concert to fine-tune mitochondrial density and activity in response to cellular energy requirements and environmental cues.

2.3 Interactions with Neural Control Circuits Implicated in Psychiatric Disorders

Each of the metabolic control circuits described above, in turn, impacts the production of neurotransmitters and other metabolites, as well as neuronal firing dynamics and axon conduction velocity, that will affect the tuning of neural control circuits that regulate emotional and cognitive processes mechanistically implicated in psychiatric signs and symptoms. These include the *prefrontal-limbic circuit*, which regulates perception of threat and risk (Mujica-Parodi et al. 2017), the *corticostriatal circuit* in which dopamine mediates learning as well as perception of pleasure and reward (Ashby and Crossley 2011; Frank 2005; Humphries et al. 2006), the *thalamo-cortical-cerebellar circuit* that regulates agency to enable distinction between self- and other-generated sensations (Abram et al. 2022; Giraldo-Chica et al. 2018), and the *ventral stream* (DiCarlo et al. 2012), which regulates perception of uncertainty and ambiguity.

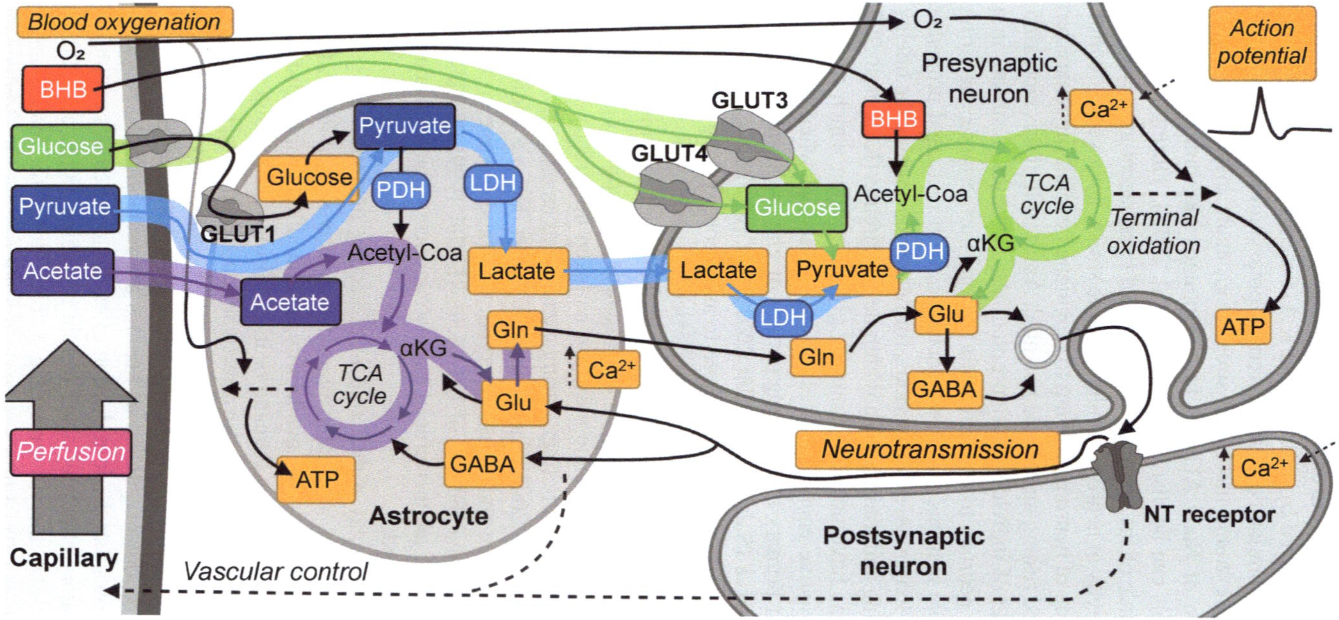

Figure 6.2 Measurement and modulation of metabolic pathways in the neuron. Magnetic resonance spectroscopy (MRS) can be used to measure key metabolites and neurotransmitters in the pathways shown as an *in vivo* modality in humans (Rothman et al. 2011). For example, ^{1}H MRS can measure glucose, BHB, lactate, glutamate/glutamine, and GABA; ^{13}C MRS can measure these as well as acetate and pyruvate; and ^{31}P MRS can measure ATP (Rothman et al. 2011). Created in BioRender. Mujica-Parodi, L. (2025) https://BioRender.com/cc5ff0r

2.3.1 Metabolic Dysregulation and Neural Circuit Destabilization

The ability of neurons to fire action potentials relies on the maintenance of ion gradients across the neuronal cell membrane (Wang et al. 2012). This crucial task is primarily managed by neuronal Na/K ATPases, which pump sodium ions out of the cell while simultaneously transferring potassium ions back in. However, under conditions of extreme metabolic stress (e.g., acute insulin resistance) or through the inhibition of Na/K ATPases by compounds like ouabain, the balance of ion concentrations can be disrupted (Wang et al. 2012). During mild stress, K^+ accumulates outside of the neuron but is captured by adjacent astrocytes, preserving the electrical activity of neurons. Under extreme stress, however, K^+ continues to accumulate outside of neurons, leading to heightened excitability characterized, in part, by an elevated resting membrane potential. As a result of this increased excitability, the amplitude of individual action potentials decreases, resulting in a diminished capacity for neurons to transmit information. Furthermore, the firing patterns of hyperexcitable neurons are more stochastic, with increased propensities both for bursting and longer quiescent periods, further compromising the fidelity of information processing and transmission within neural circuits (Kula et al. 2024).

As neurons have no local stores of glycogen upon which to draw, the provision of alternative energy sources becomes essential when they are deprived of glucose, their primary fuel. During periods of glucose scarcity, a limited astrocytic glycogen reserve (Saltiel and Kahn 2001) may prioritize glucose availability for neurons, a phenomenon often referred to as glucose sparing (Rothman et al. 2022). In addition, other energy substrates, particularly astrocytic lactate and pyruvate as well as blood-borne ketones, become crucial in supplying this deficit (Magistretti et al. 1999). Lactate has often been tied to astrocytic support of neuronal metabolism via the ANLS, but this lactate may also be taken up directly from the blood when there is a plentiful supply (e.g., during exercise) as an alternative circulating energy source (Smith et al. 2003; van Hall et al. 2009). Ketone bodies, available in the body as acetoacetate and beta-hydroxybutyrate (and acetone, to a lesser extent) are produced by the metabolism of fatty acids (Laffel 1999; Puchalska and Crawford 2017) and provide another crucial circulating energy source when neuronal access to glucose is limited.

Ketone bodies play a dual role in protecting neurons during acute metabolic stress. They serve as a valuable alternative energy source for neurons, particularly in situations where glucose availability is limited or compromised, such as during periods of high metabolic demand or in conditions like insulin resistance (Mujica-Parodi et al. 2020). Ketones also directly modulate neuronal firing by facilitating the opening of ATP-sensitive K^+ channels (Yellen 2008). This effect is due to ATP compartmentation, where glycolytic ATP that is normally sensed by the ATP-sensitive K^+ channels is downregulated as ketones are metabolized (Yellen 2008). Recent work also shows that the introduction of exogenous ketones counteracts several of the dysregulatory effects of acute insulin resistance (Kula et al. 2024) and indeed bypasses GLUT4, thereby providing a "back door" for fueling insulin resistance neurons (Figure 6.2). While this study did not measure K^+ ion dynamics directly, the established role of ketones in regulating K^+ as well as subsequent modeling efforts suggests that exogenous ketones counteract the dysregulation of potassium dynamics induced by metabolic stressors, contributing substantially to the stabilization and synchronization of brain-wide neuronal dynamics.

2.3.2 Spatial Heterogeneity of Metabolic Dysregulation

Each of the metabolic pathways discussed thus far can have an additional component contributing to its effects on neural circuitry: the heterogeneous sensitivity of brain regions to specific metabolic insults. For example, the distribution of insulin-sensitive GLUT4 transporter is highly nonuniform throughout the brain, with particularly concentrated expression in the medial temporal lobe and temporal pole (Gryglewski et al. 2018), making these regions with relatively higher sensitivity to insulin uniquely vulnerable to metabolic dysfunction in conditions such as diabetes. The use of these novel spatial transcriptomic tools (Hansen et al. 2022) is discussed further below (see Section 5).

2.4 The Challenge of Identifying the Driving Mechanism within a Set of Mutually Interacting Control Circuits

Because of multiple feedback loops within the system, disruption to any one part of the system will affect every other part of the system. This makes it nontrivial to dissociate driving mechanisms from downstream consequences. In particular, such systems are impervious to standard statistical approaches because correlations and differences in mean values cannot consider the role of compensatory mechanisms and feedback (Mujica-Parodi et al. 2017). Returning to our HVAC analogy, if a house is too cold, measuring the temperature cannot distinguish between the driving mechanism being a breakdown in the thermostat or a breakdown in the heating. Likewise, if a cohort's brains are hypometabolic due to a breakdown in cellular energy sensing, their brains may show statistically significant differences from comparison groups in both insulin resistance and mitochondria *without either insulin resistance or mitochondria being the driving mechanism* (for a compelling example of this in mitochondrial defects, see Sturm et al. 2023). This becomes a trap for confirmation bias when hypothesis-based research motivates one to test only a single mechanism against a null hypothesis. Measuring dysregulation within and across multiple homeostatic control circuits requires alternative ways of formulating questions, experimental design, analytic strategies, and diagnostic-therapeutic approaches. While determining how best to do so is currently an active area of investigation, we touch on these challenges below (see Section 5).

3 What Is Emergence?

A rigorous examination of how changes in metabolic regulatory pathways can exhibit observable changes at different scales requires the concept of *emergence*. Originating in the study of complex systems, emergence describes how intricate collective patterns and behaviors manifest from simple interaction rules among a population of individual agents. The wave-like murmuration phenomenon is a textbook example of an emergent property: each individual bird flies according to its own rules, but the collective flocking dynamics exhibit a structure that transcends individual interactions.

This concept of emergence is of special relevance to neuroscience, evidenced by phenomena like the oscillatory patterns observed in EEG experiments and the regional control circuits that underpin much of our understanding of behavior (Deco et al. 2017; Mujica-Parodi et al. 2017). These behaviors depend not on the capabilities of individual neurons, but rather on the strength and nature of their interactions, forming a whole that exceeds the sum of its parts. In this framework, cognition itself can be considered an emergent phenomenon: no single neuron or brain region gives rise to cognitive capabilities, but collectively they create cognition. The study of neuropsychiatric disorders is therefore the study of how changes in interaction rules at different scales alter the emergent phenomena of interest in a way that deviates from healthy to disorder.

The utility of emergent phenomena comes with a caveat: a delicate balance exists between order and chaos (Weistuch et al. 2021a). Minute changes in the strength of inter-component interactions can trigger cascading effects that spread throughout the entire system. During phase transitions, such as the transformation of a liquid to a gas or from a synchronized to a desynchronized brain state, the strength of these interactions crosses a critical threshold, resulting in the abrupt appearance of distinct macroscopic properties. Criticality often appears in systems that exhibit emergent properties, and the brain is no exception.

3.1 Metabolic Regulation As an Emergent Property of Bioenergetic Pressures

Some of the most studied effects in metabolism arise as emergent properties from simple changes in glycolytic regulation. For example, recent work has shown that the shift toward aerobic glycolysis (leading to production of lactate despite oxygenated conditions)—originally termed the Warburg effect and often seen in cells with metabolic dysregulation—emerges as a consequence of cells employing glycolysis over respiration to optimize for speed of ATP generation instead of ATP yield (Kukurugya et al. 2024). Crucially, this shift is driven by a need for speed rather than efficiency of ATP production, as this occurs in the setting of sufficient oxygen for mitochondrial respiration to occur (DeBerardinis and Chandel 2020; Potter et al. 2016). In this case, the effect is that any state that pressures a cell to generate ATP more quickly drives the cell toward less efficient energy production, leading to a metabolically unfavorable state and an observable increase in lactate production as a potential biomarker.

At the most fundamental level, cells within the body serve their functions by converting energy into outputs; thus, alterations in bioenergetic stresses can propagate up to the macroscale by affecting many different cellular processes (Yang et al. 2021). At the level of protein interactions, recent work has been devoted to combining active-matter physics approaches with an explicit consideration of energy costs associated with different protein functions. While this work has largely focused on physical processes that occur within the cell (e.g., transport via motor proteins, DNA replication), the field requires neural models, as neuron functions—particularly the maintenance of reversal potentials—are some of the most energetically expensive processes within the body (Yellen 2008).

3.2 Criticality in Signaling and Response

Multiple independent studies have suggested that the brain operates at a critical point between desynchronized randomness and complete synchrony, as seen in epilepsy (Schneidman et al. 2006). It has been theorized that the brain operates near criticality in order to maximize sensitivity and response to perturbations, allowing for flexibility in the face of different neurometabolic stressors. Recent work has shown that brain aging causes the brain to deviate from this critical point, perhaps explaining the loss of cognitive capacity associated with advanced aging (Weistuch et al. 2021b). Small changes in the strength of region-to-region correlations lead to significant alterations in the distribution of brain states (Chesebro et al. 2023). Consequently, the sensitivity of the brain to disruptions in the microscale interaction rules between its regions contributes to the pronounced phenotypic changes observed in brain aging.

This shift away from criticality in aging is also associated with significant alterations in neuronal network activities, as measured by both fMRI and EEG. Both network stability, quantified by the average similarity in functional connectivity across regions over time, and neuronal synchrony, measured as the average instantaneous co-alignment of these regions, were found to decrease with age (Mujica-Parodi et al. 2020; Weistuch et al. 2021b). The observed decreases in network stability and synchrony in older brains imply a greater propensity to switch between cognitive states. This heightened switching behavior is likely a consequence of the brain's decreased ability to sustain high-energy functional states over time, as indicated by the opposing effect of the ketogenic diet. Indeed, given that normal glucose utilization in the brain is linked to the number of communication pathways used, the reduced synchrony observed in older brains suggests a preference for lower-energy states (Mujica-Parodi et al. 2020).

Interestingly, both a ketogenic diet and the administration of an exogenous ketone bolus countered these effects in normal healthy adults, suggesting that metabolic modulation may play a role in preserving large-scale brain behaviors, potentially serving a neuroprotective function (Antal et al. 2025; Mujica-Parodi et al. 2020). Taken together, these findings indicate that while aging shifts the brain away from criticality, metabolic intervention can cause the reversal of some of these large-scale changes, highlighting the potential for minute interventions to restore homeostasis. More broadly, despite the complex nature of neuropsychiatric illnesses, if their principal features can be attributed to emergent brain behaviors, small system-wide interventions may hold promise for their reversal.

3.3 Metastability in Neural Fields and Traveling Waves

One of the first emergent phenomena discovered and studied within the brain are the electrical fields generated by synchronous neural activity. Classic EEG experiments determined different core frequency bands associated with specific regions and processes (alpha, beta, gamma, delta, and theta waves), establishing an emergent macroscale metric that was observable noninvasively. Since these initial studies, however, the properties of these emergent waves have been found to contain more utility than simply being observable in humans. Recent work has shown that traveling waves in the cortex encode specific information during learning tasks, which can provide feedback as neurons alter synaptic activity during the task, and that CNS-active drugs used in anesthesia and psychiatric treatments fundamentally alter these traveling wave dynamics.

EEG and fMRI studies of traveling waves within the brain, combined with computational models that capture neural field dynamics, reveal that metastable dynamics are a common feature of brain activity (Roberts et al. 2019). Related to the concept of criticality, metastable dynamics show the ability to visit multiple patterns of activity (e.g., spiraling vs. traveling wave dynamics) in a predictable temporal sequence. The ability to exhibit metastable brain dynamics is closely related to the existence of underlying neuronal parameters at critical points near bifurcations to allow for switching between these macroscopic states within a physiologically controllable parameter space (Chesebro et al. 2023; Deco et al. 2017; Roberts et al. 2019).

4 Emergent Phenomena of Metabolic (Dys)Regulation

Since emergent properties are observable at many different scales in the brain, it is useful to examine how different neuronal phenomena are explained as emerging from metabolic manipulations. In this section, we summarize several recent approaches that leverage metabolic perturbations to observe effects at both the micro- and macroscale through the lens of emergence.

4.1 Axonal Conduction Velocity

The maintenance of ion gradients across the neuron cell membrane allows for the neuron's most central activity: the firing of action potentials and their traversal through an axon to communicate electrical signals to other cells. The propagation of electric current along an axon is described by classical cable theory, a mathematical framework that captures how factors like the axon's diameter and resistance influence the conduction velocity of the action potential (Kula et al. 2024). Remarkably, this theory achieves this without delving into the detailed biophysics of the neuron. Instead, it leverages how individual axonal units synergize to produce a collective propagation pattern.

Recent work using patch-clamp recordings of hippocampal neurons has revealed that metabolic factors significantly influence conduction velocity (Kula et al. 2024). In acute insulin resistance, there is a decrease in axonal conductance velocity, which is reversible with the administration of ketones (Kula et al. 2024). Concurrent with these experiments, recent work has demonstrated that metabolic influences on conduction velocity can be modeled through a detailed metabolic reformulation of cable theory that explicitly includes the action of the Na/K ATPase. This theory was then used to connect increases in neuronal excitability following acute insulin resistance and decreases in Schaffer collateral conduction velocity. This new modeling method provided a mechanistic explanation of how insulin resistance decreased conduction velocity and how the introduction of alternative energy in the form of exogenous ketones reversed these effects (Kula et al. 2024).

4.2 Ion Gradient Bifurcations Delineate Physical Limits

Emergent and critical phenomena at the macroscale are often predictable through mathematical modeling. Criticality in neural systems is often a product of lying between multiple bifurcations in phase space, and bifurcation theory provides the analytical techniques and computational tools for describing such sudden changes. For example, applying bifurcation theory to an established neural mass model revealed that minor alterations in neuronal ion gradients can lead to abrupt shifts in multi-regional oscillatory dynamics (Chesebro et al. 2023). In particular, all ion gradients exhibited flip bifurcations (allowing full use of phase space), and some also demonstrated torus bifurcations (allowing for subthreshold voltage changes as well as spiking dynamics).

Both the Chesebro et al. (2023) and other studies (e.g., Deco et al. 2017) have shown that the most realistic neural and neural mass dynamics were observed near bifurcation points, meaning changes within this narrow region of ion gradient parameter space lead to a large effect on interregional synchrony. This further supports the criticality hypothesis and shows that changes in ion dynamics—known to be dysregulated by aging and regulated by ketones—can induce large-scale changes akin to those observed experimentally. Therefore, existing models of neuronal dynamics could theoretically forecast macroscopic brain behaviors consistent with both known and unknown phenotypes of other brain disorders.

4.3 Microscale Models of Neurometabolic Changes Explain Features Emerging at the EEG Scale

Computational models designed to capture emergent macroscopic properties from microscale changes can connect human-imaging experiments to bench-side work. For example, prior work has shown that the consumption of exogenous ketones confers beneficial effects (measured by fMRI network stability) (Mujica-Parodi and Strey 2020). Separate prior work has shown that ketones have several potential microscale mechanistic effects; prominent among these is the regulation of ATP-sensitive potassium channels. It would be impossible to simultaneously observe channel activity and record EEG/fMRI in humans; however, we can use modeling to simulate these changes. Building on the bifurcation analyses discussed before, recent work using the same model has shown that ion gradient dysregulation near the bifurcation points previously identified leads to global synchrony loss (measured by EEG) similar to that seen in aging, and this effect is reversed by ketone administration (van Nieuwenhuizen et al. 2024). Importantly, the model showed that the only microscale parameter that demonstrated changes in the simulated EEG signal that matched the age and pharmacological results was potassium, giving insight for future microscale exploration of these effects.

5 A Way Forward for Metabolic Neuropsychiatry

5.1 Measuring Dysregulation

In Section 2.3, we ended by suggesting that measuring dysregulation within and across multiple homeostatic control circuits requires alternative ways of formulating questions, experimental design, analytic strategies, and diagnostic-therapeutic approaches. These follow from three considerations. *First*, control circuits can "break" in many ways. There are basic conceptual distinctions between a circuit's sensor, excitatory, and inhibitory components. There are, however, additional potential points of failure that result from the fact that metabolic regulation requires many stages, including access to energy, utilization of energy, catalyzing downstream physiological processes, and replenishing the precursors required for each of these. Based on which of many potential points of failure is responsible, the same control circuit can be dysregulated in many ways, with distinct etiologies, resulting in divergent clinical features and thus disorders. *Second*, dysregulation is most sensitively characterized by dynamics, not amplitudes; therefore, it is most clearly seen in response to perturbation, as the system components work to regain homeostasis. *Third*, clinical symptoms typically do not mark the beginnings of dysregulation but rather the end stage of a chronic degenerative process. As such, it can sometimes be more informative to measure the impact of mechanistic targets at shorter timescales on biomarkers rather than at longer timescales on clinical symptoms, particularly in high-risk clinical populations before clinical symptoms emerge.

5.2 Accounting for Spatial and Temporal Specificity in Experimental Perturbations

To discover the mechanisms driving emergent results, neural circuits must receive *several simultaneous perturbations* that are *measured longitudinally.* By orienting ourselves to this experimental design, we move away from a paradigm where single mechanism perturbations are studied in isolation, leading to emergent properties that can be difficult to distinguish from each other and toward a way of testing multiple competing hypotheses in parallel. With the emergence of higher resolution spatial transcriptomic data sets paired with large PET-imaging cohorts, it is now possible to examine the spatial heterogeneity pathways throughout the brain (Hansen et al. 2022). The regional expression of certain proteins (e.g., insulin-sensitive glucose transporter overexpression in the temporal lobe) show patterns of regional vulnerabilities to metabolic processes that have traditionally been thought of as system-wide effects. While the rich heterogeneity of neurotransmitter receptors—and gene expression, in general—throughout the brain has been known for some time, it has been difficult to quantitatively analyze in a meaningful way due to a lack of data. The emergence of new tools quantifying the different levels of neurotransmitter density throughout the brain, and the work of the Allen Human Brain Atlas and subsequent spatial transcriptomic toolboxes based on it (Hansen et al. 2022), have opened the doors to whole-brain analyses of spatial variability, ushering in an exciting new era of neuroimaging.

5.3 Computational Modeling Has Important Roles in a Priori Experimental Design and Post Hoc Analyses

Given an appropriate model that bridges the metabolic and neural circuit scales, we can predict macroscale changes that occur in response to a microscale perturbation. For example, the administration of ketones is associated with macroscale effects in EEG observed experimentally analogous to those seen by manipulation K^+ reversal potentials *in silico* (van Nieuwenhuizen et al. 2024). Future work should therefore focus on the development of models that can predict which microscale perturbations are most capable of achieving a desired macroscale brain state. As these models capture increasing microscale complexity and are fit to human-imaging studies, they open up a promising frontier of increased predictive power where models can be used to rank potential therapeutic targets in order of likelihood.

5.3.1 Exploiting the Physics that Transcends Scales

While traditional neural mass models have proven useful in capturing certain emergent phenomena, they are all phenomenological to some degree and thus unable to capture some properties at all. For example, the Larter-Breakspear neural mass model captures emergent changes in synchrony due to metabolic stress but lacks the ability to predict true EEG power spectra. It also trades biophysical detail for computational efficiency in inhibitory neuron populations, meaning that effects specific to this population cannot be adequately modeled at the microscale. Other models suffer similar constraints; Jansen-Rit neural masses, for example, offers a realistic power spectrum at the cost of microscale details (e.g., ion dynamics).

Recent work bringing approaches from physics and complex systems has demonstrated a promising new technique to analytically derive neural mass models from biophysically detailed neurons (Chen and Campbell 2022; Montbrio et al. 2015). The approach, typically termed next-generation neural mass models, draws on the Ott-Antonsen ansatz approach (Montbrio et al. 2015) to simplify Fokker-Planck equation analyses to compute an analytical form of the average voltage and firing rate (and synaptic dynamics, if modeled) of an arbitrarily-sized population of neurons that preserves any detailed modeling of membrane voltage when simplified to the population average. This approach has been demonstrated to capture microscale effects of several standard neurons (QIF, Izhikevich, theta), while also capturing emergent properties at the LFP and EEG scale (interregional synchrony, realistic power spectrum, beta-rebound in motor cortex) (Chen and Campbell 2022).

5.3.2 Optimizing Experimental Design through a Priori Simulations

A viable computational model that bridges scales can also predict what emergent changes would potentially be observable in an experiment, leading to more targeted experimental design. Recent work has shown that, given an arbitrarily complex model and parameter hyperspace, the parameters identifiable from a specific observer function can be condensed into a minimal and sufficient latent representation (Antal et al. 2024). For example, if an entire brain is simulated as an ensemble of coupled neural mass models and observed via fMRI (i.e., the neural mass LFP signal is convolved with a hemodynamic model), then the microscale parameters that are observable as macroscale effects are identified as a minimal latent representation. This representation serves two functions: it can be used a priori to inform experimental designs as it can predict which parameters a given experimental modality can resolve, and it can serve to provide a space for fitting unique parameters from experimentally collected data. When used in combination with modern data fitting techniques like simulation-based inference, this pipeline provides a robust method for solving the inverse problem (Antal et al. 2024). Pairing simulations of different experimental modalities with this approach can be used to see which combinations will provide the greatest ability to distinguish between experimental hypotheses.

5.4 Designing Future Multiscale Experiments

Taking these approaches together, we next illustrate how to design experiments aimed at identifying the most likely microscopic mechanisms of neuropsychiatric disorders and their potential treatments. This approach requires

- an initial *multiscale computational model* to explain how the dynamics propagate across scales,
- a *targeted set of perturbations* to the circuit(s) of interest, and
- *measures of microscale dysregulation* that have been identified as potentially driving mechanisms by the first two parts.

To illustrate this approach, we consider two motivating examples: the circuit dynamics of glutamate/GABA imbalance, and the alterations to circuits with neurodegenerative disorders.

5.4.1 *Glutamate/GABA Imbalance*

Consider an experiment designed to probe the metabolic components of glutamate/GABA balance in the frontal cortex circuitry—dynamics that are dysregulated in many conditions, ranging from schizophrenia (Moghaddam and Javitt 2012; Trujillo et al. 2022) to Parkinson disease (McGregor and Nelson 2019). First, one would use viable computational models that can explain how the circuit changes during alterations to excitatory-inhibitory balance. Models that address criticality (Weistuch et al. 2021b) and capture neurotransmitter dynamics (Pathak et al. 2024) or metabolic driven ion gradient changes (van Nieuwenhuizen et al. 2024) would be sufficient to provide initial models for testing multiple mechanistic hypotheses. Second, there would need to be perturbations of the system in ways that probe the metabolic dynamics, such as dietary ketogenesis (Mujica-Parodi and Strey 2020), versus circuit dynamics through, for example, circuit-specific tasks (Pathak et al. 2024). These studies can be augmented with perturbations driven by psychoactive agents with known behavioral and circuit effects, such as ketamine challenges to study predictive coding deficits (Kort et al. 2017). One could then use these data to test the dynamics simulated by a range of potential mechanistic parameters in the original models, which would then be used to identify which mechanisms are valid explanations of macroscale dynamics (for a complete example of this approach, see van Nieuwenhuizen et al. 2024). Finally, one would define metabolic variables that can be measured peripherally, such as blood and salivary cell-free mitochondrial DNA (Sturm et al. 2023), as well as in the CNS, such as lactate and glutamate/GABA levels from MRS (Hone-Blanchet et al. 2023). These would provide a validation that the microscale mechanisms implicated by the computational model are indeed altered in the ways predicted by the macroscale observations.

5.4.2 *Neurodegenerative Processes*

Alzheimer disease is characterized by macroscale changes in brain structure (e.g., hippocampal atrophy) and function, including globally decreased glucose metabolism, abnormal brain regional communication, and cognitive decline (Antal et al. 2025; Arnold et al. 2018; Baker et al. 2011; Cunnane et al. 2011; Giorgio et al. 2024). One could follow the same approach outlined above to identify key components of disordered metabolic processes in Alzheimer disease compared to healthy aging. First, one would begin from a computational model that could identify key components that are dysregulated by Alzheimer processes, such as regional deposition of amyloid dysregulating circuit dynamics (Giorgio et al. 2024). Then, one would design metabolic perturbations that would tie directly into these dynamics, (e.g., simultaneous fMRI/FDG-PET) to quantify overlapping changes in circuit activity and glucose uptake (Hahn et al. 2020). Finally, one would identify key microscale biomarkers (e.g., β-amyloid and tau samples in cerebrospinal fluid) and additional imaging modalities for regions unable to be sampled (e.g., tau-PET to assess regional accumulation in the brain) to validate the leading candidate mechanisms identified in the macroscale data set by the computational models.

5.4.3 Looking to Future Experiments

Experimental designs in this vein provide two important transitions in how we think about future work. First, by designing an experiment with multiscale metabolism in mind from the beginning, there is a guaranteed level of explainability given by the models for which data are collected directly, and variability from unknown effects to the system can be directly quantified. Second, by creating more precise mechanistic explanations of individual data, there is a fundamental shift bringing experiments closer to *in silico* digital twins at an individual level, providing more personalized recommendations for how interventions will affect patients differently. By extending these experimental paradigms to longitudinal studies, an accurate, detailed, individualized trajectory of neurometabolic regulation can be mapped.

6 Summary

The primary point made in this chapter is that the impact of metabolism on psychiatric disorders involves one dysregulation across *multiple interconnected feedback loops*. Thus, standard statistical approaches—comparing a single snapshot of a hypothesized variable between patients and controls and/or correlating values across variables—are unlikely to be informative for three reasons. First, since all control circuits are connected, dysregulation of one component will be reflected by aberrant values across all other components and thus will fail to distinguish between driving mechanisms versus downstream effects. Second, dysregulation is characterized by altered dynamics in response to perturbation, which can include values that are higher *or* lower *or* equivalent to normal, depending on when they are measured. Third, even small disruptions at the mechanistic scale can give rise to emergent effects at the clinical scale that are not obvious and even counterintuitive. Given all three of these factors, experimental designs of complex mutually interacting control circuits (as per the role of metabolic dysregulation in neuropsychiatric disorders) will benefit from measuring variables *simultaneously across multiple control circuits* (i.e., as *continuous variables*) in response to *systematic perturbation of each control circuit*, informed by *multiscale* (*mechanistic-to-clinical scale*) *computational models*. While experimentally, analytically, and computationally challenging, methodological advances across all three areas are quickly making these types of scientific frameworks a reality, even in human clinical studies.

7 Glossary of Terms

AMPK (AMP-activated protein kinase): Cellular energy sensor that promotes catabolism while inhibits anabolism when cellular energy levels are depleted.

Astrocyte-neuron lactate shuttle (ANLS): System where astrocytes provide lactate as an energy substrate to neurons.

Bifurcation: A sudden qualitative change in the behavior of a system as a parameter is varied. *Flip bifurcation* is a bifurcation characterized by period-doubling dynamics when crossed (significantly increased cycling of the system). *Torus bifurcation* is a bifurcation characterized by additional complexity of activity (second limit cycle in phase space).
Circadian rhythm: The approximately 24-hour cycle of biological processes in living organisms.
Criticality: A state of a system poised between ordered and chaotic, often associated with optimal information processing and adaptability.
Emergence: The appearance of complex patterns or behaviors arising from simple interactions among individual components of a system.
GLUT4: Insulin-sensitive membrane glucose transporter.
Ketone bodies: Molecules produced from fatty acid oxidation during periods of low food intake or carbohydrate restriction, serving as an alternative energy source for the brain. Ketones present in the body are acetoacetate, beta-hydroxybutyrate, and small amounts of acetone.
Metastability: A state of a system characterized by the ability to transition between multiple patterns of activity in a predictable temporal sequence.
mTOR (mammalian target of rapamycin): Protein kinase that regulates cell growth, proliferation, and survival in response to nutrients and growth factors.
Neural mass model: A mathematical model that describes the average activity of a population of neurons without needing to simulate the entire population.
Parameter hyperspace: A multidimensional space where each dimension represents a parameter of a model or system. In complex systems, this space can be vast, encompassing all possible combinations of parameter values.
PGC-1α (peroxisome proliferator-activated receptor gamma coactivator 1-α): Key regulator of mitochondrial biogenesis and function.
Phase transition: A change in a feature of a physical system, often abrupt, as a result of a change in external conditions.
Respiration: In the context of cellular metabolism, respiration is the process by which cells break down molecules (e.g., glucose) to produce energy in the form of ATP. Typically occurs in the mitochondria and requires oxygen (aerobic respiration), although some cells can respire anaerobically.
Spatial transcriptomics: A technique that allows for the measurement of gene expression while preserving information about the spatial location of the measured cells within a tissue.

References

Abram S, Hua J, and Ford J (2022) Consider the Pons: Bridging the Gap on Sensory Prediction Abnormalities in Schizophrenia. Trends Neurosci 45 (11):798–808. https://doi.org/10.1016/j.tins.2022.08.008

Antal BB, Chesebro AG, Strey HH, Mujica-Parodi LR, and Weistuch C (2024) Achieving Occam's Razor: Deep Learning for Optimal Model Reduction. PLoS Comput Biol 20 (7):e1012283. https://doi.org/10.1371/journal.pcbi.1012283

Antal BB, van Nieuwenhuizen H, Chesebro AG, et al. (2025) Brain Aging Shows Nonlinear Transitions, Suggesting a Midlife "Critical Window" for Metabolic Intervention. PNAS 122 (10):e2416433122. https://doi.org/10.1073/pnas.2416433122

Arnold SE, Arvanitakis Z, Macauley-Rambach SL, et al. (2018) Brain Insulin Resistance in Type 2 Diabetes and Alzheimer Disease: Concepts and Conundrums. Nat Rev Neurol 14 (3):168–181. https://doi.org/10.1038/nrneurol.2017.185

Ashby FG, and Crossley MJ (2011) A Computational Model of How Cholinergic Interneurons Protect Striatal-Dependent Learning. J Cogn Neurosci 23 (6):1549–1566. https://doi.org/10.1162/jocn.2010.21523

Baker LD, Cross DJ, Minoshima S, et al. (2011) Insulin Resistance and Alzheimer-Like Reductions in Regional Cerebral Glucose Metabolism for Cognitively Normal Adults with Prediabetes or Early Type 2 Diabetes. Arch Neurol 68 (1):51–57. https://doi.org/10.1001/archneurol.2010.225

Cantó C, and Auwerx J (2009) PGC-1α, SIRT1 and AMPK, an Energy Sensing Network That Controls Energy Expenditure. Curr Opin Lipidol 20 (2):98–105. https://doi.org/10.1097/mol.0b013e328328d0a4

Chen L, and Campbell SA (2022) Exact Mean-Field Models for Spiking Neural Networks with Adaptation. J Comput Neurosci 50 (4):445–469. https://doi.org/10.1007/s10827-022-00825-9

Chesebro AG, Mujica-Parodi LR, and Weistuch C (2023) Ion Gradient-Driven Bifurcations of a Multi-Scale Neuronal Model. Chaos Solitons Fractals 167:113120. https://doi.org/10.1016/j.chaos.2023.113120

Cunnane SC, Nugent S, Roy M, et al. (2011) Brain Fuel Metabolism, Aging and Alzheimer's Disease. Nutrition 27 (1):3–20. https://doi.org/10.1016/j.nut.2010.07.021

DeBerardinis RJ, and Chandel NS (2020) We Need to Talk About the Warburg Effect. Nat Metab 2 (2):127–129. https://doi.org/10.1038/s42255-020-0172-2

Deco G, Kringelbach ML, Jirsa VK, and Ritter P (2017) The Dynamics of Resting Fluctuations in the Brain: Metastability and Its Dynamical Cortical Core. Sci Rep 7 (1):3095. https://doi.org/10.1038/s41598-017-03073-5

DiCarlo JJ, Zoccolan D, and Rust NC (2012) How Does the Brain Solve Visual Object Recognition? Neuron 73 (3):415–434. https://doi.org/10.1016/j.neuron.2012.01.010

Frank MJ (2005) Dynamic Dopamine Modulation in the Basal Ganglia: A Neurocomputational Account of Cognitive Deficits in Medicated and Nonmedicated Parkinsonism. J Cogn Neurosci 17 (1):51–72. https://doi.org/10.1162/0898929052880093

Gerich JE, Meyer C, Woerle HJ, and Stumvoll M (2001) Renal Gluconeogenesis: Its Importance in Human Glucose Homeostasis. Diabetes Care 24 (2):382–391. https://doi.org/10.2337/diacare.24.2.382

Giorgio J, Adams JN, Maass A, Jagust WJ, and Breakspear M (2024) Amyloid Induced Hyperexcitability in Default Mode Network Drives Medial Temporal Hyperactivity and Early Tau Accumulation. Neuron 112 (4):676–686.e674. https://doi.org/10.1016/j.neuron.2023.11.014

Giraldo-Chica M, Rogers BP, Damon SM, Landman BA, and Woodward ND (2018) Prefrontal-Thalamic Anatomical Connectivity and Executive Cognitive Function in Schizophrenia. Biol Psychiatry 83 (6):509–517. https://doi.org/10.1016/j.biopsych.2017.09.022

Gromada J, Franklin I, and Wollheim CB (2007) A-Cells of the Endocrine Pancreas: 35 Years of Research but the Enigma Remains. Endocr Rev 28 (1):84–116. https://doi.org/10.1210/er.2006-0007

Gryglewski G, Seiger R, James GM, et al. (2018) Spatial Analysis and High Resolution Mapping of the Human Whole-Brain Transcriptome for Integrative Analysis in Neuroimaging. Neuroimage 176:259–267. https://doi.org/10.1016/j.neuroimage.2018.04.068

Hahn A, Breakspear M, Rischka L, et al. (2020) Reconfiguration of Functional Brain Networks and Metabolic Cost Converge during Task Performance. eLife 9:e52443. https://doi.org/10.7554/elife.52443

Handschin C, and Spiegelman BM (2006) Peroxisome Proliferator-Activated Receptor Gamma Coactivator 1 Coactivators, Energy Homeostasis, and Metabolism. Endocr Rev 27 (7):728–735. https://doi.org/10.1210/er.2006-0037

Hansen JY, Shafiei G, Markello RD, et al. (2022) Mapping Neurotransmitter Systems to the Structural and Functional Organization of the Human Neocortex. Nat Neurosci 25 (11):1569–1581. https://doi.org/10.1038/s41593-022-01186-3

Hardie DG, Ross FA, and Hawley SA (2012) AMPK: A Nutrient and Energy Sensor That Maintains Energy Homeostasis. Nat Rev Mol Cell Biol 13 (4):251–262. https://doi.org/10.1038/nrm3311

Herzig S, and Shaw RJ (2018) AMPK: Guardian of Metabolism and Mitochondrial Homeostasis. Nat Rev Mol Cell Biol 19 (2):121–135. https://doi.org/10.1038/nrm.2017.95

Hone-Blanchet A, Antal B, McMahon L, et al. (2023) Acute Administration of Ketone Beta-Hydroxybutyrate Downregulates 7T Proton Magnetic Resonance Spectroscopy-Derived Levels of Anterior and Posterior Cingulate GABA and Glutamate in Healthy Adults. Neuropsychopharmacology 48 (5):797–805. https://doi.org/10.1038/s41386-022-01364-8

Humphries MD, Stewart RD, and Gurney KN (2006) A Physiologically Plausible Model of Action Selection and Oscillatory Activity in the Basal Ganglia. J Neurosci 26 (50):12921–12942. https://doi.org/10.1523/jneurosci.3486-06.2006

Jiang G, and Zhang BB (2003) Glucagon and Regulation of Glucose Metabolism. Am J Physiol Endocrinol Metab 284 (4):E671–E678. https://doi.org/10.1152/ajpendo.00492.2002

Kim J, and Guan K-L (2019) mTOR as a Central Hub of Nutrient Signalling and Cell Growth. Nat Cell Biol 21 (1):63–71. https://doi.org/10.1038/s41556-018-0205-1

Kim J, Kundu M, Viollet B, and Guan K-L (2011) AMPK and mTOR Regulate Autophagy through Direct Phosphorylation of Ulk1. Nat Cell Biol 13 (2):132–141. https://doi.org/10.1038/ncb2152

Kort NS, Ford JM, Roach BJ, et al. (2017) Role of N-Methyl-D-Aspartate Receptors in Action-Based Predictive Coding Deficits in Schizophrenia. Biol Psychiatry 81 (6):514–524. https://doi.org/10.1016/j.biopsych.2016.06.019

Kukurugya MA, Rosset S, and Titov DV (2024) The Warburg Effect Is the Result of Faster ATP Production by Glycolysis Than Respiration. PNAS 121 (46):e2409509121. https://doi.org/10.1073/pnas.2409509121

Kula B, Antal B, Weistuch C, et al. (2024) D-β-Hydroxybutyrate Stabilizes the Hippocampal CA3-CA1 Circuit during Acute Insulin Resistance. PNAS Nexus 3 (5):196. https://doi.org/10.1093/pnasnexus/pgae196

Laffel L (1999) Ketone Bodies: A Review of Physiology, Pathophysiology and Application of Monitoring to Diabetes. Diabetes Metab Res Rev 15 (6):412–426. https://doi.org/10.1002/(sici)1520-7560(199911/12)15:6%3C412::aid-dmrr72%3E3.0.co;2-8

Magistretti PJ, Pellerin L, Rothman DL, and Shulman RG (1999) Energy on Demand. Science 283 (5401):496–497. https://doi.org/10.1126/science.283.5401.496

McGregor MM, and Nelson AB (2019) Circuit Mechanisms of Parkinson's Disease. Neuron 101 (6):1042–1056. https://doi.org/10.1016/j.neuron.2019.03.004

Moghaddam B, and Javitt D (2012) From Revolution to Evolution: The Glutamate Hypothesis of Schizophrenia and Its Implication for Treatment. Neuropsychopharmacology 37 (1):4–15. https://doi.org/10.1038/npp.2011.181

Montbrio E, Pazo D, and Roxin A (2015) Macroscopic Description for Networks of Spiking Neurons. Phys Rev X 5 (2):021028. https://doi.org/10.1103/PhysRevX.5.021028

Mujica-Parodi LR, Amgalan A, Sultan SF, et al. (2020) Diet Modulates Brain Network Stability, a Biomarker for Brain Aging, in Young Adults. PNAS 117 (11):6170–6177. https://doi.org/10.1073/pnas.1913042117

Mujica-Parodi LR, Cha J, and Gao J (2017) From Anxious to Reckless: A Control Systems Approach Unifies Prefrontal-Limbic Regulation across the Spectrum of Threat Detection. Front Syst Neurosci 11:00018. https://doi.org/10.3389/fnsys.2017.00018

Mujica-Parodi LR, and Strey HH (2020) Making Sense of Computational Psychiatry. Int J Neuropsychopharmacol 23 (5):339–347. https://doi.org/10.1093/ijnp/pyaa013

Pathak A, Brincat SL, Organtzidis H, et al. (2024) Biomimetic Model of Corticostriatal Micro-Assemblies Discovers New Neural Code. https://doi.org/10.1101/2023.11.06.565902

Pilkis SJ, and Granner DK (1992) Molecular Physiology of the Regulation of Hepatic Gluconeogenesis and Glycolysis. Annu Rev Physiol 54:885–909. https://doi.org/10.1146/annurev.ph.54.030192.004321

Potter M, Newport E, and Morten KJ (2016) The Warburg Effect: 80 Years on. Biochem Soc Trans 44 (5):1499–1505. https://doi.org/10.1042/bst20160094

Puchalska P, and Crawford PA (2017) Multi-Dimensional Roles of Ketone Bodies in Fuel Metabolism, Signaling, and Therapeutics. Cell Metab 25 (2):262–284. https://doi.org/10.1016/j.cmet.2016.12.022

Puigserver P, and Spiegelman BM (2003) Peroxisome Proliferator-Activated Receptor-γ Coactivator 1α (PGC-1α): Transcriptional Coactivator and Metabolic Regulator. Endocr Rev 24 (1):78–90. https://doi.org/10.1210/er.2002-0012

Roberts JA, Gollo LL, Abeysuriya RG, et al. (2019) Metastable Brain Waves. Nat Commun 10 (1):1056. https://doi.org/10.1038/s41467-019-08999-0

Rodgers JT, Lerin C, Haas W, et al. (2005) Nutrient Control of Glucose Homeostasis through a Complex of PGC-1α and SIRT1. Nature 434 (7029):113–118. https://doi.org/10.1038/nature03354

Rorsman P, and Braun M (2013) Regulation of Insulin Secretion in Human Pancreatic Islets. Annu Rev Physiol 75:155–179. https://doi.org/10.1146/annurev-physiol-030212-183754

Rothman DL, De Feyter HM, de Graaf RA, Mason GF, and Behar KL (2011) ^{13}C MRS Studies of Neuroenergetics and Neurotransmitter Cycling in Humans. NMR Biomed 24 (8):943–957. https://doi.org/10.1002/nbm.1772

Rothman DL, Dienel GA, Behar KL, et al. (2022) Glucose Sparing by Glycogenolysis (Gsg) Determines the Relationship between Brain Metabolism and Neurotransmission. J Cereb Blood Flow Metab 42 (5):844–860. https://doi.org/10.1177/0271678X211064399

Saltiel AR, and Kahn CR (2001) Insulin Signalling and the Regulation of Glucose and Lipid Metabolism. Nature 414 (6865):799–806. https://doi.org/10.1016/j.jpsychires.2011.09.018

Sapolsky RM (1986) Glucocorticoid Toxicity in the Hippocampus: Reversal by Supplementation with Brain Fuels. J Neurosci 6 (8):2240–2244. https://doi.org/10.1523/jneurosci.06-08-02240.1986

Scarpulla RC (2011) Metabolic Control of Mitochondrial Biogenesis through the PGC-1 Family Regulatory Network. Biochim Biophys Acta 1813 (7):1269–1278. https://doi.org/10.1016/j.bbamcr.2010.09.019

Schneidman E, Berry MJ, Segev R, and Bialek W (2006) Weak Pairwise Correlations Imply Strongly Correlated Network States in a Neural Population. Nature 440 (7087):1007–1012. https://doi.org/10.1038/nature04701

Smith D, Pernet A, Hallett WA, et al. (2003) Lactate: A Preferred Fuel for Human Brain Metabolism *in Vivo*. J Cereb Blood Flow Metab 23 (6):658–664. https://doi.org/10.1097/01.wcb.0000063991.19746.11

Sturm G, Karan KR, Monzel AS, et al. (2023) Oxphos Defects Cause Hypermetabolism and Reduce Lifespan in Cells and in Patients with Mitochondrial Diseases. Commun Biol 6 (1):22. https://doi.org/10.1038/s42003-022-04303-x

Szendroedi J, Phielix E, and Roden M (2011) The Role of Mitochondria in Insulin Resistance and Type 2 Diabetes Mellitus. Nat Rev Endocrinol 8 (2):92–103. https://doi.org/10.1038/nrendo.2011.138

Taniguchi CM, Emanuelli B, and Kahn CR (2006) Critical Nodes in Signalling Pathways: Insights into Insulin Action. Nat Rev Mol Cell Biol 7 (2):85–96. https://doi.org/10.1038/nrm1837

Trujillo P, Song AK, Hay KR, et al. (2022) Dopamine-Induced Changes to Thalamic GABA Concentration in Impulsive Parkinson Disease Patients. NPJ Parkinsons Dis 8 (1):37. https://doi.org/10.1038/s41531-022-00298-8

van Hall G, Strømstad M, Rasmussen P, et al. (2009) Blood Lactate Is an Important Energy Source for the Human Brain. J Cereb Blood Flow Metab 29 (6):1121–1129. https://doi.org/10.1038/jcbfm.2009.35

van Nieuwenhuizen H, Chesebro AG, Polizu C, et al. (2024) Ketosis Regulates K^+ Ion Channels, Strengthening Brain-Wide Signaling Disrupted by Age. Imaging Neurosci 2:10.1162. https://doi.org/10.1162/imag_a_00163

Wang F, Smith NA, Xu Q, et al. (2012) Astrocytes Modulate Neural Network Activity by Ca^{2+}-Dependent Uptake of Extracellular K^+. Sci Signal 5 (218):ra26. https://doi.org/10.1126/scisignal.2002334

Weistuch C, Mujica-Parodi LR, and Dill K (2021a) The Refractory Period Matters: Unifying Mechanisms of Macroscopic Brain Waves. Neural Comput 33 (5):1145–1163. https://doi.org/10.1162/neco_a_01371

Weistuch C, Mujica-Parodi LR, Razban RM, et al. (2021b) Metabolism Modulates Network Synchrony in the Aging Brain. PNAS 118 (40):e2025727118. https://doi.org/10.1073/pnas.2025727118

Wu Z, Puigserver P, Andersson U, et al. (1999) Mechanisms Controlling Mitochondrial Biogenesis and Respiration through the Thermogenic Coactivator PGC-1. Cell 98 (1):115–124. https://doi.org/10.1016/s0092-8674(00)80611-x

Yang X, Heinemann M, Howard J, et al. (2021) Physical Bioenergetics: Energy Fluxes, Budgets, and Constraints in Cells. PNAS 118 (26):e2026786118. https://doi.org/10.1073/pnas.2026786118

Yellen G (2008) Ketone Bodies, Glycolysis, and K_{ATP} Channels in the Mechanism of the Ketogenic Diet. Epilepsia 49 (Suppl 8):80–82. https://doi.org/10.1111/j.1528-1167.2008.01843.x

Open Access This chapter is licensed under the terms of the Creative Commons Attribution-NonCommercial-NoDerivatives 4.0 International License (http://creativecommons.org/licenses/by-nc-nd/4.0/), which permits any noncommercial use, sharing, distribution and reproduction in any medium or format, as long as you give appropriate credit to the original author(s) and the source, provide a link to the Creative Commons license and indicate if you modified the licensed material. You do not have permission under this license to share adapted material derived from this chapter or parts of it.

The images or other third party material in this chapter are included in the chapter's Creative Commons license, unless indicated otherwise in a credit line to the material. If material is not included in the chapter's Creative Commons license and your intended use is not permitted by statutory regulation or exceeds the permitted use, you will need to obtain permission directly from the copyright holder.

7

Systemic Mitochondrial Alterations in Alzheimer Disease Dementia

Allie Amick and Anthony J. A. Molina

Abstract Alzheimer disease (AD) is a progressive neurodegenerative disease characterized by pathophysiological changes that start years before the onset of cognitive decline. First observed by Alois Alzheimer in 1906, symptomatic AD is characterized by extracellular amyloid-beta (Aβ) plaques, intracellular neurofibrillary tau tangles, and neurodegeneration within the brain in conjunction with cognitive decline. In addition to the traditional hallmark signs, mitochondrial dysfunction occurs in the brain at the earliest stages of disease progression and is also apparent in the peripheral cells and tissues of patients with AD. This review examines the role of mitochondria in AD and the specific alterations that have been described in the literature. Furthermore, evidence is reviewed for mitochondrial dysfunction across various cell types and tissues and insights provided into the factors that can drive systemic bioenergetic decline in AD.

Keywords Alzheimer disease, genetic risk, mitochondrial dysfunction, bioenergetic decline, mitochondrial cascade hypothesis

---

Anthony J. A. Molina (✉)
University of California San Diego School of Medicine, Stein Institute for Research on Aging, Department of Medicine, Division of Geriatrics, Gerontology, and Palliative Care, La Jolla, CA 92093, USA
Email: ajmolina@health.ucsd.edu

Allie Amick
Section on Gerontology and Geriatrics, Sticht Center for Healthy Aging and Alzheimer's Prevention, Department of Internal Medicine, Wake Forest Baptist Medical Center, Winston-Salem, NC 27104, USA

© Frankfurt Institute for Advanced Studies (FIAS) 2025

D. Öngür, J. M. Ford (eds.), *Metabolic Neuropsychiatry*, Strüngmann Forum Reports,
https://doi.org/10.1007/978-3-032-05630-6_7

1 The Mitochondrial Cascade Hypothesis

In 2004, Russell Swerdlow and Shaharyar Khan proposed the mitochondrial cascade hypothesis as an alternative to the amyloid cascade hypothesis to better explain the onset and risk of developing sporadic Alzheimer disease (sAD) (Swerdlow and Khan 2004). The hypothesis states that in sporadic, late-onset AD, mitochondrial function affects amyloid precursor protein (APP) expression, APP processing, or beta amyloid (Aβ) accumulation, and if an amyloid cascade truly exists, mitochondrial function triggers it. The mitochondrial cascade hypothesis addresses the limitations of the amyloid cascade hypothesis by explaining how individuals without mutations that induce aberrant APP processing develop sAD. The hypothesis maintains the following points and possible outcomes due to mitochondrial damage and inefficiency. First, genetic differences set the foundation for mitochondrial function and durability. Variations in nuclear DNA and mitochondrial DNA (mtDNA) genes that encode the subunits of the electron transport chain (ETC) complexes determine the baseline electron transfer system (ETS) efficiency and mitochondrial reactive oxygen species (ROS) production. Therefore, individuals with genetic predispositions for aberrant mitochondrial function would be more likely to develop AD later in life. Both maternal and paternal nuclear DNA contribute to baseline mitochondrial function and AD risk; however, as mtDNA is inherited only from the mother, maternal inheritance has been proposed to contribute more to differences in mitochondrial function and AD risk than paternal inheritance (Bassett et al. 2006). In one study, adult children with mothers affected by sAD had decreased brain glucose metabolism in brain regions affected in individuals with AD (Mosconi et al. 2007). Children of fathers affected by sAD and children of parents without dementia did not show metabolic abnormalities (Mosconi et al. 2007). Moreover, it has been reported that individuals with a maternal, rather than paternal, family history of AD had a three to nine times higher incidence of developing AD (Gómez-Tortosa et al. 2007). Mechanistically, decreased complex IV activity is apparent in platelets from individuals with a maternal family history of AD but not in individuals with a paternal or no parental history of AD (Mosconi et al. 2011). Altogether, these results suggest that maternal inheritance and mtDNA play a key role in mitochondrial dysfunction related to AD.

Changes in brain mitochondrial function with advancing age are determined by the combined influences of genetics and extrinsic factors. Measures of glucose metabolism in the brains of young adults carrying the AD risk factor allele APOE4 reveal decreased glucose uptake in the posterior cingulate, parietal, temporal, and prefrontal cortices (Reiman et al. 2004). A study of young APOE4 carriers found reduced complex IV activity but no evidence of elevated soluble Aβ, histologic Aβ, or tau pathology in the posterior cingulate cortex (Valla et al. 2010). These results suggest that mitochondrial dysfunction can precede Aβ deposition in regions susceptible to AD due to genetic factors. Over time, behaviors such as poor diet can impair mitochondrial function and increase the risk of developing AD (Więckowska-Gacek et al. 2021).

The mitochondrial cascade hypothesis posits that sAD pathology accumulates when mitochondrial dysfunction exceeds a threshold. Early pathological changes include alterations in APP, sAPPα, and Aβ homeostasis. For example, inhibiting complex IV shifts APP processing to produce more Aβ (Gabuzda et al. 1994; Gasparini et al. 1997). Moreover, impairments in both oxidative phosphorylation and glycolysis can affect APP processing (Mueller et al. 2005). These findings suggest that mitochondrial dysfunction can initiate processes that lead to Aβ deposition. Other proposed effects of mitochondrial dysfunction in the mitochondrial cascade hypothesis include tau phosphorylation and tangle formation, cell cycle reentry, and synaptic loss. Notably, Aβ and tau pathology have been reported to affect mitochondrial function, potentially creating a feedback loop that accelerates neurodegeneration (see Figure 7.1).

In summary, differences in mitochondrial function driven by genetic factors and the cumulative impacts of behaviors and lifetime exposures can impact the likelihood of developing sAD. When a threshold of mitochondria dysfunction is reached, downstream processes drive the production of Aβ and plaque deposition, tau tangle formation, synaptic loss, and neurodegeneration. We posit that the mitochondrial cascade hypothesis also suggests that mitochondrial dysfunction associated with the development of AD is systemic. Indeed, work from our group and others indicates that mitochondrial alterations apparent in peripheral cells are associated with AD dementia. These studies and their potential implications will be discussed in greater detail later in this review.

2 The Electron Transport Chain and the Alzheimer Disease Brain

The mitochondrial bioenergetics research is commonly focused on the ETC and the passage of electrons that drive the mitochondrial membrane potential required for oxidative phosphorylation. The ETC is responsible for generating energy as well as ROS, which can facilitate cell signaling at low levels or cause oxidative damage when produced in excess (Mueller et al. 2005). Alterations to the ETS—whether in

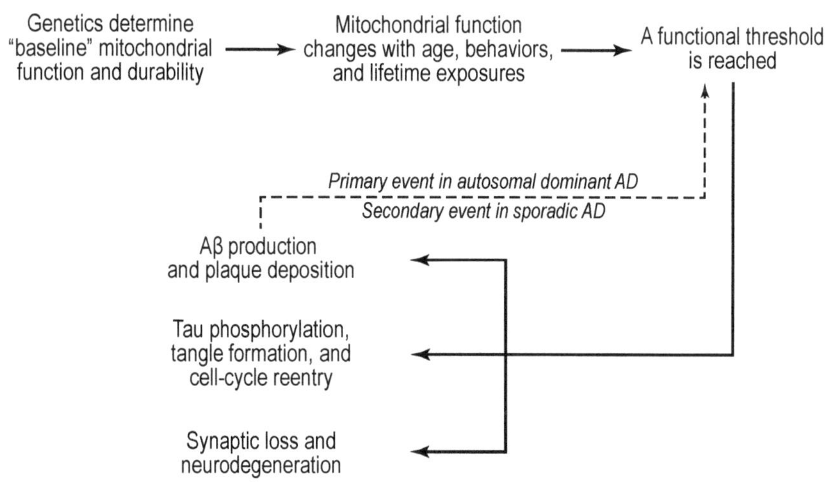

Figure 7.1 Mitochondrial cascade hypothesis. Adapted with permission from Elsevier from Swerdlow et al. (2010).

the form of disrupted ETC subunit expression, changes in the activity of the complexes, or other perturbations—can drive mitochondrial dysfunction and impact disease risk and progression.

2.1 Complex I

Complex I is a key entry point into the ETS and receives electrons from NADH produced by the tricarboxylic acid cycle. Complex I is composed of 45 subunit proteins, seven of which are encoded by mtDNA and translated in the mitochondrial matrix, making it the largest and most complex structure in the ETS (Sharma et al. 2009). More subunits increase the opportunity for errors; in fact, complex I deficiency is the most common cause of mitochondrial disorders. In AD, studies suggest that decreased complex I expression occurs in the frontal cortex and temporal lobe, especially in the transentorhinal and entorhinal cortices. In early AD and AD confirmed postmortem, expression of all complex I subunits in the frontal cortex is reduced. In particular, subunit 1 progressively decreases with the increasing severity of pathology (Manczak et al. 2004). Using postmortem human tissue, Adav et al. (2019) examined proteomes of the medial frontal gyrus in early-onset AD, late-onset AD, and nondemented aging controls. They found 13 downregulated complex I subunits in early-onset AD, 12 in late-onset AD, and only three in nondemented aging, demonstrating a distinct difference in complex I subunit expression and abundance between AD and nondemented aging. They also found that two subunits downregulated in early-onset AD were responsible for stabilizing complex I in the inner membrane, suggesting that destabilization of complex I could be significant in early-onset AD. Recently, new positron emission tomography (PET) tracers specific to complex I have enabled direct measurement of mitochondrial complex I in relation to Aβ and tau pathology. It has been reported that in early stages of AD, complex I levels are closely associated with tau levels in the transentorhinal cortex, entorhinal cortex, and hippocampus but not with levels of Aβ. There was also no relationship with Aβ in other brain regions (Terada et al. 2021). Further studies are needed to understand the relationship between Aβ and mitochondrial function earlier in disease progression, when Aβ starts to accumulate, perhaps during midlife in individuals with the APOE4 allele as a risk factor or who are the children of mothers that developed AD. In addition, it is still unclear if tau leads to mitochondrial dysfunction or if mitochondrial dysfunction leads to tau tangles. New PET imaging technologies may be able to address this key question.

Complex I is generally accepted to be downregulated in AD, but this could indicate complex I dysfunction, a compensatory mechanism in response to stress, or both. Decreased expression of ETS proteins, particularly subunits of complex I, in the brain of long-lived mice can improve several aging hallmarks, such as reduced ROS production and increased bioenergetic capacity and complex I assembly (Miwa et al. 2014). In particular, the subunits that stabilize complex I in the inner membrane are downregulated with age. This evidence suggests that downregulating complex I proteins may function as a compensatory mechanism (Adav et al. 2019). Furthermore, partial inhibition of complex I is beneficial in AD models. Using primary murine neurons and the APP/PS1 mouse model, systemic partial inhibition of complex I led to key improvements such as increased biogenesis, cognitive protection, long-term potentiation, oxidative stress, and inflammation (Trushin et al. 2020). The APP/PS1 AD model is a double transgenic mouse expressing mouse/human APP and a mutant human presenilin 1 (PS1), directed to neuronal cells. Another study in APP/PS1 female mice found that systemic partial inhibition of complex I improved energy homeostasis, long-term potentiation, proteostasis, synaptic activity, and dendritic spine maturation. In addition, oxidative stress and inflammation were improved in both the brain and periphery (Stojakovic et al. 2021). Both studies were conducted in APP expression mouse models, suggesting that complex I inhibitors may have an effect when Aβ is inducing cellular stress. Thus, it is unclear whether these inhibitors would be beneficial in sAD. Complex I is also a major site of ROS production. If produced in excess, ROS, which is elevated in AD, leads to cellular damage. Therefore, partial inhibition or downregulation of complex I is likely to mediate a decrease in ROS production and may positively impact cellular health.

2.2 Complex II

Complex II donates electrons to the ETS through the oxidation of succinate and does not transfer protons from the matrix to the intermembrane space. Complex II has not been a major focus in AD research. There are reports of increased (Bubber et al. 2005), decreased (Armand-Ugon et al. 2017; de la Monte and Wands 2006), and no difference (Cottrell et al. 2001; Maurer et al. 2000) in complex II activity or expression between controls, in human and mouse models of AD. Although several studies examining complex II find decreases in gene expression and protein expression, they are inconsistent in the subunits and brain regions affected. There is also some evidence that Aβ may inhibit complex II activity (Kaneko et al. 1995). Overall, we do not have overwhelming evidence that complex II plays a significant role in AD. However, because the ETS components are functionally interconnected, there may be complex II contributions to AD pathophysiology that are not reflected in protein and activity assays.

2.3 Complex III

Coenzyme Q receives electrons from the four ETS entry points and passes them to complex III. Complex III is a major site of electron leakage out of the ETS, which results in ROS production. Despite ROS elevation in AD, complex III has not been a major focus in AD research. Some studies have found minor differences in complex III protein and gene expression. One study found increased expression of one subunit in the frontal cortex of individuals with AD (Manczak et al. 2004). A separate study found decreased expression of a different subunit in the temporal cortex in individuals with AD (Kim et al. 2000). The reduced expression of complex III genes has also been reported in five brain regions of individuals with AD (Liang et al. 2008). However, another study found no difference in complex III gene expression between AD brains and controls (de la Monte and Wands 2006). Overall, the results for complex III in AD are mixed or suggest decreased expression of individual subunits, although the literature is sparse. It should be noted that complex I inhibition can decrease complex III activity (Chen et al. 2003). Complex I activity is compromised in AD, leading to more ROS from complex I but possibly decreasing ROS generated from complex III (Rummel and Butterfield 2022). Overall, the impact of complex III on AD pathology is unclear.

2.4 Complex IV

Complex IV is one of the key mitochondrial complexes affected in AD. Substantial evidence from postmortem tissue studies indicates that complex IV subunits are downregulated and that complex IV has decreased activity in the frontal cortex, temporal lobe, and parietal lobes during symptomatic AD (Armand-Ugon et al. 2017; Chagnon et al. 1995; Kenney and Bennett 2019; Kish et al. 1992; Maurer et al. 2000, 2001; Mutisya et al. 1994; Mastroeni et al. 2017). A meta-analysis found that out of 51 studies, 41 reported a decrease, six reported an increase, and four reported no difference in complex IV activity in human tissue and mouse AD models (Morais et al. 2021). Complex IV activity is also reduced in the posterior cingulate cortex of young adults at risk of developing sAD. Expression levels are elevated in this population (Perkins et al. 2016; Valla et al. 2010), indicating that complex IV activity is impaired in brain regions susceptible to AD pathology well before disease onset and may possibly be compensatory for the reduced activity by upregulating expression. Further, a mouse model of AD with a complex IV deficiency had fewer plaques and reduced oxidative stress than controls, suggesting that a decrease in complex IV activity may reduce oxidative stress (Fukui et al. 2007).

2.5 Complex V/ATP Synthase

ATP synthase, also referred to as complex V, utilizes the proton gradient generated by the ETS to generate ATP from ADP. Complex V is generally downregulated in AD. In one cell model that used N2a neuroblastoma cells expressing APOE4, there was a significant decrease in all 12 detected subunits compared to cells expressing APOE3, whereas the other ETS complexes I–IV only had decreases in 25–47% of detected subunits. In addition, ATP-linked respiration was also decreased (Orr et al. 2019). Studies of human brain tissue found decreased expression of the entire complex in hippocampal tissue (Schägger and Ohm 1995) and decreased gene expression in the posterior cingulate cortex, medial temporal gyrus, hippocampus, entorhinal cortex, and visual cortex with validation of decreased protein expression by western blot in the posterior cingulate cortex (Liang et al. 2008). Another study described a 50–60% decrease in the β-subunit mRNA transcript in the mid-temporal cortex but not the motor cortices of AD patients compared to age-matched controls, suggesting a role in brain regions affected in AD (Chandrasekaran et al. 1997).

Postmortem analysis of brain tissue can confirm AD pathology, but such studies are typically from patients who were later in disease progression with significant neurodegeneration. A study utilizing brain tissue from individuals in Braak stages I and II (no overt symptoms of AD and at the earliest stages of tau pathology) described elevated complex V lipoxidation and decreased activity in the entorhinal and transentorhinal cortices (Terni et al. 2010). It is not possible to know if these individuals would have progressed to full AD, but the early stage of tau pathology gives insight into mitochondrial alterations during the early stages of tau tangle formation. Overall, various models and stages of AD are associated with complex V reduction or downregulation; however, the mechanisms behind dysregulation remain unclear.

2.6 Electron Transferring Flavoprotein: Ubiquinone Oxidoreductase and Fatty Acid β-Oxidation

The electron transferring flavoprotein–ubiquinone oxidoreductase (ETF–QO) transfers electrons derived from fatty acid β-oxidation, amino acid catabolism, and choline metabolism into the ETS. Studies explicitly examining ETF–QO in AD could not be found. However, ample literature reports the role of fatty acid β-oxidation in AD, the primary source of electrons entering the ETS through ETF–QO. A study that examined lipid pathology in the hippocampus in an APOE4 mouse model found decreased fatty acid β-oxidation as compared to APOE3 mice, which led to the accumulation of toxic lipid droplets and reduced bioenergetic capacity, suggesting that impairment of fatty acid β-oxidation could lead to energy deficits and greater risk of AD (Qi et al. 2021a). Reduced fatty acid degradation and buildup of lipid droplets were also noted in these mice (Qi et al. 2021b), indicating dysregulated fatty acid metabolism in the mitochondria. The control and regulation of fatty acid β-oxidation is critical because an accumulation of fatty acids can cause neurodegeneration (Schönfeld and Reiser 2017). Furthermore, several top-risk genes for sAD, including ABCA7, APOE, APOJ, PICALM, SORL1, and TREM2, are involved in lipid trafficking or metabolism (Karch and Goate 2015).

2.7 Mitochondrial Glycerol-3-Phosphate Dehydrogenase

Mitochondrial glycerol-3-phosphate dehydrogenase (GpDH) contributes to bioenergetic capacity and oxidizes glycerol-3-phosphate to dihydroxyacetone phosphate and resides in the intermembrane of the mitochondria. There is also a cytosolic glycerol-3-phosphate that converts dihydroxyacetone phosphate back to glycerol-3-phosphate. Localization of GpDH varies widely between cell types and brain regions, making comparisons between studies difficult (Nguyen et al. 2003). One study of whole-brain tissue from a triple-transgenic AD mouse model reported downregulated mitochondrial GpDH and six other mitochondrial proteins involved in the tricarboxylic acid cycle and oxidative phosphorylation (Ciavardelli et al. 2010). Another study of aging PS-1 KI transgenic mice found no difference in GpDH expression, although no designation was made between mitochondrial and cytosolic GpDH (Fu et al. 2009). There is limited information about mitochondrial GpDH in the context of AD. However, mitochondrial GpDH is a source of superoxide formation. Therefore, alterations in mitochondrial GpDH have the potential to play a role in ROS production, which is elevated in AD, although direct evidence is lacking.

3 Systemic Mitochondrial Alterations in Alzheimer Disease

An important feature of mitochondrial alterations associated with AD is that these changes are not exclusive to neuronal cells. Indeed, a growing body of work is demonstrating that the mitochondrial bioenergetic profiles of peripheral cells and tissues are also altered in patients with AD dementia. In particular, blood cell bioenergetic profiling has emerged as a promising approach for examining the systemic mitochondrial alterations associated with AD. These provide a minimally invasive approach for human mitochondrial bioenergetic profiling that is amenable for use in a wide array of patient populations and for repeated assessments. Many groups are now using these approaches to track bioenergetic changes associated with the progression of age-related diseases and responses to interventions.

3.1 Platelets

Platelets are short-lived cells (8–9 days) that originate from megakaryocytes in the bone marrow and are primarily known for their role in thrombus formation and coagulation. Although they are anuclear, they contain mitochondria and can influence their environment by releasing extracellular vesicles, growth factors, chemokines, coagulant factors, and RNA species. They also uptake and adapt to factors present in plasma by endocytosis. Platelets are considered good peripheral candidates for studying AD because they have the necessary enzymes to produce APP and generate Aβ (Kroll and Schafer 1989; Racchi and Govoni 2003). Furthermore, work from our group has shown that platelet bioenergetics correlates with FDG-PET measures of glucose uptake in the brain, linking platelet and brain metabolism (Tyrrell et al. 2017).

Complex IV activity in platelets is consistently compromised in studies examining mitochondrial components in AD. An early study found that complex IV activity was reduced by ~13% in platelets from AD patients compared to controls while also reporting no difference in complex II, complex III, and citrate synthase activity (Parker et al. 1994). Another study with similar results found that complex IV activity was reduced by 15% in AD patients, but there was no difference in complex IV levels. It also reported decreased ATP and increased ROS levels (Cardoso et al. 2004). In a study examining the inheritance of AD, platelet bioenergetics in cognitively normal adult children of mothers with a history of AD had reduced complex IV activity compared to control individuals with no maternal history of AD (Mosconi et al. 2011). Platelets' mitochondrial function is compromised in patients with AD and individuals at risk of AD.

3.2 Peripheral Blood Mononuclear Cells

As they freely circulate throughout the body, peripheral blood mononuclear cells (PBMCs) are exposed to all circulating factors present in the blood, and have an array of altered functions in AD. Work from our lab has shown that PBMC bioenergetics are related to white matter, gray matter, and total intracranial volumes in older adults, demonstrating that PBMC bioenergetics can report on the brain (Mahapatra et al. 2018). Furthermore, in a study with 53 mild to moderate AD patients and 30 age-matched controls, basal respiration was decreased in the AD compared to controls (Maynard et al. 2015). PBMCs also show alterations other than bioenergetics in AD. Lymphocytes, which make up 70–90% of PBMC cells, specifically the $CD4^+$ portion, had increased apoptosis and reduced mitochondrial membrane potential in mild cognitive impairment and AD compared to cognitively normal individuals. Lymphocytes from AD and individuals with mild cognitive impairment also had more pronounced apoptosis and decreased mitochondrial membrane potential when challenged with mitochondrial inhibitors than cells derived from cognitively normal individuals (Leuner et al. 2012). In a large-scale study examining PBMCs from participants with normal cognition, mild cognitive impairment, and dementia due to probable AD, we observed a progressive decline in bioenergetic capacity as assessed by high-resolution mitochondrial respirometry (Mahapatra et al. 2022).

3.3 Fibroblasts

Fibroblasts have many documented changes in mitochondrial function within the context of AD. For example, in one familial AD study, fibroblasts from patients with PSEN1 mutations had diminished mitochondrial content and reduced basal, maximal, spare, and ATP-linked respiration compared to controls (Gray and Quinn 2015). A separate study, however, found that fibroblasts from patients with the PSEN1 mutation had elevated respiration compared to controls (Bell et al. 2018). Fibroblasts are also valuable for studying sAD in addition to familial AD. Fibroblasts from sAD patients have impaired bioenergetic function, and the mitochondria are more fragmented (Bell et al. 2018; Pérez et al. 2017). In addition, fibroblasts from sAD patients also have downregulated genes for mitochondrial homeostasis, including genes for fission and fusion (Martín-Maestro et al. 2017). As a result, mitochondrial dynamics and membrane recovery are impaired.

4 Drivers of Systemic Bioenergetic Decline Associated with Aging and Alzheimer Disease

We posit that the systemic nature of bioenergetic alterations associated with AD is mediated by factors circulating in the blood. In studies of aging, seminal studies have demonstrated that connecting the circulatory systems of young and old animals, an approach called heterochronic parabiosis, has systemic "rejuvenating" effects for older animals and detrimental effects for younger animals (Conboy et al. 2013; Eggel and Wyss-Coray 2014). The effects of blood are evident in skeletal muscle (Brack et al. 2007; Conboy et al. 2005), liver (Conboy et al. 2005), heart (Loffredo et al. 2013), and pancreas (Salpeter et al. 2013). The effects of young blood on older parabionts include increased neurogenesis accompanied by improved synaptic plasticity and cognitive function (Katsimpardi et al. 2014; Villeda et al. 2011, 2014). A study from our lab used the parabiosis model to demonstrate that circulating factors mediate age-related bioenergetic decline (Gonzalez-Armenta et al. 2020). We examined mitochondrial morphology and respiration of skeletal muscle fiber bundles from heterochronic pairs and isochronic controls. Across all measures, young heterochronic parabionts exhibited mitochondrial deficits compared to young isochronic controls. Importantly, we observed that young and old heterochronic pairs exhibited an equalization of low bioenergetic capacities, similar to the older controls. This striking finding demonstrates that circulating factors may be sufficient to drive age-related bioenergetic decline.

In addition to mediating the age-related bioenergetic decline, we have reported that noncellular factors present in human serum can mediate mitochondrial improvements associated with interventions such as diet and exercise. Using serum collected from overweight and obese older adults, before and after resistance training or resistance training plus caloric restriction interventions, we demonstrated that circulating factors were sufficient to confer bioenergetic improvements in muscle cells *in vitro* (Gonzalez-Armenta et al. 2024).

Similar to age-related mitochondrial bioenergetic decline, blood-borne factors can mediate systemic bioenergetic differences associated with AD. Naïve neuronal cells exposed *in vitro* to dilute fibrinogen-depleted plasma from participants with normal cognition, mild cognitive impairment, and dementia due to probable AD-induced bioenergetic changes were associated with the progression of AD and cognitive decline (Amick et al. 2022). Interestingly, the bioenergetic capacity of neuronal cells exposed to serum was positively correlated with the bioenergetic capacity of PBMCs isolated from the serum donor. These results suggest that circulating factors can modulate the bioenergetics of naïve neurons according to the bioenergetic capacity of the donor.

4.1 "Mitoactive" Circulating Factors

Numerous classes of circulating factors can modify mitochondrial function across various cell types. Thus, it is unlikely that any single molecule is solely responsible for systemic mitochondrial alterations associated with the progression of AD.

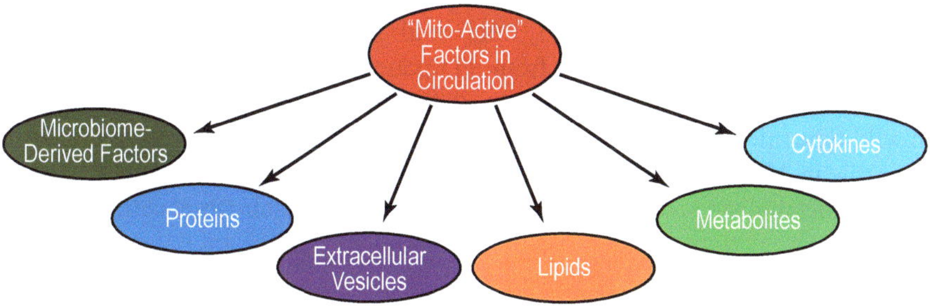

Figure 7.2 Blood contains numerous circulating factors that can directly alter mitochondrial function.

Potentially mitoactive factors are depicted in Figure 7.2, which summarizes many of the likely classes of factors that can directly impact mitochondrial function via different mechanisms.

The combined effects of these factors may underlie the systemic nature of bioenergetic changes associated with aging and AD. Below, we will highlight two of the classes listed in Figure 7.2 that have been relatively understudied with regard to potential effects on mitochondrial function.

4.1.1 Extracellular Vesicles

Extracellular vesicles (EVs) are present in all biofluids (van der Pol et al. 2012) and are loaded with unique cargo, including RNAs, proteins, and metabolites that relate to the cell of origin and the physiologic and metabolic state of the organism (Eissa et al. 2016; de Jong et al. 2012; Ramteke et al. 2015; Schlaepfer et al. 2015). EVs released by cells under stressed or pathologic states are different from those released under normal physiologic conditions (Chistiakov et al. 2016; de Jong et al. 2012; Hedlund et al. 2011; Panigrahi et al. 2018; Ramteke et al. 2015; Sheller et al. 2016). For example, small EVs secreted in response to oxidative stress are loaded with unique proteins representative of a pro-oxidative physiological state (Eissa et al. 2016; de Jong et al. 2012; Hedlund et al. 2011). For these reasons, small EVs are being extensively studied for the diagnosis and prognosis of diseases such as AD, Parkinson disease, multiple sclerosis, and cancer (Asai et al. 2015; Chistiakov et al. 2016; Eissa et al. 2016; Kalluri 2016; Panigrahi and Deep 2017; Soria et al. 2017; Winston et al. 2016). It is likely that EVs impact the function of the cells they target; however, little is known about how they affect cellular metabolism and mitochondrial bioenergetics.

In AD, seminal studies demonstrate that small EVs carry pathogenic proteins that can spread neuronal dysfunction (Asai et al. 2015; Baker et al. 2016; Kapogiannis et al. 2015; Malm et al. 2016; Ngolab et al. 2017; Rajendran et al. 2006; Soria et al. 2017; Winston et al. 2016; Yuyama and Igarashi 2017). Brain tissue-derived small EVs are present in circulation and are being examined as potential mediators of AD-related pathological changes (Fiandaca et al. 2015; Kapogiannis et al. 2015; Vella et al. 2016). For example, the levels of Aβ and P-tau in neuron-derived small EVs predict AD development (Fiandaca et al. 2015; Winston et al. 2016). Separate studies examining small EV RNA species report that miRNAs, mRNAs, and noncoding

RNAs are differentially expressed in AD patients compared to healthy individuals (Cheng et al. 2015; Liu et al. 2014; Lugli et al. 2015). Recent studies further demonstrate that small EVs isolated from AD patients, cell lines expressing APP, or P301L tau transgenic rTg4510 mice can elicit pathological responses when injected in wild-type mice (Baker et al. 2016; Winston et al. 2016; Zheng et al. 2017). While their role in AD is apparent, future studies are needed to determine how EVs affect cellular metabolism and mitochondrial bioenergetics in their target cells.

4.1.2 Extracellular Lipids

Among the various classes of circulating factors that can play a role in AD, lipids are of particular interest. Several studies comparing normo-cognitive and cognitively impaired individuals have found differences in circulating noncellular lipids. For instance, a study of 123 individuals divided into AD, mild cognitive impairment, and elderly control groups demonstrated that the abundance of just ten lipid metabolites could identify patients from healthy controls (Proitsi et al. 2015). In another study, several species of sphingomyelin were decreased, and two ceramide species were increased in early AD plasma compared to controls (Han et al. 2011). Moreover, Liu et al. (2021) reported that Aβ is contributory, if not causal, to the differences observed in peripheral lipidomes between cognitive groups.

We recently demonstrated that lipids may play a role in AD-associated changes in mitochondrial function. Using mass spectrometry, we analyzed the abundance of lipid metabolites in the serum of participants with normal cognition, mild cognitive impairment, and dementia due to probable AD. We found that the abundance of glycocholic acid was positively correlated with the bioenergetic effects of the serum samples on neuronal cells *in vitro*. Moreover, the abundance of glycocholic acid was also correlated with the bioenergetic capacity of the serum donor, as reported by respirometric analyses of PBMCs. Conversely, linoleic acid abundance was negatively correlated with the same bioenergetic measures (Amick et al. 2022). In a larger cohort of 378 participants, we found that glycocholic acid abundance was lower by 53% in the dementia group compared to the normal cognition group, and that linoleic acid abundance was higher by 56% in the dementia group compared to the normal cognition group. Future studies are required to determine the direct bioenergetic effects of glycocholic acid, linoleic acid, and other potentially mitoactive lipids that are differentially expressed in the circulation of patients with AD.

5 Conclusion

Mitochondrial dysfunction plays a central role in the pathophysiology of AD and is influenced by a combination of intrinsic (e.g., genetic) and extrinsic factors. Extending the mitochondrial cascade hypothesis, studies show that mitochondrial alterations associated with AD are systemic and apparent in peripheral cells. For this reason, it may be possible to utilize more readily accessible cells, such as those found in blood, to track systemic bioenergetic changes associated with the progression of AD. Moreover, blood cells may provide a way to evaluate the potential systemic bioenergetic benefits of promising interventions, such as diet and exercise, which are known to have broad metabolic benefits. Finally, we are beginning to understand the factors driving bioenergetic changes associated with the progression of AD. These factors include a multitude of circulating molecules that can impact mitochondrial function across many cell and tissue types. While no single factor can be solely responsible for bioenergetic changes associated with AD, efforts to identify key regulators may inform the development of therapeutic strategies for countering mitochondrial dysfunction and its consequences.

References

Adav SS, Park JE, and Sze SK (2019) Quantitative Profiling Brain Proteomes Revealed Mitochondrial Dysfunction in Alzheimer's Disease. Mol Brain 12:1–12. https://doi.org/10.1186/s13041-019-0430-y

Amick KA, Mahapatra G, Gao Z, et al. (2022) Plasma Glycocholic Acid and Linoleic Acid Identified as Potential Mediators of Mitochondrial Bioenergetics in Alzheimer's Dementia. Front Aging Neurosci 14:954090. https://doi.org/10.3389/fnagi.2022.954090

Armand-Ugon M, Ansoleaga B, Berjaoui S, and Ferrer I (2017) Reduced Mitochondrial Activity Is Early and Steady in the Entorhinal Cortex but It Is Mainly Unmodified in the Frontal Cortex in Alzheimer's Disease. Curr Alzheimer Res 14:1327–1334. https://doi.org/10.2174/1567205014666170505095921

Asai H, Ikezu S, Tsunoda S, et al. (2015) Depletion of Microglia and Inhibition of Exosome Synthesis Halt Tau Propagation. Nat Neurosci 18 (11):1584–1593. https://doi.org/10.1038/nn.4132

Baker S, Polanco JC, and Gotz J (2016) Extracellular Vesicles Containing P301L Mutant Tau Accelerate Pathological Tau Phosphorylation and Oligomer Formation but Do Not Seed Mature Neurofibrillary Tangles in Alz17 Mice. J Alzheimers Dis 54 (3):1207–1217. https://doi.org/10.3233/JAD-160371

Bassett SS, Yousem DM, Cristinzio C, et al. (2006) Familial Risk for Alzheimer's Disease Alters fMRI Activation Patterns. Brain 129 (Pt 5):1229–1239. https://doi.org/10.1093/brain/awl089

Bell SM, Barnes K, Clemmens H, et al. (2018) Ursodeoxycholic Acid Improves Mitochondrial Function and Redistributes Drp1 in Fibroblasts from Patients with Either Sporadic or Familial Alzheimer's Disease. J Mol Biol 430 (21):3942–3953. https://doi.org/10.1016/j.jmb.2018.08.019

Brack AS, Conboy MJ, Roy S, et al. (2007) Increased Wnt Signaling during Aging Alters Muscle Stem Cell Fate and Increases Fibrosis. Science 317 (5839):807–810. https://doi.org/10.1126/science.1144090

Bubber P, Haroutunian V, Fisch G, Blass JP, and Gibson GE (2005) Mitochondrial Abnormalities in Alzheimer Brain: Mechanistic Implications. Ann Neurol 57 (5):695–703. https://doi.org/10.1002/ana.20474

Cardoso SM, Proença MT, Santos S, Santana I, and Oliveira CR (2004) Cytochrome *C* Oxidase Is Decreased in Alzheimer's Disease Platelets. Neurobiol Aging 25 (1):105–110. https://doi.org/10.1016/s0197-4580(03)00033-2

Chagnon P, Bétard C, Robitaille Y, Cholette A, and Gauvreau D (1995) Distribution of Brain Cytochrome Oxidase Activity in Various Neurodegenerative Diseases. Neuroreport 6 (5):711–715. https://doi.org/10.1097/00001756-199503270-00002

Chandrasekaran K, Hatanpää K, Rapoport SI, and Brady DR (1997) Decreased Expression of Nuclear and Mitochondrial DNA-Encoded Genes of Oxidative Phosphorylation in Association Neocortex in Alzheimer Disease. Brain Res Mol Brain Res 44 (1):99–104. https://doi.org/10.1016/s0169-328x(96)00191-x

Chen Q, Vazquez EJ, Moghaddas S, Hoppel CL, and Lesnefsky EJ (2003) Production of Reactive Oxygen Species by Mitochondria: Central Role of Complex III. J Biol Chem 278 (38):36027–36031. https://doi.org/10.1074/jbc.M304854200

Cheng L, Doecke JD, Sharples RA, et al. (2015) Prognostic Serum miRNA Biomarkers Associated with Alzheimer's Disease Shows Concordance with Neuropsychological and Neuroimaging Assessment. Mol Psychiatr 20 (10):1188–1196. https://doi.org/10.1038/mp.2014.127

Chistiakov DA, Orekhov AN, and Bobryshev YV (2016) Cardiac Extracellular Vesicles in Normal and Infarcted Heart. Int J Mol Sci 17 (1):63. https://doi.org/10.3390/ijms17010063

Ciavardelli D, Silvestri E, Del Viscovo A, et al. (2010) Alterations of Brain and Cerebellar Proteomes Linked to Aβ and Tau Pathology in a Female Triple-Transgenic Murine Model of Alzheimer's Disease. Cell Death Dis 1 (10):e90. https://doi.org/10.1038/cddis.2010.68

Conboy IM, Conboy MJ, Wagers AJ, et al. (2005) Rejuvenation of Aged Progenitor Cells by Exposure to a Young Systemic Environment. Nature 433 (7027):760–764. https://doi.org/10.1038/nature03260

Conboy MJ, Conboy IM, and Rando TA (2013) Heterochronic Parabiosis: Historical Perspective and Methodological Considerations for Studies of Aging and Longevity. Aging Cell 12 (3):525–530. https://doi.org/10.1111/acel.12065

Cottrell DA, Blakely EL, Johnson MA, Ince PG, and Turnbull DM (2001) Mitochondrial Enzyme-Deficient Hippocampal Neurons and Choroidal Cells in AD. Neurology 57 (2):260–264. https://doi.org/10.1212/wnl.57.2.260

de Jong OG, Verhaar MC, Chen Y, et al. (2012) Cellular Stress Conditions Are Reflected in the Protein and RNA Content of Endothelial Cell-Derived Exosomes. J Extracell Vesicles 1:18396. https://doi.org/10.3402/jev.v1i0.18396

de la Monte SM, and Wands JR (2006) Molecular Indices of Oxidative Stress and Mitochondrial Dysfunction Occur Early and Often Progress with Severity of Alzheimer's Disease. J Alzheimers Dis 9 (2):167–181. https://doi.org/10.3233/jad-2006-9209

Eggel A, and Wyss-Coray T (2014) A Revival of Parabiosis in Biomedical Research. Swiss Med Wkly 144 (0506):w13914. https://doi.org/10.4414/smw.2014.13914

Eissa S, Matboli M, Aboushahba R, Bekhet MM, and Soliman Y (2016) Urinary Exosomal microRNA Panel Unravels Novel Biomarkers for Diagnosis of Type 2 Diabetic Kidney Disease. J Diabetes Complications 30 (8):1585–1592. https://doi.org/10.1016/j.jdiacomp.2016.07.012

Fiandaca MS, Kapogiannis D, Mapstone M, et al. (2015) Identification of Preclinical Alzheimer's Disease by a Profile of Pathogenic Proteins in Neurally Derived Blood Exosomes: A Case-Control Study. Alzheimers Dement 11 (6):600–607.e601. https://doi.org/10.1016/j.jalz.2014.06.008

Fu YJ, Xiong S, Lovell MA, and Lynn BC (2009) Quantitative Proteomic Analysis of Mitochondria in Aging PS-1 Transgenic Mice. Cell Mol Neurobiol 29 (5):649–664. https://doi.org/10.1007/s10571-009-9359-5

Fukui H, Diaz F, Garcia S, and Moraes CT (2007) Cytochrome *C* Oxidase Deficiency in Neurons Decreases Both Oxidative Stress and Amyloid Formation in a Mouse Model of Alzheimer's Disease. PNAS 104 (35):14163–14168. https://doi.org/10.1073/pnas.0705738104

Gabuzda D, Busciglio J, Chen LB, Matsudaira P, and Yankner BA (1994) Inhibition of Energy Metabolism Alters the Processing of Amyloid Precursor Protein and Induces a Potentially Amyloidogenic Derivative. J Biol Chem 269 (18):13623–13628. https://doi.org/10.1016/S0021-9258(17)36875-8

Gasparini L, Racchi M, Benussi L, et al. (1997) Effect of Energy Shortage and Oxidative Stress on Amyloid Precursor Protein Metabolism in COS Cells. Neurosci Lett 231 (2):113–117. https://doi.org/10.1016/s0304-3940(97)00536-3

Gómez-Tortosa E, Barquero MS, Barón M, et al. (2007) Variability of Age at Onset in Siblings with Familial Alzheimer Disease. Arch Neurol 64 (12):1743–1748. https://doi.org/10.1001/archneur.64.12.1743

Gonzalez-Armenta JL, Bergstrom J, Lee J, et al. (2024) Serum Factors Mediate Changes in Mitochondrial Bioenergetics Associated with Diet and Exercise Interventions. Geroscience 46 (1):349–365. https://doi.org/10.1007/s11357-023-00855-w

Gonzalez-Armenta JL, Li N, Lee RL, Lu B, and Molina AJA (2020) Heterochronic Parabiosis: Old Blood Induces Changes in Mitochondrial Structure and Function of Young Mice. J Gerontol A Biol Sci Med Sci 76 (3):434–439. https://doi.org/10.1093/gerona/glaa299

Gray NE, and Quinn JF (2015) Alterations in Mitochondrial Number and Function in Alzheimer's Disease Fibroblasts. Metab Brain Dis 30 (5):1275–1278. https://pubmed.ncbi.nlm.nih.gov/25862550/

Han X, Rozen S, Boyle SH, et al. (2011) Metabolomics in Early Alzheimer's Disease: Identification of Altered Plasma Sphingolipidome Using Shotgun Lipidomics. PLoS One 6 (7):e21643. https://doi.org/10.1371/journal.pone.0021643

Hedlund M, Nagaeva O, Kargl D, Baranov V, and Mincheva-Nilsson L (2011) Thermal- and Oxidative Stress Causes Enhanced Release of NKG2D Ligand-Bearing Immunosuppressive Exosomes in Leukemia/Lymphoma T and B Cells. PLoS One 6 (2):e16899. https://doi.org/10.1371/journal.pone.0016899

Kalluri R (2016) The Biology and Function of Exosomes in Cancer. J Clin Invest 126 (4):1208–1215. https://doi.org/10.1172/JCI81135

Kaneko I, Yamada N, Sakuraba Y, Kamenosono M, and Tutumi S (1995) Suppression of Mitochondrial Succinate Dehydrogenase, a Primary Target of β-Amyloid, and Its Derivative Racemized at Ser Residue. J Neurochem 65 (6):2585–2593. https://doi.org/10.1046/j.1471-4159.1995.65062585.x

Kapogiannis D, Boxer A, Schwartz JB, et al. (2015) Dysfunctionally Phosphorylated Type 1 Insulin Receptor Substrate in Neural-Derived Blood Exosomes of Preclinical Alzheimer's Disease. FASEB J 29 (2):589–596. https://doi.org/10.1096/fj.14-262048

Karch CM, and Goate AM (2015) Alzheimer's Disease Risk Genes and Mechanisms of Disease Pathogenesis. Biol Psychiatry 77 (1):43–51. https://doi.org/10.1016/j.biopsych.2014.05.006

Katsimpardi L, Litterman NK, Schein PA, et al. (2014) Vascular and Neurogenic Rejuvenation of the Aging Mouse Brain by Young Systemic Factors. Science 344 (6184):630–634. https://doi.org/10.1126/science.1251141

Kenney PM, and Bennett JP, Jr. (2019) Alzheimer's Disease Frontal Cortex Mitochondria Show a Loss of Individual Respiratory Proteins but Preservation of Respiratory Supercomplexes. Int J Alzheimers Dis 2019:4814783. https://doi.org/10.1155/2019/4814783

Kim SH, Vlkolinsky R, Cairns N, and Lubec G (2000) Decreased Levels of Complex III Core Protein 1 and Complex V β Chain in Brains from Patients with Alzheimer's Disease and Down Syndrome. Cell Mol Life Sci 57 (12):1810–1816. https://doi.org/10.1007/pl00000661

Kish SJ, Bergeron C, Rajput A, et al. (1992) Brain Cytochrome Oxidase in Alzheimer's Disease. J Neurochem 59 (2):776–779. https://doi.org/10.1111/j.1471-4159.1992.tb09439.x

Kroll M, and Schafer A (1989) Biochemical Mechanisms of Platelet Activation. Blood 74 (4):1181–1195. https://doi.org/10.1182/blood.V74.4.1181.1181

Leuner K, Schulz K, Schutt T, et al. (2012) Peripheral Mitochondrial Dysfunction in Alzheimer's Disease: Focus on Lymphocytes. Mol Neurobiol 46 (1):194–204. https://doi.org/10.1007/s12035-012-8300-y

Liang WS, Reiman EM, Valla J, et al. (2008) Alzheimer's Disease Is Associated with Reduced Expression of Energy Metabolism Genes in Posterior Cingulate Neurons. PNAS 105 (11):4441–4446. https://doi.org/10.1073/pnas.0709259105

Liu CG, Song J, Zhang YQ, and Wang PC (2014) MicroRNA-193b Is a Regulator of Amyloid Precursor Protein in the Blood and Cerebrospinal Fluid Derived Exosomal microRNA-193b Is a Biomarker of Alzheimer's Disease. Mol Med Rep 10 (5):2395–2400. https://doi.org/10.3892/mmr.2014.2484

Liu Y, Thalamuthu A, Mather KA, et al. (2021) Plasma Lipidome Is Dysregulated in Alzheimer's Disease and Is Associated with Disease Risk Genes. Transl Psychiatry 11 (1):344. https://doi.org/10.1038/s41398-021-01362-2

Loffredo FS, Steinhauser ML, Jay SM, et al. (2013) Growth Differentiation Factor 11 Is a Circulating Factor That Reverses Age-Related Cardiac Hypertrophy. Cell 153 (4):828–839. https://doi.org/10.1016/j.cell.2013.04.015

Lugli G, Cohen AM, Bennett DA, et al. (2015) Plasma Exosomal miRNAs in Persons with and without Alzheimer Disease: Altered Expression and Prospects for Biomarkers. PLoS One 10 (10):e0139233. https://doi.org/10.1371/journal.pone.0139233

Mahapatra G, Gao Z, Bateman JR, 3rd, et al. (2022) Blood-Based Bioenergetic Profiling Reveals Differences in Mitochondrial Function Associated with Cognitive Performance and Alzheimer's Disease. Alzheimers Dement 19:1466–1478. https://doi.org/10.1002/alz.12731

Mahapatra G, Smith SC, Hughes TM, et al. (2018) Blood-Based Bioenergetic Profiling Is Related to Differences in Brain Morphology in African Americans with Type 2 Diabetes. Clin Sci 132 (23):2509–2518. https://doi.org/10.1042/CS20180690

Malm T, Loppi S, and Kanninen KM (2016) Exosomes in Alzheimer's Disease. Neurochem Int 97:193–199. https://doi.org/10.1016/j.neuint.2016.04.011

Manczak M, Park BS, Jung Y, and Reddy PH (2004) Differential Expression of Oxidative Phosphorylation Genes in Patients with Alzheimer's Disease: Implications for Early Mitochondrial Dysfunction and Oxidative Damage. Neuromolecular Med 5 (2):147–162. https://doi.org/10.1385/nmm:5:2:147

Martín-Maestro P, Gargini R, García E, et al. (2017) Slower Dynamics and Aged Mitochondria in Sporadic Alzheimer's Disease. Oxid Med Cell Longev 2017:9302761. https://doi.org/10.1155/2017/9302761

Mastroeni D, Khdour OM, Delvaux E, et al. (2017) Nuclear but Not Mitochondrial-Encoded Oxidative Phosphorylation Genes Are Altered in Aging, Mild Cognitive Impairment, and Alzheimer's Disease. Alzheimers Dement 13 (5):510–519. https://doi.org/10.1016/j.jalz.2016.09.003

Maurer I, Zierz S, and Möller HJ (2000) A Selective Defect of Cytochrome *C* Oxidase Is Present in Brain of Alzheimer Disease Patients. Neurobiol Aging 21 (3):455–462. https://doi.org/10.1016/s0197-4580(00)00112-3

Maynard S, Hejl AM, Dinh TS, et al. (2015) Defective Mitochondrial Respiration, Altered DNTP Pools and Reduced AP Endonuclease 1 Activity in Peripheral Blood Mononuclear Cells of Alzheimer's Disease Patients. Aging 7 (10):793–815. https://doi.org/10.18632/aging.100810

Miwa S, Jow H, Baty K, et al. (2014) Low Abundance of the Matrix Arm of Complex I in Mitochondria Predicts Longevity in Mice. Nat Commun 5 (1):3837. https://doi.org/10.1038/ncomms4837

Morais FM, Ribeiro AM, Moreira FA, and Silva PVG (2021) Systematic Review and Meta-Analysis on the Role of Mitochondrial Cytochrome *C* Oxidase in Alzheimer's Disease. Acta Neuropsychiatr 33 (2):55–64. https://doi.org/10.1017/neu.2020.43

Mosconi L, Brys M, Switalski R, et al. (2007) Maternal Family History of Alzheimer's Disease Predisposes to Reduced Brain Glucose Metabolism. PNAS 104 (48):19067–19072. https://doi.org/10.1073/pnas.0705036104

Mosconi L, de Leon M, Murray J, et al. (2011) Reduced Mitochondria Cytochrome Oxidase Activity in Adult Children of Mothers with Alzheimer's Disease. J Alzheimers Dis 27 (3):483–490. https://doi.org/10.3233/jad-2011-110866

Mueller CFH, Laude K, McNally JS, and Harrison DG (2005) ATVB in Focus: Redox Mechanisms in Blood Vessels. Arterioscler Thromb Vasc Biol 25 (2):274–278. https://doi.org/10.1161/01.atv.0000149143.04821.eb

Mutisya EM, Bowling AC, and Beal MF (1994) Cortical Cytochrome Oxidase Activity Is Reduced in Alzheimer's Disease. J Neurochem 63 (6):2179–2184. https://doi.org/10.1046/j.1471-4159.1994.63062179.x

Ngolab J, Trinh I, Rockenstein E, et al. (2017) Brain-Derived Exosomes from Dementia with Lewy Bodies Propagate α-Synuclein Pathology. Acta Neuropathol Commun 5 (1):46. https://doi.org/10.1186/s40478-017-0445-5

Nguyen NH, Bråthe A, and Hassel B (2003) Neuronal Uptake and Metabolism of Glycerol and the Neuronal Expression of Mitochondrial Glycerol-3-Phosphate Dehydrogenase. J Neurochem 85 (4):831–842. https://doi.org/10.1046/j.1471-4159.2003.01762.x

Orr AL, Kim C, Jimenez-Morales D, et al. (2019) Neuronal Apolipoprotein E4 Expression Results in Proteome-Wide Alterations and Compromises Bioenergetic Capacity by Disrupting Mitochondrial Function. J Alzheimers Dis 68 (3):991–1011. https://doi.org/10.3233/jad-181184

Panigrahi GK, and Deep G (2017) Exosomes-Based Biomarker Discovery for Diagnosis and Prognosis of Prostate Cancer. Front Biosci (Landmark Ed) 22:1682–1696. https://doi.org/10.2741/4565

Panigrahi GK, Praharaj PP, Peak TC, et al. (2018) Hypoxia-Induced Exosome Secretion Promotes Survival of African-American and Caucasian Prostate Cancer Cells. Sci Rep 8 (1):3853. https://doi.org/10.1038/s41598-018-22068-4

Parker WD, Jr., Mahr NJ, Filley CM, et al. (1994) Reduced Platelet Cytochrome *C* Oxidase Activity in Alzheimer's Disease. Neurology 44 (6):1086–1090. https://doi.org/10.1212/wnl.44.6.1086

Pérez MJ, Ponce DP, Osorio-Fuentealba C, Behrens MI, and Quintanilla RA (2017) Mitochondrial Bioenergetics Is Altered in Fibroblasts from Patients with Sporadic Alzheimer's Disease. Front Neurosci 11:553. https://doi.org/10.3389/fnins.2017.00553

Perkins M, Wolf AB, Chavira B, et al. (2016) Altered Energy Metabolism Pathways in the Posterior Cingulate in Young Adult Apolipoprotein E ε4 Carriers. J Alzheimers Dis 53 (1):95–106. https://doi.org/10.3233/jad-151205

Proitsi P, Kim M, Whiley L, et al. (2015) Plasma Lipidomics Analysis Finds Long Chain Cholesteryl Esters to Be Associated with Alzheimer's Disease. Transl Psychiatry 5 (1):e494. https://doi.org/10.1038/tp.2014.127

Qi G, Mi Y, Shi X, et al. (2021a) ApoE4 Impairs Neuron-Astrocyte Coupling of Fatty Acid Metabolism. Cell Rep 34:108572. https://doi.org/10.1016/J.CELREP.2020.108572

Qi G, Mi Y, and Yin F (2021b) Characterizing Brain Metabolic Function *ex Vivo* with Acute Mouse Slice Punches. STAR Protoc 23:100559. https://doi.org/10.1016/J.XPRO.2021.100559

Racchi M, and Govoni S (2003) The Pharmacology of Amyloid Precursor Protein Processing. Exp Gerontol 38 (1–2):145–157. https://doi.org/10.1016/s0531-5565(02)00158-4

Rajendran L, Honsho M, Zahn TR, et al. (2006) Alzheimer's Disease β-Amyloid Peptides Are Released in Association with Exosomes. PNAS 103 (30):11172–11177. https://doi.org/10.1073/pnas.0603838103

Ramteke A, Ting H, Agarwal C, et al. (2015) Exosomes Secreted under Hypoxia Enhance Invasiveness and Stemness of Prostate Cancer Cells by Targeting Adherens Junction Molecules. Mol Carcinog 54 (7):554–565. https://doi.org/10.1002/mc.22124

Reiman EM, Chen K, Alexander GE, et al. (2004) Functional Brain Abnormalities in Young Adults at Genetic Risk for Late-Onset Alzheimer's Dementia. PNAS 101 (1):284–289. https://doi.org/10.1073/pnas.2635903100

Rummel NG, and Butterfield DA (2022) Altered Metabolism in Alzheimer Disease Brain: Role of Oxidative Stress. Antioxid Redox Signal 36:16–18. https://doi.org/10.1089/ARS.2021.0177

Salpeter SJ, Khalaileh A, Weinberg-Corem N, et al. (2013) Systemic Regulation of the Age-Related Decline of Pancreatic β-Cell Replication. Diabetes 62 (8):2843–2848. https://doi.org/10.2337/db13-0160

Schägger H, and Ohm TG (1995) Human Diseases with Defects in Oxidative Phosphorylation. Eur J Biochem 227 (3):916–921. https://doi.org/10.1111/j.1432-1033.1995.0916p.x

Schlaepfer IR, Nambiar DK, Ramteke A, et al. (2015) Hypoxia Induces Triglycerides Accumulation in Prostate Cancer Cells and Extracellular Vesicles Supporting Growth and Invasiveness Following Reoxygenation. Oncotarget 6 (26):22836–22856. https://doi.org/10.18632/oncotarget.4479

Schönfeld P, and Reiser G (2017) Brain Energy Metabolism Spurns Fatty Acids as Fuel Due to Their Inherent Mitotoxicity and Potential Capacity to Unleash Neurodegeneration. Neurochem Int 109:68–77. https://doi.org/10.1016/j.neuint.2017.03.018

Sharma LK, Lu J, and Bai Y (2009) Mitochondrial Respiratory Complex I: Structure, Function and Implication in Human Diseases. Curr Med Chem 16 (10):1266–1277. https://doi.org/10.2174/092986709787846578

Sheller S, Papaconstantinou J, Urrabaz-Garza R, et al. (2016) Amnion-Epithelial-Cell-Derived Exosomes Demonstrate Physiologic State of Cell under Oxidative Stress. PLoS One 11 (6):e0157614. https://doi.org/10.1371/journal.pone.0157614

Soria FN, Pampliega O, Bourdenx M, et al. (2017) Exosomes, an Unmasked Culprit in Neurodegenerative Diseases. Front Neurosci 11:26. https://doi.org/10.3389/fnins.2017.00026

Stojakovic A, Trushin S, Sheu A, et al. (2021) Partial Inhibition of Mitochondrial Complex I Ameliorates Alzheimer's Disease Pathology and Cognition in APP/PS1 Female Mice. Commun Biol 4 (1):61. https://doi.org/10.1038/s42003-020-01584-y

Swerdlow RH, Burns JM, and Khan SM (2010) The Alzheimer's Disease Mitochondrial Cascade Hypothesis. J Alzheimers Dis 20 (Suppl 2):S265–S279. https://doi.org/10.3233/jad-2010-100339

Swerdlow RH, and Khan SM (2004) A "Mitochondrial Cascade Hypothesis" for Sporadic Alzheimer's Disease. Med Hypotheses 63 (1):8–20. https://doi.org/10.1016/j.mehy.2003.12.045

Terada T, Therriault J, Kang MSP, et al. (2021) Mitochondrial Complex I Abnormalities Is Associated with Tau and Clinical Symptoms in Mild Alzheimer's Disease. Mol Neurodegener 16 (1):28. https://doi.org/10.1186/s13024-021-00448-1

Terni B, Boada J, Portero-Otin M, Pamplona R, and Ferrer I (2010) Mitochondrial ATP-Synthase in the Entorhinal Cortex Is a Target of Oxidative Stress at Stages I/II of Alzheimer's Disease Pathology. Brain Pathol 20 (1):222–233. https://doi.org/10.1111/j.1750-3639.2009.00266.x

Trushin S, Stojakovic A, Chang S-Y, and Trushina E (2020) Partial Mitochondrial Complex I Inhibitors as Disease-Modifying Therapeutics for Alzheimer's Disease. Alzheimers Dement 16 (S9):e045529. https://doi.org/10.1002/alz.045529

Tyrrell DJ, Bharadwaj MS, Jorgensen MJ, et al. (2017) Blood-Based Bioenergetic Profiling Reflects Differences in Brain Bioenergetics and Metabolism. Oxid Med Cell Longev 2017:7317251. https://doi.org/10.1155/2017/7317251

Valla J, Yaari R, Wolf AB, et al. (2010) Reduced Posterior Cingulate Mitochondrial Activity in Expired Young Adult Carriers of the ApoE ε4 Allele, the Major Late-Onset Alzheimer's Susceptibility Gene. J Alzheimers Dis 22 (1):307–313. https://doi.org/10.3233/jad-2010-100129

van der Pol E, Boing AN, Harrison P, Sturk A, and Nieuwland R (2012) Classification, Functions, and Clinical Relevance of Extracellular Vesicles. Pharmacol Rev 64 (3):676–705. https://doi.org/10.1124/pr.112.005983

Vella LJ, Hill AF, and Cheng L (2016) Focus on Extracellular Vesicles: Exosomes and Their Role in Protein Trafficking and Biomarker Potential in Alzheimer's and Parkinson's Disease. Int J Mol Sci 17 (2):173. https://doi.org/10.3390/ijms17020173

Villeda SA, Luo J, Mosher KI, et al. (2011) The Ageing Systemic Milieu Negatively Regulates Neurogenesis and Cognitive Function. Nature 477 (7362):90–94. https://doi.org/10.1038/nature10357

Villeda SA, Plambeck KE, Middeldorp J, et al. (2014) Young Blood Reverses Age-Related Impairments in Cognitive Function and Synaptic Plasticity in Mice. Nat Med 20 (6):659–663. https://doi.org/10.1038/nm.3569

Więckowska-Gacek A, Mietelska-Porowska A, Wydrych M, and Wojda U (2021) Western Diet as a Trigger of Alzheimer's Disease: From Metabolic Syndrome and Systemic Inflammation to Neuroinflammation and Neurodegeneration. Ageing Res Rev 70:101397. https://doi.org/10.1016/J.ARR.2021.101397

Winston CN, Goetzl EJ, Akers JC, et al. (2016) Prediction of Conversion from Mild Cognitive Impairment to Dementia with Neuronally Derived Blood Exosome Protein Profile. Alzheimers Dement 3:63–72. https://doi.org/10.1016/j.dadm.2016.04.001

Yuyama K, and Igarashi Y (2017) Exosomes as Carriers of Alzheimer's Amyloid-ß. Front Neurosci 11:229. https://doi.org/10.3389/fnins.2017.00229

Zheng T, Pu J, Chen Y, et al. (2017) Exosomes Secreted from HEK293-APP Swe/Ind Cells Impair the Hippocampal Neurogenesis. Neurotox Res 32 (1):82–93. https://doi.org/10.1007/s12640-017-9713-1

Open Access This chapter is licensed under the terms of the Creative Commons Attribution-NonCommercial-NoDerivatives 4.0 International License (http://creativecommons.org/licenses/by-nc-nd/4.0/), which permits any noncommercial use, sharing, distribution and reproduction in any medium or format, as long as you give appropriate credit to the original author(s) and the source, provide a link to the Creative Commons license and indicate if you modified the licensed material. You do not have permission under this license to share adapted material derived from this chapter or parts of it.

The images or other third party material in this chapter are included in the chapter's Creative Commons license, unless indicated otherwise in a credit line to the material. If material is not included in the chapter's Creative Commons license and your intended use is not permitted by statutory regulation or exceeds the permitted use, you will need to obtain permission directly from the copyright holder.

8

Metabolic Abnormalities in Psychiatric Disorders

A Transdiagnostic, Whole-Body Approach

Sabina Berretta, Ana C. Andreazza, Dorit Ben-Shachar, Javier Gilbert-Jaramillo, Jill R. Glausier, Margaret Hahn, Iris-Tatjana Kolassa, R. Nehir Mavioglu, Anthony J. A. Molina, Martin Picard, Zoltan Sarnyai, Robert E. McCullumsmith, Johann Steiner, and Kim Q. Do

Abstract Psychiatric disorders have traditionally been conceptualized as conditions primarily affecting the brain. However, mounting evidence challenges this view, revealing complex interactions between systemic energy metabolism, immune function, and brain physiology. This chapter examines the critical role of metabolic abnormalities in psychiatric disorders, advocating for a "whole-body" framework of investigation. Recent research demonstrates that individuals with severe psychiatric disorders experience significantly reduced life expectancy, largely due to metabolic and cardiovascular conditions that are not fully attributable to medication side effects. Evidence from genetic as well as peripheral and central nervous system studies reveals multifaceted metabolic dysregulation across psychiatric disorders, including altered glucose metabolism, mitochondrial function, and oxidative stress. Notably, specific metabolic signatures may correspond to distinct clinical phenotypes that cut across traditional diagnostic boundaries, suggesting the need for a transdiagnostic approach to both research and treatment strategies. The relationship between metabolic dysfunction and psychiatric symptoms is bidirectional and complex. Although metabolic abnormalities may not directly cause psychiatric disorders, they likely represent significant risk factors that interact with genetic, physiological, environmental, and socioeconomic conditions. Understanding these interactions is crucial for developing more effective, personalized therapeutic

Sabina Berretta (✉)
Basic Neuroscience Division, McLean Hospital, Belmont, MA; Harvard Medical School, Boston, MA; Stanley Center for Psychiatric Research at Broad Institute, Cambridge MA, USA
Email: sberretta@mclean.harvard.edu

Affiliations for the coauthors are available in the List of Contributors.

© Frankfurt Institute for Advanced Studies (FIAS) 2025

D. Öngür, J. M. Ford (eds.), *Metabolic Neuropsychiatry*, Strüngmann Forum Reports,
https://doi.org/10.1007/978-3-032-05630-6_8

Group photos (top left to bottom right) Sabina Berretta, Ana Andreazza, Kim Do, Robert McCullumsmith, Dorit Ben-Shachar, Margaret Hahn, Ana Andreazza, Javier Gilbert-Jaramillo, Nehir Mavioglu, Martin Picard, Kim Do, Anthony Molina, Zoltan Sarnyai, Jill Glausier, Johann Steiner, Martin Picard, Sabina Berretta, Iris-Tatjana Kolassa strategies. Future investigations focused on identifying biology-based symptom constructs may be better able to guide individualized treatment approaches than traditional diagnostic categories.

Keywords Metabolic syndrome, mitochondria, insulin resistance, oxidative stress

1 Considering a "Whole-Body" Framework for Brain Disorders

Central nervous system (CNS) disorders, particularly psychiatric disorders, have long been conceived as uniquely "brain disorders," in relative isolation with respect to the rest of the body. However, a growing recognition of the interplay between brain physiology and factors, such as the immune system, gut microbiome, stress and trauma exposure, and metabolism, is challenging this perspective. Thus, it is not surprising that system metabolic abnormalities are being linked to an increasing number of CNS disorders, including major depressive and psychosis spectrum disorders (Alagiakrishnan and Halverson 2024; Karabatsiakis et al. 2014; Kuang et al. 2018; Montanari et al. 2025; Yuksel et al. 2021) as well as neurodegenerative disorders such as Alzheimer and Parkinson diseases (Chang et al. 2025; Ding et al. 2025; Elzinga et al. 2025; Fakih et al. 2022). Indeed, if these disorders are not solely confined to the brain, but result from complex "whole-body" interactions, the very concept of "brain disorders" may need to be reconsidered. At a minimum, the "whole-body" framework warrants careful consideration.

Here, we use the term *severe psychiatric conditions* (SPCs) to denote illnesses marked by significant severity and chronicity, resulting in substantial functional impairment, limitation to major life activities (e.g., independent living), and increased risk to life (e.g., suicide, self-harm, incarceration, substance overdose). These include schizophrenia (SZ), bipolar disorder (BD), major depressive disorder (MDD), and severe instances of conditions such as personality disorders, obsessive-compulsive disorder, post-traumatic stress disorder (PTSD), eating disorders and other severe anxiety disorders. The impact of systemic illnesses on individuals with SPCs is starkly evidenced by a significant reduction of their life expectancy by 10 to 20 years (Chang et al. 2011; Correll et al. 2017; Hoang et al. 2013; Lawrence et al. 2013; Nordentoft et al. 2013). While increased risk of death by suicide contributes to excess mortality, communicable and noncommunicable diseases also account for premature mortality in persons with SPCs (Ali et al. 2022; Hoang et al. 2013; Walker et al. 2015). Relative to those without, persons with SPCs are at higher risk for cancer, cardiovascular and respiratory disease, as well as medication side effects, and this risk is worsened by poor access to health care, "diagnostic overshadowing," and ensuing delays in diagnosis and treatment (Ali et al. 2022; Chang et al. 2011; Correll et al. 2017; Hoang et al. 2013; Lawrence et al. 2013; Nordentoft et al. 2013; Walker et al. 2015). The *Journal of the American College of Cardiology* reports that individuals with SPCs have twice the cardiovascular mortality rate observed in the general population, with notably higher rates of conditions like metabolic syndrome and diabetes (Goldfarb et al. 2022). Supporting these findings, a large French cohort study on SPCs revealed a high prevalence of metabolic syndrome among psychiatric patients: over two-thirds do not receive adequate treatment (Brouwer et al. 2024; Leboyer et al. 2022). These considerations highlight the urgency to invest in research on psychiatric disorders based on a whole-body framework focused on understanding the mechanistic role of metabolic abnormalities and directed at novel treatments.

In asking "whether" and "how" metabolic abnormalities play a pathophysiological role in CNS disorders, it is important to recognize that this term, "metabolic abnormalities," encompasses a broad range of conditions. These abnormalities are likely multifarious, including several different systemic and central mechanisms, potentially distinct in different clusters of people and conditions. A narrower focus on the metabolic syndrome and its relationship with brain disorders may offer a more specific framework. The definition of metabolic syndrome includes a constellation of interrelated metabolic risk factors related to insulin resistance, including central obesity, high blood sugar, elevated blood pressure, elevated triglycerides, and reduced high-density lipoprotein cholesterol. In turn, these factors heighten the risk of developing cardiovascular diseases and diabetes (Foley et al. 2010; Grundy et al. 2005; Pan et al. 2012; Skilton et al. 2007; Vogelzangs et al. 2011). This general definition has been put forth in different versions and is not without controversy, perhaps reflecting different emphasis in its clinical aspects and health consequences (Expert Panel on Detection 2001; Grundy et al. 2005; Oda 2012; Reaven 2004, 2006; Zimmet and Alberti 2005). Compelling evidence supports the association of a large number of brain disorders with the metabolic syndrome and dysregulation of appetite and eating, posing important questions on the reciprocal relationship between systemic and CNS disorders, in general, and psychiatric disorders, in particular. Yet, metabolic abnormalities in CNS disorders do not consistently match the definition of metabolic syndrome. Notably, growing evidence also points to the disruption of mitochondrial function and morphology in a broad range of CNS disorders, detected both systemically and in the brain. Although it is plausible to postulate that mitochondrial dysfunction, on the one hand, and systemic metabolic abnormalities (e.g., dysregulated blood sugar, insulin signaling, dyslipidemia), on the other, may impact and perhaps cause each other, the mechanisms underlying their relationships are not well understood.

The emerging landscape suggests that unique patterns of metabolic changes may be shared across diagnoses while also being specific to distinct transdiagnostic phenotypes. Furthermore, the time frame of mechanisms driving these changes, along with their interactions, may vary across prenatal, postnatal, and adult life, and may affect different brain regions and their associated functions. This complexity poses both a daunting challenge and an exciting opportunity, offering the potential for new insights into the biological underpinnings of psychiatric conditions. Here, we discuss evidence for metabolic dysregulation in SPCs, focusing on genetic, systemic, and human brain studies. We highlight some examples in the context of an extensive body of knowledge on this topic, examine the potential impact of these abnormalities on synaptic functions, and review potential cause-effect relationships between them. Thereafter, we ask whether distinct patterns of metabolic abnormalities may be specific to distinct clinical phenotypes and discuss strategies for transdiagnostic and stratification approaches.

2 Evidence for Metabolic Abnormalities from Blood/Peripheral Studies

2.1 Insulin Signaling

Given the presence of metabolic abnormalities in SZ, a complex relationship between SZ and insulin resistance has long been suspected (Agarwal et al. 2020; Barrett and Serre 1924; Kasanin 1926; Kooy 1919). For instance, glucose tolerance curves were reported to differ in persons with psychiatric disorders with respect to unaffected controls (Raphael and Parsons 1921). Notably, evidence for insulin dysregulation was reported in drug-naïve first-episode SZ (Fernandez and Torres-Alemán 2012). This latter study showed elevated fasting glucose and insulin levels, higher insulin resistance, and lower glucose tolerance in first-episode patients. Post-oral glucose tolerance testing showed greater insulin resistance in first-episode patients as well as in their nonaffected siblings (Enez Darcin et al. 2015; Pillinger et al. 2017). Notably, dysregulation of fasting glucose and insulin levels in SZ have been reported in conjunction with increased levels of immune and stress factors including cortisol and interleukins, highlighting potential interactions between metabolic, immune, and stress signaling (Chouinard et al. 2019; Fernandez-Egea et al. 2009; Ryan et al. 2003).

2.2 Glucose Metabolism

Elevated blood glucose levels, repeatedly shown in drug-naïve first-episode SZ (Chen et al. 2016; Pillinger et al. 2017), provide ample substrate for glycolysis in the periphery (Figure 8.1). Consistent with these observations, elevated peripheral glycolysis in first-episode SZ has been reported (Herberth et al. 2011). Proteome analysis of peripheral blood mononuclear cells (PBMCs) from first-episode, antipsychotic-naïve patients revealed a significant number of differentially expressed proteins involved in glycolysis pathway and the tricarboxylic acid (TCA) cycle, a central metabolic pathway that plays a critical role in cellular respiration. Others have identified multiple upregulated glycolytic and TCA cycle enzymes in circulating leukocytes in first-episode SZ (Jiang et al. 2019; Martins-de-Souza et al. 2011). Metabolome analyses of PBMCs revealed higher levels of octanoic acid and lower levels of maltose, valine, inositol, sorbitol, creatinine, and fumaric acid, indicative of increased carbon flux through the TCA cycle, in patients as compared to healthy controls (Gomez-Casati et al. 2016; Liu et al. 2014). A follow-up study showed a widespread increase of multiple glycolytic metabolites, including glucose-6-phosphate, ribose-5-phosphate, and fructose-6-phosphate, along with a decrease in the TCA cycle metabolite in PBMCs of patients (Liu et al. 2015). Comparing transcriptomic data from first-episode psychosis and individuals with early dysglycemia, a meta-analysis identified 221 common gene expression signatures (Lee and Chung 2024; Weske et al. 2001). Together, these findings suggest an intrinsic link between psychosis spectrum disorders and dysglycemic states, independent of the disease chronicity and bioenergetic effects of antipsychotic drugs.

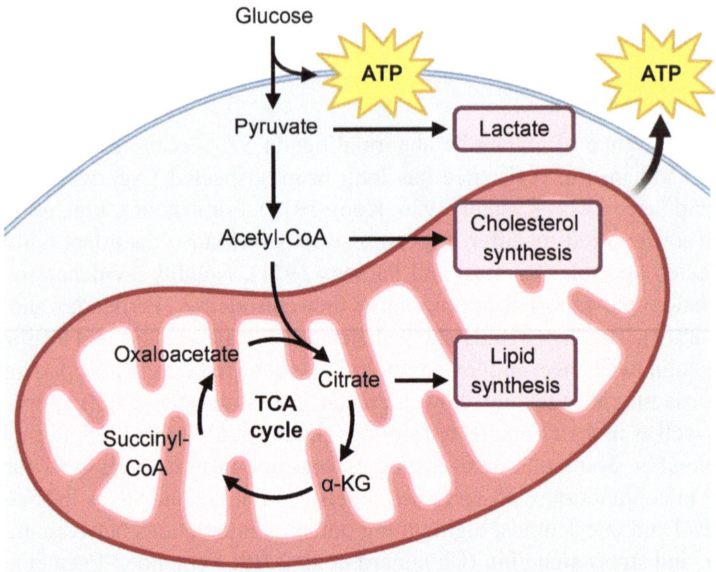

Figure 8.1 Bioenergetic glucose metabolism in eukaryotic cells. This diagram illustrates the primary metabolic pathways derived from glucose, emphasizing its conversion to ATP, lactate, and entry into the TCA cycle via acetyl-CoA. Downstream products include intermediates for cholesterol and lipid synthesis, known to be dysregulated in SPCs. Created in BioRender by A. C. Andreazza (2025) https://BioRender.com/r2bgipk

Lactate is produced from pyruvate as an end product of glycolysis. Increased extracellular lactate could indicate a heightened activity of glycolysis or a decreased lactate clearance. The former, when in the presence of adequate oxygen concentration, is known as aerobic glycolysis (Warburg effect). Thus, levels of lactate and pyruvate have been used as an index of glycolytic activity. A series of studies suggests that lactate/pyruvate changes may become progressively altered in the course of psychosis spectrum disorders. In a group of young individuals at clinical high risk for psychosis, serum levels of L-lactate were lower in comparison to healthy controls and negatively correlated with negative symptoms using PANSS. Although no information was available on how many converted to psychosis, this study suggests that glycolytic activity may predate its onset (Onozato et al. 2020). In first-episode, antipsychotic-naïve SZ, lactate dehydrogenase B, which catalyzes pyruvate transformation to lactate, was increased in the peripheral blood mononuclear cells and coincided with elevated serum lactate level (Herberth et al. 2011).

BD is increasingly recognized as a condition marked by significant metabolic and mitochondrial dysfunction. Elevated lactate levels have been consistently observed among persons with BD, particularly in those with metabolic syndrome (Zachos et al. 2024). These findings align with earlier evidence of impaired mitochondrial function (Kuang et al. 2018), suggesting that increased reliance on glycolysis due to mitochondrial inefficiency may lead to lactate accumulation in this disorder. Metabolomic analysis revealed distinct metabolic profiles in BD patients stratified by lactate levels and metabolic syndrome. For instance, elevated citrate and alpha-ketoglutarate levels were associated with high lactate and metabolic syndrome, whereas increased kynurenine and tryptophan levels were noted in those with low lactate and metabolic syndrome, suggesting alterations in key metabolic pathways like the TCA cycle and kynurenine pathway. These insights emphasize the complex interplay between mitochondrial dysfunction and metabolic disruptions and highlight the need for personalized therapeutic approaches targeting mitochondrial health.

2.3 Association of Obesity with Psychiatric Disorders

Obesity has been postulated to increase risk for psychiatric conditions. Analyses of the Austrian national registry data of inpatient services from 1997 to 2014 revealed significant associations between obesity and psychiatric conditions such as depression, anxiety, eating disorders, SZ, and somatization disorders across all age groups (Leutner et al. 2023). These analyses, including ~45 million hospital stays of nine million individuals, show that obesity often precedes psychiatric conditions, with time-order ratios indicating that obesity was frequently the first diagnosis for most people (e.g., time-order ratio ~2 for recurrent depression). Gender and age differences were prominent: women exhibited higher risks for depressive and anxiety disorders, whereas men had elevated risks for psychosis spectrum disorders and nicotine addiction. Support for the idea that obesity may represent a risk factor for psychiatric symptoms, or at least precede their onset, comes from imaging studies, which demonstrated increased visceral fat accumulation in individuals with SZ prior to antipsychotic exposure (Thakore 2005; Thakore et al. 2002). Evidence discussed above, showing impaired glucose tolerance and insulin resistance in first-episode psychosis, also supports the hypothesis that metabolic abnormalities may be intrinsic to the disease process and not uniquely treatment-induced (Perry et al. 2016; Pillinger et al. 2017). Together, these findings indicate that obesity commonly precedes SPCs and highlight the urgency of implementing screening for psychiatric symptoms in persons with obesity, critically needed for prevention and early treatment. Notably, the growing use of glucagon-like peptide-1 (GLP-1) receptor agonists, typically used to treat diabetes II and for weight management, shows promises to benefit psychiatric conditions (Breit and Hubl 2025; Ebrahimi et al. 2025; Gholami et al. 2025; Xiang and Peng 2025).

Distinct aspects of the metabolic syndrome may be considered in the context of its role in brain disorders. Visceral fat (measured by waist to hip ratio) is strongly associated with adverse brain changes, including increased white matter hyperintensities, reduced gray and white matter volumes, and altered white matter microstructure (Dolatshahi et al. 2024; Lampe et al. 2019; Raji et al. 2024). These effects may be mediated by inflammation, with elevated pro-inflammatory cytokines (e.g., IL-6, CRP) playing a central role. The predictive value of visceral fat is consistent across sexes and age groups, although it is particularly pronounced in women, highlighting its significance as a modifiable risk factor for brain health.

2.4 Shared Environmental Factors for Obesity and Mental Disorders

Environmental risk factors, such as physical inactivity, poor diet, socioeconomic status, and chronic stress, contribute significantly to both obesity and psychiatric disorders like MDD and SZ (Flores-Dorantes et al. 2020). Physical inactivity and high-calorie diets lead to obesity and are also associated with increased inflammation and neuroendocrine dysregulation, which exacerbate psychiatric symptoms. Socioeconomic deprivation intensifies these risks by limiting access to nutritious food and health care resources. Additionally, maternal early-life adversity, obesity, and stress have been linked to an increased risk of psychiatric conditions in offsprings, including MDD, anxiety disorders, SZ and attention-deficit/hyperactivity disorder (ADHD), mediated by inflammatory and hormonal changes (Flores-Dorantes et al. 2020). These factors highlight the intertwined pathways of environmental and biological influences in obesity and mental health disorders.

3 Effects of Systemic Metabolic Abnormalities on Brain Functions

The evidence reviewed thus far offers strong support for the association of metabolic abnormalities with SPCs, but do systemic abnormalities in fact affect brain functions? In recent years, overwhelming evidence for the powerful impact of immune activation in brain disorders has provided a clear answer to this question (Flux and Lowry 2023; Gallego Deike et al. 2023; Gangadin et al. 2024; Neupane et al. 2023; Réus et al. 2023; Sørensen and Benros 2023; Zakia et al. 2023; Zhang et al. 2023). A clear example of the effects of systemic metabolic changes on the CNS is the impact of systemic diseases (e.g., kidney or liver diseases) and systemic immune activation during an infection. These conditions impose significantly higher energetic requirements with respect to healthy baseline states. However, cells and organisms have access only to a finite energy budget (Pontzer 2018; Urlacher et al. 2019). Disease-related high-energy states "steal" energy from other systems, including potentially the brain (Bobba-Alves et al. 2022; Sercel et al. 2024). This instance may also occur physiologically, such as during the postprandial phase (after a meal), when the digestive system consumes 10–20% of the energy budget, leading to sluggishness and fatigue (wanting to nap or postprandial "coma"). Several mechanisms exist to alert the brain that other organs in the body are in need of energy. One emerging factor is growth differentiation factor 15 (GDF15). GDF15 is produced by energy-stressed cells, which release it in the bloodstream, eventually reaching its receptors on neurons in the brainstem (area postrema) (Lockhart et al. 2020). GDF15 signaling to the brain triggers nausea, energy conservation behaviors, and hyperglycemia-promoting endocrine signals involving the sympathetic nervous system and the hypothalamus-pituitary axis (Cimino et al. 2021; Engström Ruud et al. 2024). Notably, GDF15 is elevated in primary mitochondrial disorders and correlates strongly with fatigue (Huang et al. 2024). It has been postulated that systemic cell energy deficit signals reaching the brain may modulate subjective experience and aging trajectories (Shaulson et al. 2024). The contribution of these pathways to psychiatric conditions is virtually unexplored, although some evidence points to their impact on depression (Milaneschi et al. 2020; Sørensen and Benros 2023).

Systemic metabolic abnormalities affect the brain in significant ways but may not be sufficient to cause psychiatric disorders. Therefore, it is reasonable to postulate that metabolic abnormalities may represent significant risk factors that combine with life stressors, lack of resources and social support, genetic vulnerabilities, and other environmental factors. Given the heterogeneity of systemic metabolic abnormalities reported in CNS disorders and their occurrence at different times in life, it is plausible to speculate that their impact on the brain may be equally diversified. This consideration raises the question of how systemic changes in energy utilization might affect specific brain circuits yet leave others unaffected. Instances of functional specificity within the brain, such as neuronal insulin resistance and glucocorticoid-related neurotoxicity, may offer some clues.

The effects of neuronal insulin resistance may be selective for GLUT4-dependent brain areas, and glucocorticoid neurotoxicity impacts glucocorticoid-receptor-rich brain regions. This selectivity may affect specific circuit dynamics. For instance, insulin resistance may slow axon conduction velocity (Akın et al. 2016; Kula et al. 2024), potentially causing a lag in GLUT4-dependent neural circuit relays and shifting the frequency of circuit-generated oscillations and the effectiveness of circuit responsivity to perturbations. Notably, insulin may, at least in part, be synthesized in the brain (Appel and Farr 1929; Havrankova et al. 1978; Lemche et al. 2024), with important implications regarding region specificity of insulin resistance and signaling in psychiatric disorders. Another example highlighting the potential for cell/brain region specificity of systemic metabolic abnormalities is the contribution of cytokines, which originate from nonimmune cells and organs as well as the immune system, in signaling systemic metabolic dysregulation to the brain (Dantzer et al. 2008; Shaulson et al. 2024). The cell/brain region specificity of cytokine receptors in the brain is likely to mediate the differential impact of these signals. These considerations illustrate a more general point: cognitive and affective symptoms may depend critically not just on whether the circuit is affected but on how it is affected. Distinct mechanisms may affect the same circuit differently by targeting specific points of failure. An obvious example is the distinction between type 1 and type 2 diabetes and postprandial hypoglycemia. While all three conditions involve the same control circuit regulating glucose, they implicate different points of failure: a distinction most clearly revealed by perturbing the system (i.e., glucose bolus) and measuring the resulting dynamics as the system attempts to regain homeostasis.

4 Effects of Cumulative Trauma on Metabolic Abnormalities

Do other risk factors for SPCs, such as exposure to trauma, impact metabolic abnormalities? In survivors of the Rwandan genocide and the Ugandan civil wars, as well as in emergency rescue workers, Kolassa and colleagues observed that traumatic stress has cumulative lifetime effects and increases, in a dose-response manner, the risk of somatic symptoms and psychiatric disorders such as PTSD and MDD (Behnke et al. 2020; Kolassa et al. 2010a, b). Other studies have also described a cumulative impact of lifetime stress load and childhood maltreatment load at the biomolecular level (Gola et al. 2014). In women with a history of childhood maltreatment, blood cells exhibited increased pro-inflammatory cytokine secretion, which correlated with the severity of maltreatment. Similar findings were observed in trauma survivors with PTSD and MDD (Gola et al. 2014), where the risk of these disorders increases with more severe traumatic stress load. Higher levels of inflammation were associated with elevated mitochondrial activity and increased reactive oxygen species (ROS) production, DNA damage, and telomere attrition (Bergholz et al. 2017; Boeck et al. 2018; Morath et al. 2014). These changes increase energy requirements needed for antioxidative mechanisms and cell repair. Over time, they may contribute to greater cell wear-and-tear, resulting in a state of allostatic load. Together, these studies indicate that early-life adversity and adult life trauma increase energetic requirements systemically and in the brain.

5 Genetic Evidence for Metabolic Abnormalities in Psychiatric Disorders

The possibility that abnormal glucose metabolism is an inherent part of SPC pathophysiology was initially supported by initial linkage studies in SZ. These identified three genes involved in glycolysis and two variants in its regulation (Olsen et al. 2008; Stone et al. 2004; Tanner et al. 2018), respectively:

1. 6-phosphofructo-2-kinase/fructose-2,6-bisphosphatase 2 (Pfkfb2; 1q32.2),
2. hexokinase 3 (HK3; 5q35.3),
3. pyruvate kinase 3 (PK3; 15q23),
4. phosphoenolpyruvate carboxykinase 1 (PCK1), and
5. fructose-1,6-biphosphatase (FBP1).

This suggests that abnormal glucose metabolism is an inherent part of the disease pathophysiology.

The Psychiatric Genomics Consortium Schizophrenia Phase 2 (PGC-SCZ2) genome-wide association study (GWAS), which consisted of 35,476 cases and 46,839 control subjects, reported evidence for 22 nuclear-encoded mitochondrial genes (Schizophrenia Working Group of the Psychiatric Genomics Consortium 2014). However, mitochondria-related pathways showed no significant association with the disease in other large GWAS studies of SZ (Trubetskoy et al. 2022). Using a gene-based approach, one study directly tested the identified nuclear-encoded mitochondrial and mitochondria-related (nucleus-mitochondria crosstalk) genes associated with SZ but did not detect significant gene sets. It did, however, provide evidence for PGC-SCZ2 variants' association analysis of individual single nucleotide polymorphisms (Gonçalves et al. 2018). Further evidence includes enrichment of expression quantitative trait loci and an excess of large (> 500 Kb) deletions and rare copy number variants in nuclear-encoded mitochondrial genes in individuals with SZ (Szatkiewicz et al. 2014).

Similarly, direct GWAS evidence specifically linking nuclear-encoded mitochondrial genes to BD and MDD is not as robust as for other genetic loci. Some associations, however, have been noted, and genes involved in the mitochondrial energy metabolism pathways have been implicated in these disorders. Beyond individual gene associations, analyses of GWAS data have shown that pathways involving mitochondrial functions are enriched in BD and MDD. This suggests that while individual mitochondrial-related genes might not reach genome-wide significance alone, their collective role may be significant (Alcocer-Gómez et al. 2016; Clay et al. 2011; Klinedinst and Regenold 2015; Scaini et al. 2016).

5.1 Rare Genetic Mitochondrial Diseases: Association with Psychiatric Symptoms

Rare genetic mitochondrial diseases are caused by mutations in mitochondrial or nuclear DNA that impair mitochondrial ability to generate energy (Figure 8.1). The functional severity of genetic mitochondrial disease is greatest in organs with particularly high-energy demands, such as brain, heart, and muscle. Persons affected by these diseases experience more hospital stays, heightened susceptibility to health problems, and reduced quality of life. Approximately 35% of persons with rare mitochondrial genetic diseases have psychiatric diagnoses, supporting an association between mitochondrial deficits and psychiatric symptoms. Consistent with these observations, a recent study reported a notably higher occurrence of psychiatric illness in persons with mitochondrial disease (18%) than in those with multiple sclerosis (9%) (Rosella et al. 2025). Healthcare costs for mitochondrial disease patients with psychiatric diseases surpassed those of multiple sclerosis. These clinical observations point to the possibility that mitochondrial diseases may increase susceptibility to psychiatric disorders, suggesting that some mitochondrial pathology could contribute to brain pathophysiology in the context of psychiatric disorders.

5.2 Peripheral Mitochondrial Abnormalities

Investigations on mitochondrial abnormalities in blood and other peripheral tissue of persons with SPCs offer the potential to identify targets for therapeutic strategies and efficacy indices for treatments targeting mitochondrial functions. Several studies have reported mitochondrial abnormalities in SZ and BD, including reduced mitochondrial DNA copy numbers, complex I function, and altered intracellular mitochondria location (Akarsu et al. 2015; Bergman and Ben-Shachar 2016; Cataldo et al. 2010; Clay et al. 2011; Fizíková et al. 2023; Gubert et al. 2013; Valiente-Pallejà et al. 2020; Wang et al. 2022). While these abnormalities may not have been consistently replicated, several findings highlighting the specificity of these changes underscore the importance of stratification and transdiagnostic approaches. For instance, mitochondrial DNA copy number reductions may positively correlate with psychosis severity and be affected by specific antipsychotic medications (Chestkov et al. 2018; Kumar et al. 2018). Although this finding was not replicated in a recent study, a significantly higher prevalence of mitochondrial DNA abnormalities was reported in persons with SZ compared to unaffected controls (Bulduk et al. 2024).

In MDD, mitochondrial functioning in PBMCs was found to be significantly impaired (Karabatsiakis et al. 2014). Specifically, measures of routine and uncoupled respiration, spare respiratory capacity, coupling efficiency, and ATP turnover-related respiration were reduced in persons with MDD. These findings led to the suggestion that mitochondrial respiration in PBMCs may be a promising target for therapeutic treatment of MDD (Karabatsiakis et al. 2014). Although these results were not replicated in two follow-up studies (Fernström et al. 2021; Gumpp et al. 2021), indices of mitochondrial functions, including mitochondrial content factors and complex I activity, were significantly higher in SSRI responders as compared to nonresponders (Fernström et al. 2021), potentially pointing to the usefulness of indices of mitochondrial activity as biomarkers for SSRI responsivity.

6 Evidence for Metabolic Abnormalities in Brain Disorders

Altered immuno-oxidative mechanisms in the brain, including oxidative stress, mitochondrial dysfunction, neuroinflammation, and cell-mediated immune response, have been shown to severely disrupt neural functions in CNS disorders. Balance between the redox pair of nicotinamide adenine dinucleotides (oxidized NAD^+ and reduced NADH) reflects the oxidative state of cells and the ability of biological systems to carry out energy production. Significant reduction of the NAD^+ to NADH ratio has been reported in chronic SZ compared to a matched healthy control group as well as in first-episode SZ compared to both a matched first-episode BD and matched healthy control groups (Kim et al. 2017). Impaired glucose metabolism and mitochondrial functions have been identified by using a variety of approaches and techniques. Briefly, decreased expression of genes encoding different glycolytic enzymes, such as the rate-limiting hexokinase, have been reported, along with decreased amount of enzyme proteins and activity, specifically in glutamatergic neurons in the dorsolateral prefrontal cortex in SZ (Henkel et al. 2022; Sahay et al. 2023; Sullivan et al. 2018, 2019b, c; Swerdlow and Wilkins 2020). Furthermore, multiple mitochondrial impairments, including altered expression of electron transport chain (ETC) enzymes such as complex I and complex V (ATP synthase), have been shown in the brains of people with SZ (Andreazza et al. 2010; Cuenod et al. 2022; Karry et al. 2004; O'Brien et al. 2024; Shan et al. 2014; Wang et al. 2009). Brain imaging studies have shown decreased glucose utilization in the frontal cortex (hypofrontality) and a switch to more glycolytic ATP production, associated with elevated brain lactate levels (Du et al. 2018; Duarte and Xin 2019; Rowland et al. 2016; Sullivan et al. 2019a; Yuksel et al. 2015). *In vivo* and postmortem studies have consistently reported lower brain pH in people with SPCs, particularly SZ (Stein et al. 2023). If less ATP is required from oxidative phosphorylation (OXPHOS), there might be a relative increase in the glycolysis to OXPHOS ratio (Figure 8.1). In turn, this predicts the observed slightly more acidic brain tissue and increased lactate. These findings led to the conceptualization of SZ as a disease of impaired brain bioenergetics (Henkel et al. 2022).

Energy dysregulation is also considered to be a central feature of BD pathophysiology (Cataldo et al. 2010; Clay et al. 2011; Du et al. 2018; Duarte and Xin 2019; Sullivan et al. 2019a). Mania may represent a condition of heightened cerebral energy metabolism facilitated by hyperglycolysis and glutaminolysis (Campbell and Campbell 2024). When oxidative glucose metabolism becomes impaired in the brain, neurons can utilize glutamate as an alternative substrate to generate energy through oxidative phosphorylation. In this way, it has been hypothesized that the upregulation of glycolysis and glutaminolysis causes the brain to enter a state of heightened metabolism and excitatory activity, which may underlie the subjective experience of mania (Campbell and Campbell 2024). Supporting this hypothesis, recent studies have identified mitochondrial abnormalities and resulting elevation of deleterious free oxygen radicals in BD (Andreazza et al. 2008; Cingi Yirun et al. 2016; Kotzaeroglou and Tsamesidis 2022). Taken together, recent progress in the neurobiological understanding of SZ and BD points toward the mechanistically important role of abnormal brain bioenergetics in their pathophysiology.

6.1 Oxidative Stress and Redox Dysregulation

Oxidative stress and redox dysregulation play crucial roles in psychiatric disorders through their interaction with key pathophysiological processes. Research has identified a convergence of oxidative stress, NMDAR hypofunction, and neuroinflammation during neurodevelopment that affects, in particular, neural connectivity and cognitive function in conditions like SZ (Cabungcal et al. 2014; Do et al. 2009; Khadimallah et al. 2022; Steullet et al. 2016). Of particular interest is the impact of oxidative stress on parvalbumin-positive interneurons (PVIs), a specific class of neurons crucial for neural synchronization. Indeed, with high firing discharge, PVIs utilize high amounts of energy provided by mitochondrial oxidative phosphorylation; this makes them particularly vulnerable to redox imbalance and the ensuing oxidative stress (Kann 2016). In these cells, oxidative stress triggers a cascade of molecular events, beginning with the upregulation of microRNA miR-137, a significant genetic risk factor for SZ. This upregulation leads to decreased expression of COX6A2, a component of cytochrome C oxidase complex IV, and results in impaired mitochondrial function and the accumulation of damaged mitochondria. This mechanism creates a self-perpetuating cycle of cellular dysfunction that may progressively affect neural circuit operation, especially in the prefrontal cortex. Clinical studies have shown promising results with targeted antioxidant interventions. The mitochondria-targeted antioxidant MitoQ, for instance, has demonstrated effectiveness in protecting prefrontal PVI circuits from oxidative stress-induced damage. In individuals with early psychosis, exosome-based measurements of miR-137 and COX6A2 plasma levels serve as biomarkers of cellular dysfunction, correlating with reduced gamma oscillations, cognitive impairment, and poorer functional outcomes (Kann 2016; Khadimallah et al. 2022). Investigating whether circulating brain-derived exosomal microRNAs (miRNAs) could serve as mechanism-based biomarkers for predicting psychosis transition in clinical high-risk individuals, a recent study identified a panel of four microRNAs (miR-132, miR-9, miR-941, miR-34a) that predict psychosis transition in clinical high-risk individuals with remarkable accuracy (AUC = 0.97). A study including 344 participants, who were followed for up to five years, showed that these blood biomarkers (involved in redox function, neuroinflammation, and blood-brain barrier integrity) substantially outperformed current assessment methods, detecting 90.6% patients with psychosis. These findings, validated in an independent cohort (NAPLS3, AUC = 0.92), suggest that blood measures of redox function could provide impactful psychosis prediction and offer new therapeutic targets for this vulnerable population (Do 2025). These findings have significant therapeutic implications, offering potential for both targeted interventions and measurable outcomes in prevention and treatment development. The identification of specific biomarkers enables better patient stratification and treatment monitoring, potentially improving the precision of therapeutic approaches for psychiatric disorders (Cuenod et al. 2022).

6.2 Is Synaptic Pathology in Psychiatric Disorders the Cause or Effect of Mitochondrial Alterations?

A brief summary of the relationships between synaptic activity and mitochondria may give insight into this question. Neuronal synaptic signaling requires substantial energy, and the majority of this energy is in the form of ATP synthesized via OXPHOS within mitochondria localized near synaptic structures (Acin-Perez and Enriquez 2014; Harris et al. 2012). Persistent increases or decreases in levels of synaptic signaling raise or lower ATP synthesis, respectively (Hall et al. 2012; Harris et al. 2012; Khatri and Man 2013; Li and Sheng 2022; Wong-Riley 2012). To meet changes in energy use, neurons make coordinated adjustments in the expression of OXPHOS-related transcripts, especially ETC complex subunits (Barshad et al. 2018; Devin and Rigoulet 2007; van Waveren and Moraes 2008; Williams et al. 1987; Wong-Riley 2012). Indeed, the coordinated expression of ETC complexes appears to be a conserved feature across many cell and tissue types (Barshad et al. 2018; van Waveren and Moraes 2008). This relationship reflects the fact that the coordinated expression and cooperative activity of ETC components is required for ATP synthesis via OXPHOS within mitochondria (Hall et al. 2012; Harris et al. 2012; Whittaker et al. 2011). As such, levels of nuclear DNA-encoded ETC complex subunits (nETC) are positively correlated, as are mitochondria DNA-encoded ETC complex subunits (mETC), across a variety of cell types and tissues (Barshad et al. 2018; van Waveren and Moraes 2008). Thus, persistent reductions in synaptic neurotransmission are expected to result in a synchronized lowering of levels of ETC components.

Genetic studies identify glutamatergic synaptic alterations as a core etiological mechanism of SPCs, particularly SZ (Bryois et al. 2020; Genovese et al. 2016; Hall et al. 2015; Kirov et al. 2012; Schizophrenia Working Group of the Psychiatric Genomics Consortium 2014; Trubetskoy et al. 2022). Consistent with these findings, *in vivo* and postmortem brain studies have identified altered expression of molecular, structural, and physiological markers reflective of deficient cortical glutamate synaptic neurotransmission (Glausier and Lewis 2013; Howes et al. 2024;

Schoonover et al. 2020). Further supporting the presence of reduced cortical glutamatergic synaptic signaling in SPCs, accumulating *in vivo* studies and transcriptomic, proteomic, and enzymatic postmortem studies of energetic measures suggest lower OXPHOS in cortical regions in these conditions (Arion et al. 2017; Bralet et al. 2016; Enwright and Lewis 2021; Fujimoto et al. 2007; Gandal et al. 2019; Glausier et al. 2020; Hazlett et al. 2019; Kraguljac et al. 2012; Minzenberg et al. 2009; Mitelman et al. 2018; Park et al. 2006; Sukumar et al. 2020). For example, in SZ, transcriptomic analyses in the dorsolateral prefrontal cortex total gray matter and glutamatergic pyramidal neurons showed the stereotypic coordinated reduction of OXPHOS gene expression that occurs in response to chronic reductions in neuronal firing and synaptic neurotransmission. Furthermore, targeted analysis of a gene set that indexed the diversity of mitochondrial functions (including OXPHOS, Ca^{2+} buffering, and reactive species generation) showed a specific and selective reduction in functional pathways related to OXPHOS. Weighted gene co-expression network analysis further revealed that OXPHOS-related network structure was completely preserved, as would be predicted if neurons were making coordinated adjustments to respond to lower ATP consumption (Glausier et al. 2020).

Mitochondrial abundance, size, and morphology also change in predictable ways in direct response to altered energetic demand at synapses (Cserép et al. 2018; Hackenbrock 1981; Justs et al. 2022; Li and Sheng 2022; Rossi and Pekkurnaz 2019; Scalettar et al. 1991). Studies using electron microscopy, an approach that permits direct visualization of mitochondria at the nanometer resolution level, show mitochondrial alterations consistent with these experimental findings in the cortex of individuals with SPCs, including smaller size or number with generally preserved ultrastructural integrity (Roberts 2017; Roberts et al. 2015; Somerville et al. 2011). These electron microscopy findings differ from those identified in people with neurodegenerative disorders (e.g., Parkinson disease) or primary mitochondrial disorders (e.g., Leigh syndrome), which find ultrastructural evidence of mitochondrial dysfunction (Uittenbogaard et al. 2021; Weidling and Swerdlow 2020).

6.2.1 Evidence Supporting an Upstream Synaptic Pathology

Several considerations are consistent with the idea that mitochondrial abnormalities in SPCs may be secondary to synaptic deficits. First, as discussed above, GWAS studies do not strongly support a genetic basis for mitochondrial dysfunction, but they do provide compelling evidence for synaptic abnormalities, as risk genes encoding for synaptic components are robustly represented in several psychiatric disorders (Bigdeli et al. 2024; Borcuk et al. 2024; Owen et al. 2023; Trubetskoy et al. 2022). Second, these genes are enriched in concerted synaptic neuron-astrocyte gene expression programs, which are decreased in psychiatric disorders (Ling et al. 2024). Third, synapses are high-energy consumers, and experimental evidence shows that lower synaptic functions correspond to lower mitochondrial functions and nuclear mitochondrial genes (Cserép et al. 2018; Dhar and Wong-Riley 2009; Hevner and Wong-Riley 1991; Justs et al. 2022; Li and Sheng 2022; Liang et al. 2006; Rossi and Pekkurnaz 2019; Scalettar et al. 1991).

In neurons, the most energetically "demanding" processes involve action potential generation and synaptic signaling. Accordingly, chronically lower rates of neuronal firing cause a coordinated reduction in the expression of ETC complex subunits. For example, *in vitro* and *in vivo* manipulations that cause persistent reductions in neuronal firing result in less transcription of all subunits comprising ETC complex IV. Further, mitochondrial ultrastructure reflects less energy usage, and the direct relationship between mitochondrial function and ultrastructure, including at synaptic complexes, is well supported. Ultrastructural measures of synaptic mitochondria directly inform the relative level of activity at individual synapses. Mitochondrial abundance, size, and morphology change in direct response to altered energetic usage at synapses. For example, mitochondrial abundance and/or size are downregulated at less active synapses. Thus, if chronic reductions in neuronal firing are upstream of lower measures of OXPHOS in SPCs, then (a) expression of ETC complex subunit genes is predicted to be broadly and comparably lower, (b) the physiologically correlated expression of ETC complex subunit transcripts would be retained, and (c) mitochondrial ultrastructural integrity would be preserved. Thus, many studies of brain from individuals with SPCs, and perhaps particularly SZ, converge on the interpretation that mitochondrial OXPHOS may not be dysfunctional or defective per se, but that lower measures of OXPHOS reflect the expected consequence of long-term reductions in synaptic signaling. Therapeutics that augment synaptic signaling may prove beneficial in restoring cortical function in SPCs.

6.2.2 Evidence Supporting an Upstream Mitochondrial Pathology

Evidence for an upstream mitochondrial pathology may be equally compelling. Strong evidence for an inherent mitochondrial dysfunction, primarily of OXPHOS, in SPCs can be found in the numerous reports of mitochondrial functional deficits in a variety of peripheral cells, including blood and muscle cells, keratinocytes, and iPSCs, which are not impacted by synaptic activity (Chouinard et al. 2017; Du et al. 2018; Renshaw et al. 2001; Volz et al. 1998). In addition, transplanted mitochondria may either improve or disrupt host mitochondrial function, and thereby neuronal structure, synaptic formation and activity, both in iPSCs differentiated into glutamatergic neurons and in an animal model of SZ and their healthy controls (Robicsek et al. 2018). Mitochondrial transplantation ameliorated behavioral motor and cognitive symptoms in animal models of several CNS disorders (Chang et al. 2016; Ene et al. 2023; Nitzan et al. 2019; Wang et al. 2019). Glutamate negatively impacts mitochondria, and its toxicity in SPCs may underlie the observed mitochondrial dysfunction (Attwell and Laughlin 2001; Devine and Kittler 2018; Glancy et al. 2020; Hackenbrock 1981; Hall et al. 2012; Harris et al. 2012; Mendelsohn et al. 2022). Redox dysregulation leads to mitochondrial damage (autosomal mitophagy) (Garza-Lombó et al. 2020; Lee et al. 2011; Redmann et al. 2016; Zhang 2013).

Impairments of ETC complex that result from impaired expression of ETC complex accessory subunits, assembly factors, or ROS-induced damage negatively impact the OXPHOS process (Diaz 2010; Guo et al. 2013; Reinecke et al. 2009). Impaired ETC differentially affects the expression of nETC, mETC, and ETC complex assembly factors (Coleman et al. 2018; Dogan et al. 2018; Gohil et al. 2010; Mansilla et al. 2018; Reinecke et al. 2009; Viscomi et al. 2011). For example, downregulating the expression of ETC complex core subunits, accessory subunits, or assembly factors to impair OXPHOS does not result in a coordinated reduction in the expression of ETC complexes (Grünewald et al. 2016; Lax et al. 2016; Reinecke et al. 2009; Richter-Dennerlein et al. 2016; Stroud et al. 2016; Viscomi et al. 2011). Moreover, impaired ETC complex functioning typically disrupts other mitochondrial processes in addition to OXPHOS, such as the Krebs cycle (Garmier et al. 2008; Reinecke et al. 2009), as well as the balance between fission and fusion events (Benard et al. 2007; Chernivec et al. 2018; Duvezin-Caubet et al. 2006; Morán et al. 2010; Peng et al. 2017; Silva Ramos et al. 2016). Cells with high firing and energy demands are highly vulnerable to impaired OXPHOS. For example, PVIs have substantially higher firing rates and ATP demand than projection neurons (Kann et al. 2014). As such, perturbations of mitochondrial energetic capacity, including induction of oxidative stress by a variety of experimental manipulations, more strongly affect the firing of, and gene expression in, PVIs than in other cell types (Kann et al. 2014; Steullet et al. 2016; Whittaker et al. 2011). Finally, defects to the ETC, or damage due to ROS generation, are reflected in impaired mitochondrial ultrastructure integrity, including broken double membranes, swollen dysmorphic mitochondrial shape, and disorganized cristae (Gong et al. 2011; Guo et al. 2013; Quintana et al. 2010; Vincent et al. 2016). Thus, if the ETC is impaired and limits mitochondrial OXPHOS capacity in SPCs, then ETC complex subunit genes would show variable patterns of altered expression—with a corresponding loss of the positive correlation among ETC complex subunit transcripts and stronger effects in PVIs relative to pyramidal neurons—and mitochondrial ultrastructural integrity would be impaired.

It is worth noting, however, that although primary mitochondrial diseases and experimental manipulations in model systems robustly impair mitochondrial function and OXPHOS, the possibility that SPCs, including SZ, may be associated with more moderate impairments of OXPHOS that are not accurately reflected in these studies cannot be excluded. Current evidence of cytosolic glucose dysregulation in parallel to mitochondrial metabolic impairment stands from *in vitro* studies in a diversity of models, including iPSCs. These "snapshots," however, reflect a link that may be mediated by the artifactual conditions of these systems. Thus, the strongest evidence for the role of mitochondria in SZ may be derived from *in vivo* and *in vitro* observations in which mitochondrial transplantation and transfer, respectively, are capable of rescuing or inducing disease phenotypes (Robicsek et al. 2018). This raises the possibility that some combinatorial effect of reduced energetic demand and impaired energetic capacity may be present in SZ.

Finally, evidence linking early-life adversity—a significant risk factor for SPCs (Gerke et al. 2018)—to impaired (peripheral) mitochondrial function (Boeck et al. 2016; Karabatsiakis et al. 2014) suggests the possibility of a causal mitochondrial deficit, which may manifest as fatigue or energy depletion, as seen in depression. Patients with childhood maltreatment often develop depression later in life after a second traumatic event or due to cumulative effects of trauma and stress (Boeck et al. 2016; Gumpp et al. 2023; Nold et al. 2019).

In summary, several lines of evidence support seemingly contradictory mechanisms underlying the relationship between mitochondrial deficits and synaptic functions in SPCs. In the context of the significant heterogeneity of these disorders in virtually all dimensions (genetic, environmental, clinical), it is plausible to consider that these two mechanisms may not in fact be mutually exclusive. One possible scenario is that in subsets of SPCs, upstream mechanisms in synaptic pathology may reduce energy demand or increase DNA damage, leading to a downregulation of energy production, whereas in a different group, mitochondrial deficits may cause synaptic deficits. Another possibility is that these two mechanisms may co-exist across a spectrum, with one or the other predominant in some individuals, brain regions, or temporal stages of the disease. Future studies designed to stratify individuals with SPCs according to their metabolic and synaptic features may prove informative for understanding their etiology and identifying novel therapeutic targets (see Section 8).

6.2.3 Potential Role of Mitochondrial, Synaptic, and Astrocyte Interactions

Reciprocal negative effects of synaptic pathology and mitochondrial damage support the idea that synaptic and mitochondrial deficits may compound each other. On one hand, as discussed above, synaptic deficits lead to reduced demands on mitochondria. On the other, mitochondrial damage is expected to have extensive repercussions on neurotransmission, neuron structural plasticity, and immune factors, which in turn impact synaptic functions. Mitochondria supply not only ATP but also nucleotides and other key molecules, such as cholesterol and neurosteroids. Notably, mitochondria can be donated and received by cells (Khadimallah et al. 2022; Swerdlow and Wilkins 2020). Transcellular mitophagy plays an important role in local cell-cell signaling and neuroprotection, indicating that brain mitochondrial dysfunction may also disrupt this key function. When mitochondria are received, they are digested (mitophagy) and the components used by the cell in a beneficial manner. They may also be carrying information to the receiving cells. Some evidence indicates that cholesterol may be particularly important among the mitochondrial components transferred from astrocytes to neurons. Recent work has shown that cholesterol synthesis in astrocytes is an important factor in the synaptic neuron and astrocyte program found to be reduced in psychiatric disorders (Ling et al. 2024). It is tempting to speculate that transcellular mitophagy may mediate cell-cell effects of mitochondrial deficits in brain disorders.

Another potential mechanism mediating astrocyte-neuron effects of metabolic abnormalities may be based on the astrocyte-neuron lactate shuttle hypothesis. This model proposes that astrocytes support neuronal metabolism by supplying neurons with lactate, as an energy substrate, especially during periods of increased neuronal activity. Increased glutamate release at synapses during high neuronal activation results in higher glutamate uptake by astrocytes, signaling increased neuronal activity and triggering glycolysis and lactate production. According to this hypothesis, astrocytes would then supply neurons with an alternative energy source to meet their high metabolic demands (Pellerin and Magistretti 1994, 1996). This mechanism may allow neurons to focus on oxidative metabolism, which is more efficient for ATP generation, whereas astrocytes would take on glucose metabolism and have neuroprotective effects, as lactate would represent an immediate substrate for ATP production during energy stress. We note that the astrocyte-neuron lactate shuttle hypothesis has received some criticism. For instance, it has been argued that this hypothesis may not account for critical stoichiometric requirements and ultimately may not significantly contribute to brain activation energy demands (Dienel 2017; Rothman et al. 2012). However, its potential role in metabolic abnormalities in brain disorders has not been assessed and deserves consideration.

7 Are Metabolic Abnormalities Predominant in Distinct Clinical Phenotypes?

The question of whether specific metabolic patterns are predominantly associated with distinct clinical phenotype may be particularly relevant to approaches to stratification and efforts to understand their pathophysiological role in brain disorders. Although not extensively investigated at this time, some examples in the literature support this possibility. MDD is a highly heterogenous disorder, with a broad range of clinical presentations, responsivity to treatment, and intricate, multifaceted, and bidirectional interactions between CNS and systemic conditions (Penninx et al. 2013). This disorder is strongly associated with metabolic syndrome (Foley et al. 2010; Grundy et al. 2005; Pan et al. 2012; Skilton et al. 2007; Vogelzangs et al. 2011). Notably, metabolic abnormalities in MDD were reported selectively in a distinct clinical phenotype, presenting with a combination of immune and metabolic dysregulations (e.g., dysregulated levels of leptin, insulin, and immune markers) and atypical depression, defined as predominant mood reactivity and interpersonal rejection sensitivity in combination with hypersomnia, increased appetite, and weight gain (Milaneschi et al. 2020, 2021; Vreijling et al. 2023). These studies also show that persons with energy-related depressive symptoms are associated with more severe course of illness, such as younger age of onset and more frequent, longer depressive episodes as well as greater incidence of comorbid PTSD and eating and personality disorders (Vreijling et al. 2023).

8 A Transdiagnostic Approach

The evidence reviewed here supports the notion that distinct metabolic phenotypes may be shared among multiple brain disorders while not being consistently associated with any of them. Conversely, the clinical heterogeneity within each categorical psychiatric disorder may contribute to the challenges encountered in identifying replicable metabolic markers in clinical populations. As reviewed above, evidence for metabolic abnormalities is well documented in a growing number of CNS disorders, including SZ, MDD, BD, PTSD, and several neurodegenerative disorders. This evidence is consistent with the growing recognition that causal and vulnerability factors (e.g., genetic risk, trauma, immune and metabolic factors) may converge across categorical brain disorders in different combinations, resulting in a variety of transdiagnostic clinical phenotypes. In this context, it may be important to consider that the once broadly accepted distinction between neurodegenerative and psychiatric disorders has become increasingly blurred in recent years. Symptoms of depression, anxiety, cognitive decline, substance use, and even psychosis are shared across many "categorical" psychiatric and neurodegenerative disorders, inviting a dimensional, transdiagnostic clinical framework of investigation (Balthazar et al. 2014; Zhou and Seeley 2014; Chen et al. 2021; Creese and Lunnon 2022; Davis et al. 2023; Dolphin et al. 2023; Jay et al. 2023; Jenkins et al. 2022; Mendez 2021; Schneider et al. 2008; Tabrizi et al. 2022; Weintraub and Mamikonyan 2019; Wingo et al. 2022). As an example, it may be meaningful to ask whether a distinct metabolic signature may be associated with a specific syndrome (e.g., depression or psychosis) across disorders, with careful stratification by demographic, systemic, genetic, and environmental/socioeconomic factors (Table 8.1). This approach may better address the challenges of connecting a specific biological measure (e.g., insulin resistance) to a clinically diverse disorder (e.g., PTSD or SZ) and could facilitate the identification of symptom constructs rooted in biological evidence.

Along the same lines, it is plausible to postulate that distinct patterns of metabolic abnormalities may arise from diverse mechanisms (e.g., genetic, immune or stress-driven, neurodegenerative) and yet give rise to similar, overlapping transdiagnostic clinical phenotypes. As an example, it may be informative to consider the extent of metabolic similarities between Alzheimer disease and psychiatric

Table 8.1 Biomarkers of metabolic abnormalities in psychiatric disorders.

Metabolic Marker	Associated Psychiatric Symptoms	Clinical Implications	Emerging Research Directions
Insulin resistance	Cognitive deficits, mood instability	Early screening for metabolic syndrome in psychiatric populations	Explore the role of insulin resistance in cognitive and emotional regulation across psychiatric conditions, with a focus on prevention and early intervention
Elevated inflammatory markers	Fatigue, anhedonia, psychomotor slowing	Anti-inflammatory strategies for symptom modulation	Investigate the impact of systemic inflammation on symptom progression and treatment response across diverse psychiatric disorders
Mitochondrial dysfunction	Cognitive impairment, energy dysregulation	Targeted mitochondrial enhancers (e.g., CoQ10, creatine)	Examine mitochondrial health as a universal biomarker for mental health and resilience, and as a potential target for novel therapeutic interventions
Dysregulated lipid metabolism	Anxiety, irritability, psychosis	Integration of lipid-lowering medications in treatment	Understand lipid metabolism as a bridge between metabolic health and psychiatric conditions, with implications for both prevention and treatment strategies
Gut microbiome dysbiosis	Mood swings, cognitive decline	Probiotic and dietary interventions to restore gut microbiome balance	Investigate the microbiota-gut-brain axis as a unifying pathway influencing neurodevelopment, stress response, and emotional regulation across the lifespan

conditions, perhaps considering whether elements of the metabolic syndrome associated with MDD may be present in the prodromal depression that often precedes Alzheimer disease. At the cellular level, ultrastructural studies on Alzheimer disease have shed new light into the previous observations reporting morphological differences. By applying three-dimensional electron microscopy to visualize mitochondrial structure in brain tissue from persons with Alzheimer disease and mouse models, a new mitochondrial fission arrest phenotype has been observed, resulting in elongated structures referred to as "mitochondria-on-a-string" (Zhang et al. 2016). Exploring whether similar structures exist in other brain disorders and identifying any shared pathophysiological or clinical features could provide valuable insights. For instance, severe psychiatric disorders represent a vulnerability factor for dementia, and significant psychiatric symptoms are common in a large number of neurodegenerative disorders (Balthazar et al. 2014; Chen et al. 2021; Creese and Lunnon 2022; Davis et al. 2023; Dolphin et al. 2023; Jay et al. 2023; Jenkins et al. 2022; Schneider et al. 2008; Tabrizi et al. 2022; Weintraub and Mamikonyan 2019; Wingo et al. 2022; Zhou and Seeley 2014). Distinct systemic/CNS metabolic abnormality patterns shared by transdiagnostic (psychiatric and neurodegenerative) clinical endophenotypes may shed light on these relationships.

Implementing transdiagnostic approaches in research and clinical practice requires overcoming methodological challenges, including:

1. Defining meaningful clusters: Identify shared symptom domains across disorders that are biologically grounded.
2. Longitudinal analysis: Establish causal links between metabolic abnormalities and evolving symptom patterns.
3. Personalized interventions: Develop treatments tailored to individual metabolic profiles rather than diagnosis-specific protocols.
4. Clinical implications and integration of metabolic markers: The integration of metabolic markers into clinical practice offers a promising avenue for improving psychiatric care. By identifying specific metabolic profiles, clinicians can move toward a precision-medicine approach that tailors interventions to individual patient needs.
5. Screening: Use routine assessment of metabolic markers (e.g., HbA1c, inflammatory cytokines) in psychiatric populations to identify high-risk individuals.
6. Intervention: Incorporate metabolic-targeted treatments, such as insulin sensitizers, anti-inflammatory agents, or mitochondrial enhancers, alongside psychotropic medications.
7. Outcome tracking: Monitor changes in metabolic and psychiatric symptoms over time to refine treatment protocols.

9 Conclusions

Compelling findings presented throughout this volume underscore the role of metabolic abnormalities in CNS disorders. This evidence leads us to the following key conclusions:

1. Conditions historically referred to as "brain disorders" arise from "whole-body" interactions and should thus be investigated within this framework.
2. "Metabolic abnormalities"—a broad term encompassing heterogeneous, interacting factors and mechanisms—occur in a large number of brain disorders. Distinct metabolic signatures may correspond to specific clinical endophenotypes, such as metabolic syndrome in a subgroup of patients with atypical depression (Vreijling et al. 2023), and may be observed across multiple categorical diagnoses.
3. There is strong evidence that metabolic abnormalities significantly contribute to psychiatric disorders, and that addressing these issues may serve as a valuable complement to psychiatric care. However, metabolic abnormalities are not in themselves a sufficient cause. They may instead act as vulnerability factors or triggers for psychiatric symptoms, akin to cancer "hallmarks."
4. There is compelling evidence for the role of demographic and environmental factors (e.g., sex, age, diet, socioeconomic status, exposure to trauma, substance use) in metabolic abnormalities related to SPCs, perhaps in combination with more subtle genetic vulnerabilities.

Together, these considerations offer a path for investigations on the role of metabolic abnormalities in psychiatric disorders. Specifically, they point to the importance of transdiagnostic studies, with clearly defined clinical phenotypes, symptom constructs, and disease stages, that are sufficiently powered to allow for stratification according to demographic and environmental factors, and well-validated measures of specific metabolic functions (Table 8.1). These investigations may benefit from experimental designs focused on the association and causal relationships between focused metabolic metrics (e.g., pyruvate/lactate ratios, mitochondria ultrastructural and DNA integrity and functions, body fat measures, insulin signaling) and categorical diagnoses deconstructed into biology-based symptoms (anhedonia, loss of appetite, hypersomnia). Key outcomes of these studies may be the identification of transdiagnostic biology-based symptom constructs—clinical endophenotypes defined by their biological underpinnings rather than traditional descriptive observations. We put forth that focus on biology-based symptom constructs may be a critical requirement for individualized, evidence-based therapeutic approaches.

References

Acin-Perez R, and Enriquez JA (2014) The Function of the Respiratory Supercomplexes: The Plasticity Model. Biochim Biophys Acta Bioenerg 1837 (4):444–450. https://doi.org/10.1016/j.bbabio.2013.12.009

Agarwal SM, Caravaggio F, Costa-Dookhan KA, et al. (2020) Brain Insulin Action in Schizophrenia: Something Borrowed and Something New. Neuropharmacology 163:107633. https://doi.org/10.1016/j.neuropharm.2019.05.010

Akarsu S, Torun D, Erdem M, et al. (2015) Mitochondrial Complex I and III mRNA Levels in Bipolar Disorder. J Affect Disord 184:160–163. https://doi.org/10.1016/j.jad.2015.05.060

Akın O, Eker İ, Arslan M, et al. (2016) Association of Nerve Conduction Impairment and Insulin Resistance in Children with Obesity. Childs Nerv Syst 32 (11):2219–2224. http://dx.doi.org/10.1007/s00381-016-3210-3

Alagiakrishnan K, and Halverson T (2024) Role of Peripheral and Central Insulin Resistance in Neuropsychiatric Disorders. J Clin Med 13 (21):6607. https://doi.org/10.3390/jcm13216607

Alcocer-Gómez E, Núñez-Vasco J, Casas-Barquero N, et al. (2016) Gene Expression Profile in Major Depressive Disorder Shows Reduced Mitochondrial Biogenesis. CNS Neurosci Ther 22 (7):636–638. https://doi.org/10.1111/cns.12568

Ali S, Santomauro D, Ferrari AJ, and Charlson F (2022) Excess Mortality in Severe Mental Disorders: A Systematic Review and Meta-Regression. J Psychiatr Res 149:97–105. https://doi.org/10.1016/j.jpsychires.2022.02.036

Andreazza AC, Kauer-Sant'Anna M, Frey BN, et al. (2008) Oxidative Stress Markers in Bipolar Disorder: A Meta-Analysis. J Affect Disord 111 (2–3):135–144. https://doi.org/10.1016/j.jad.2008.04.013

Andreazza AC, Shao L, Wang JF, and Young LT (2010) Mitochondrial Complex I Activity and Oxidative Damage to Mitochondrial Proteins in the Prefrontal Cortex of Patients with Bipolar Disorder. Arch Gen Psychiatry 67 (4):360–368. https://doi.org/10.1001/archgenpsychiatry.2010.22

Appel KE, and Farr CB (1929) The Blood Sugar Reaction to Insulin in Psychoses. Arch Neurol Psychiatry 21 (1):145–148. https://doi.org/10.1001/archneurpsyc.1929.02210190148009

Arion D, Huo Z, Enwright JF, et al. (2017) Transcriptome Alterations in Prefrontal Pyramidal Cells Distinguish Schizophrenia from Bipolar and Major Depressive Disorders. Biol Psychiatry 82 (8):594–600. https://doi.org/10.1016/j.biopsych.2017.03.018

Attwell D, and Laughlin SB (2001) An Energy Budget for Signaling in the Grey Matter of the Brain. J Cereb Blood Flow Metab 21 (10):1133–1145. https://doi.org/10.1097/00004647-200110000-00001

Balthazar MLF, Pereira FRS, Lopes TM, et al. (2014) Neuropsychiatric Symptoms in Alzheimer's Disease Are Related to Functional Connectivity Alterations in the Salience Network. Hum Brain Mapp 35 (4):1237–1246. https://doi.org/10.1002/hbm.22248

Barrett TM, and Serre P (1924) Blood Analysis and Sugar Tolerance Tests in Mental Disease. J Nerv Ment Dis 59 (6):561–570

Barshad G, Blumberg A, Cohen T, and Mishmar D (2018) Human Primitive Brain Displays Negative Mitochondrial-Nuclear Expression Correlation of Respiratory Genes. Genome Res 28 (7):952–967. https://doi.org/10.1101/gr.226324.117

Behnke A, Rojas R, Karabatsiakis A, and Kolassa I-T (2020) Childhood Maltreatment Compromises Resilience against Occupational Trauma Exposure: A Retrospective Study among Emergency Medical Service Personnel. Child Abuse Negl 99:104248. https://doi.org/10.1016/j.chiabu.2019.104248

Benard G, Bellance N, James D, et al. (2007) Mitochondrial Bioenergetics and Structural Network Organization. J Cell Sci 120 (Pt 5):838–848. https://doi.org/10.1242/jcs.03381

Bergholz LM, Subic-Wrana C, Heylmann D, et al. (2017) DNA Damage in Lymphocytes of Patients Suffering from Complex Traumatization. DNA Repair 52:103–109. https://doi.org/10.1016/j.dnarep.2017.02.006

Bergman O, and Ben-Shachar D (2016) Mitochondrial Oxidative Phosphorylation System (OXPHOS) Deficits in Schizophrenia: Possible Interactions with Cellular Processes. Can J Psychiatry 61 (8):457–469. https://doi.org/10.1177/0706743716648290

Bigdeli TB, Chatzinakos C, Bendl J, et al. (2024) Biological Insights from Schizophrenia-Associated Loci in Ancestral Populations. Aug. 28, 2024. https://doi.org/10.1101/2024.08.27.24312631

Bobba-Alves N, Juster R-P, and Picard M (2022) The Energetic Cost of Allostasis and Allostatic Load. Psychoneuroendocrinology 146:105951. https://doi.org/10.1016/j.psyneuen.2022.105951

Boeck C, Koenig AM, Schury K, et al. (2016) Inflammation in Adult Women with a History of Child Maltreatment: The Involvement of Mitochondrial Alterations and Oxidative Stress. Mitochondrion 30:197–207. https://doi.org/10.1016/j.mito.2016.08.006

Boeck C, Salinas-Manrique J, Calzia E, et al. (2018) Targeting the Association between Telomere Length and Immuno-Cellular Bioenergetics in Female Patients with Major Depressive Disorder. Sci Rep 8 (1):9419. https://doi.org/10.1038/s41598-018-26867-7
Borcuk C, Parihar M, Sportelli L, et al. (2024) Network-Wide Risk Convergence in Gene Co-Expression Identifies Reproducible Genetic Hubs of Schizophrenia Risk. Neuron 112 (21):3551–3566.e3556. https://doi.org/10.1016/j.neuron.2024.08.005
Bralet M-C, Buchsbaum MS, DeCastro A, Shihabuddin L, and Mitelman SA (2016) FDG-PET Scans in Patients with Kraepelinian and Non-Kraepelinian Schizophrenia. Eur Arch Psychiatry Clin Neurosci 266 (6):481–494. https://doi.org/10.1007/s00406-015-0633-x
Breit S, and Hubl D (2025) The Effect of GLP-1RAs on Mental Health and Psychotropics-Induced Metabolic Disorders: A Systematic Review. Psychoneuroendocrinology 176:107415. https://doi.org/10.1016/j.psyneuen.2025.107415
Brouwer JMJL, Wardenaar KJ, Liemburg EJ, et al. (2024) High Persistence and Low Treatment Rates of Metabolic Syndrome in Patients with Mood and Anxiety Disorders: A Naturalistic Follow-up Study. J Affect Disord 354:451-462. https://doi.org/10.1016/j.jad.2024.03.042
Bryois J, Skene NG, Hansen TF, et al. (2020) Genetic Identification of Cell Types Underlying Brain Complex Traits Yields Insights into the Etiology of Parkinson's Disease. Nat Genet 52 (5):482–493. https://doi.org/10.1038/s41588-020-0610-9
Bulduk BK, Tortajada J, Valiente-Pallejà A, et al. (2024) High Number of Mitochondrial DNA Alterations in Postmortem Brain Tissue of Patients with Schizophrenia Compared to Healthy Controls. Psychiatry Res 337:115928. https://doi.org/10.1016/j.psychres.2024.115928
Cabungcal JH, Counotte DS, Lewis E, et al. (2014) Juvenile Antioxidant Treatment Prevents Adult Deficits in a Developmental Model of Schizophrenia. Neuron 83 (5):1073–1084. https://doi.org/10.1016/j.neuron.2014.07.028
Campbell IH, and Campbell H (2024) The Metabolic Overdrive Hypothesis: Hyperglycolysis and Glutaminolysis in Bipolar Mania. Mol Psychiatr 29 (5):1521–1527. https://doi.org/10.1038/s41380-024-02431-w
Cataldo AM, McPhie DL, Lange NT, et al. (2010) Abnormalities in Mitochondrial Structure in Cells from Patients with Bipolar Disorder. Am J Pathol 177 (2):575–585. https://doi.org/10.2353/ajpath.2010.081068
Chang C-K, Hayes RD, Perera G, et al. (2011) Life Expectancy at Birth for People with Serious Mental Illness and Other Major Disorders from a Secondary Mental Health Care Case Register in London. PLoS One 6 (5):e19590. https://doi.org/10.1371/journal.pone.0019590
Chang J-C, Wu S-L, Liu K-H, et al. (2016) Allogeneic/Xenogeneic Transplantation of Peptide-Labeled Mitochondria in Parkinson's Disease: Restoration of Mitochondria Functions and Attenuation of 6-Hydroxydopamine-Induced Neurotoxicity. Transl Res 170:40–56.e43. https://doi.org/10.1016/j.trsl.2015.12.003
Chang Y, Park J, Yun J, and Song T (2025) The Association between the Triglyceride-Glucose Index and the Incidence Risk of Parkinson's Disease: A Nationwide Cohort Study. J Mov Disord 18 (2):138-148. https://doi.org/10.14802/jmd.24131
Chen DC, Du XD, Yin GZ, et al. (2016) Impaired Glucose Tolerance in First-Episode Drug-Naïve Patients with Schizophrenia: Relationships with Clinical Phenotypes and Cognitive Deficits. Psychol Med 46 (15):3219–3230. https://doi.org/10.1017/s0033291716001902
Chen Y, Dang M, and Zhang Z (2021) Brain Mechanisms Underlying Neuropsychiatric Symptoms in Alzheimer's Disease: A Systematic Review of Symptom-General and -Specific Lesion Patterns. Mol Neurodegener 16 (1):38. https://doi.org/10.1186/s13024-021-00456-1
Chernivec E, Cooper J, and Naylor K (2018) Exploring the Effect of Rotenone-A Known Inducer of Parkinson's Disease-on Mitochondrial Dynamics in *Dictyostelium Discoideum*. Cells 7 (11):201. https://doi.org/10.3390/cells7110201

Chestkov IV, Jestkova EM, Ershova ES, et al. (2018) ROS-Induced DNA Damage Associates with Abundance of Mitochondrial DNA in White Blood Cells of the Untreated Schizophrenic Patients. Oxid Med Cell Longev 2018:8587475. https://doi.org/10.1155/2018/8587475

Chouinard V-A, Henderson DC, Dalla Man C, et al. (2019) Impaired Insulin Signaling in Unaffected Siblings and Patients with First-Episode Psychosis. Mol Psychiatr 24 (10):1513–1522. https://doi.org/10.1038/s41380-018-0045-1

Chouinard V-A, Kim S-Y, Valeri L, et al. (2017) Brain Bioenergetics and Redox State Measured by ^{31}P Magnetic Resonance Spectroscopy in Unaffected Siblings of Patients with Psychotic Disorders. Schizophr Res 187:11–16. https://doi.org/10.1016/j.schres.2017.02.024

Cimino I, Kim H, Tung YCL, et al. (2021) Activation of the Hypothalamic–Pituitary–Adrenal Axis by Exogenous and Endogenous GDF15. PNAS 118 (27):e2106868118. https://doi.org/10.1073/pnas.2106868118

Cingi Yirun M, Unal K, Altunsoy Sen N, et al. (2016) Evaluation of Oxidative Stress in Bipolar Disorder in Terms of Total Oxidant Status, Total Antioxidant Status, and Oxidative Stress Index. Arch Neuropsychiatr 53 (3):194–198. https://doi.org/10.5152/npa.2015.10123

Clay HB, Sillivan S, and Konradi C (2011) Mitochondrial Dysfunction and Pathology in Bipolar Disorder and Schizophrenia. Int J Dev Neurosci 29 (3):311–324. https://doi.org/10.1016/j.ijdevneu.2010.08.007

Coleman V, Sa-Nguanmoo P, Koenig J, et al. (2018) Partial Involvement of Nrf2 in Skeletal Muscle Mitohormesis as an Adaptive Response to Mitochondrial Uncoupling. Sci Rep 8 (1):2446. https://doi.org/10.1038/s41598-018-20901-4

Correll CU, Solmi M, Veronese N, et al. (2017) Prevalence, Incidence and Mortality from Cardiovascular Disease in Patients with Pooled and Specific Severe Mental Illness: A Large-Scale Meta-Analysis of 3,211,768 Patients and 113,383,368 Controls. World Psychiatry 16 (2):163–180. https://doi.org/10.1002/wps.20420

Creese B, and Lunnon K (2022) Neuropsychiatric Symptoms in AD: The Search for Mechanisms. Nat Rev Neurol 18 (11):639–640. https://doi.org/10.1038/s41582-022-00721-x

Cserép C, Pósfai B, Schwarcz AD, and Dénes Á (2018) Mitochondrial Ultrastructure Is Coupled to Synaptic Performance at Axonal Release Sites. eNeuro 5 (1):ENEURO.0390–0317.2018. https://doi.org/10.1523/eneuro.0390-17.2018

Cuenod M, Steullet P, Cabungcal J-H, et al. (2022) Caught in Vicious Circles: A Perspective on Dynamic Feed-Forward Loops Driving Oxidative Stress in Schizophrenia. Mol Psychiatr 27 (4):1886–1897. https://doi.org/10.1038/s41380-021-01374-w

Dantzer R, O'Connor JC, Freund GG, Johnson RW, and Kelley KW (2008) From Inflammation to Sickness and Depression: When the Immune System Subjugates the Brain. Nat Rev Neurosci 9 (1):46–56. https://doi.org/10.1038/nrn2297

Davis M-C, Hill AT, Fitzgerald PB, et al. (2023) Neurophysiological Correlates of Non-Motor Symptoms in Late Premanifest and Early-Stage Manifest Huntington's Disease. Clin Neurophysiol 153:166–176. https://doi.org/10.1016/j.clinph.2023.06.021

Devin A, and Rigoulet M (2007) Mechanisms of Mitochondrial Response to Variations in Energy Demand in Eukaryotic Cells. Am J Physiol Cell Physiol 292 (1):C52–C58. https://doi.org/10.1152/ajpcell.00208.2006

Devine MJ, and Kittler JT (2018) Mitochondria at the Neuronal Presynapse in Health and Disease. Nat Rev Neurol 19 (2):63–80. https://doi.org/10.1038/nrn.2017.170

Dhar SS, and Wong-Riley MTT (2009) Coupling of Energy Metabolism and Synaptic Transmission at the Transcriptional Level: Role of Nuclear Respiratory Factor 1 in Regulating Both Cytochrome C Oxidase and NMDA Glutamate Receptor Subunit Genes. J Neurosci 29 (2):483–492. https://doi.org/10.1523/jneurosci.3704-08.2009

Diaz F (2010) Cytochrome C Oxidase Deficiency: Patients and Animal Models. Biochim Biophys Acta 1802 (1):100–110. https://doi.org/10.1016/j.bbadis.2009.07.013

Dienel GA (2017) Lack of Appropriate Stoichiometry: Strong Evidence against an Energetically Important Astrocyte–Neuron Lactate Shuttle in Brain. J Neurosci Res 95 (11):2103–2125. https://doi.org/10.1002/jnr.24015
Ding J, Quan M, Zang P, and Jia J (2025) Associations between Serum Metabolic Syndrome Indicators Levels and Cerebrospinal Fluid Pathological Protein in Dementia and Pre-Dementia Patients. J Alzheimers Dis 104 (2):537-546. https://doi.org/10.1177/13872877251318298
Do KQ (2025) Decoding Psychosis Risk: Global Perspectives on Clinical and Molecular Predictors. In: Annual Congress of the Schizophrenia Intl Res Soc 2025. pp 117–121
Do KQ, Cabungcal JH, Frank A, Steullet P, and Cuenod M (2009) Redox Dysregulation, Neurodevelopment, and Schizophrenia. Curr Opin Neurobiol 19 (2):220–230. https://doi.org/10.1016/j.conb.2009.05.001
Dogan SA, Cerutti R, Benincá C, et al. (2018) Perturbed Redox Signaling Exacerbates a Mitochondrial Myopathy. Cell Metab 28 (5):764–775.e765. https://doi.org/10.1016/j.cmet.2018.07.012
Dolatshahi M, Commean PK, Rahmani F, et al. (2024) Relationships between Abdominal Adipose Tissue and Neuroinflammation with Diffusion Basis Spectrum Imaging in Midlife Obesity. Obesity 33 (1):41–53. https://doi.org/10.1002/oby.24188
Dolphin H, Dyer AH, McHale C, O'Dowd S, and Kennelly SP (2023) An Update on Apathy in Alzheimer's Disease. Geriatrics 8 (4):75. https://doi.org/10.3390/geriatrics8040075
Du F, Yuksel C, Chouinard V-A, et al. (2018) Abnormalities in High-Energy Phosphate Metabolism in First-Episode Bipolar Disorder Measured Using ^{31}P-Magnetic Resonance Spectroscopy. Biol Psychiatry 84 (11):797–802. https://doi.org/10.1016/j.biopsych.2017.03.025
Duarte JMN, and Xin L (2019) Magnetic Resonance Spectroscopy in Schizophrenia: Evidence for Glutamatergic Dysfunction and Impaired Energy Metabolism. Neurochem Res 44 (1):102–116. https://doi.org/10.1007/s11064-018-2521-z
Duvezin-Caubet S, Jagasia R, Wagener J, et al. (2006) Proteolytic Processing of OPA1 Links Mitochondrial Dysfunction to Alterations in Mitochondrial Morphology. J Biol Chem 281 (49):37972–37979. https://doi.org/10.1074/jbc.m606059200
Ebrahimi P, Batlle JC, Ayati A, et al. (2025) Suicide and Self-Harm Events with GLP-1 Receptor Agonists in Adults with Diabetes or Obesity: A Systematic Review and Meta-Analysis. JAMA Psychiatry. https://doi.org/10.1001/jamapsychiatry.2025.0091
Elzinga SE, Guo K, Turfah A, et al. (2025) Metabolic Stress and Age Drive Inflammation and Cognitive Decline in Mice and Humans. Alzheimers Dement 21 (3):e70060. https://doi.org/10.1002/alz.70060
Ene HM, Karry R, Farfara D, and Ben-Shachar D (2023) Mitochondria Play an Essential Role in the Trajectory of Adolescent Neurodevelopment and Behavior in Adulthood: Evidence from a Schizophrenia Rat Model. Mol Psychiatr 28 (3):1170–1181. https://doi.org/10.1038/s41380-022-01865-4
Enez Darcin A, Yalcin Cavus S, Dilbaz N, Kaya H, and Dogan E (2015) Metabolic Syndrome in Drug-Naïve and Drug-Free Patients with Schizophrenia and in Their Siblings. Schizophr Res 166 (1):201-206. https://doi.org/10.1016/j.schres.2015.05.004
Engström Ruud L, Font-Gironès F, Zajdel J, et al. (2024) Activation of GFRAL^{+} Neurons Induces Hypothermia and Glucoregulatory Responses Associated with Nausea and Torpor. Cell Rep 43 (4):113960. https://doi.org/10.1016/j.celrep.2024.113960
Enwright JF, and Lewis DA (2021) Similarities in Cortical Transcriptome Alterations between Schizophrenia and Bipolar Disorder Are Related to the Presence of Psychosis. Schizophr Bull 47 (5):1442–1451. https://doi.org/10.1093/schbul/sbaa195
Expert Panel on Detection E, and Treatment of High Blood Cholesterol in Adults (2001) Executive Summary of the Third Report of the National Cholesterol Education Program (NCEP) Expert Panel on Detection, Evaluation, and Treatment of High Blood Cholesterol in Adults (Adult Treatment Panel III). JAMA 285 (19):2486–2497. https://doi.org/10.1001/jama.285.19.2486

Fakih W, Zeitoun R, AlZaim I, et al. (2022) Early Metabolic Impairment as a Contributor to Neurodegenerative Disease: Mechanisms and Potential Pharmacological Intervention. Obesity 30 (5):982-993. https://doi.org/10.1002/oby.23400

Fernandez-Egea E, Bernardo M, Donner T, et al. (2009) Metabolic Profile of Antipsychotic-Naive Individuals with Non-Affective Psychosis. British Journal of Psychiatry 194 (5):434–438. https://doi.org/10.1192/bjp.bp.108.052605

Fernandez AM, and Torres-Alemán I (2012) The Many Faces of Insulin-Like Peptide Signalling in the Brain. Nat Rev Neurosci 13 (4):225–239. https://doi.org/10.1038/nrn3209

Fernström J, Mellon SH, McGill MA, et al. (2021) Blood-Based Mitochondrial Respiratory Chain Function in Major Depression. Transl Psychiatry 11 (1):593. https://doi.org/10.1038/s41398-021-01723-x

Fizíková I, Dragašek J, and Račay P (2023) Mitochondrial Dysfunction, Altered Mitochondrial Oxygen, and Energy Metabolism Associated with the Pathogenesis of Schizophrenia. Int J Mol Sci 24 (9):7991. https://doi.org/10.3390/ijms24097991

Flores-Dorantes MT, Díaz-López YE, and Gutiérrez-Aguilar R (2020) Environment and Gene Association with Obesity and Their Impact on Neurodegenerative and Neurodevelopmental Diseases. Front Neurosci 14:863. https://doi.org/10.3389/fnins.2020.00863

Flux MC, and Lowry CA (2023) Inflammation as a Mediator of Stress-Related Psychiatric Disorders. In: Zigmond MJ, Chesselet M-F, Wiley CA (eds) Neurobiology of Brain Disorders: Biological Basis of Neurological and Psychiatric Disorders. Elsevier, London, pp 885–911. doi:https://doi.org/10.1016/C2020-0-01841-4

Foley DL, Morley KI, Madden PAF, et al. (2010) Major Depression and the Metabolic Syndrome. Twin Res Hum Genet 13 (4):347–358. https://doi.org/10.1375/twin.13.4.347

Fujimoto T, Takeuch K, Matsumoto T, et al. (2007) Abnormal Glucose Metabolism in the Anterior Cingulate Cortex in Patients with Schizophrenia. Psychiatry Res 154 (1):49–58. https://doi.org/10.1016/j.pscychresns.2006.04.002

Gallego Deike L, Barreiro P, and Reneses B (2023) The New Profile of Psychiatric Disorders in Patients with HIV Infection. AIDS Rev 25 (1):41–53. https://doi.org/10.24875/aidsrev.22000030

Gandal MJ, Haney JR, Parikshak NN, et al. (2019) Shared Molecular Neuropathology across Major Psychiatric Disorders Parallels Polygenic Overlap. Focus 17 (1):66–72. https://doi.org/10.1176/appi.focus.17103

Gangadin SS, Enthoven AD, van Beveren NJM, Laman JD, and Sommer IEC (2024) Immune Dysfunction in Schizophrenia Spectrum Disorders. Annu Rev Clin Psychol 20 (1):229–257. https://doi.org/10.1146/annurev-clinpsy-081122-013201

Garmier M, Carroll AJ, Delannoy E, et al. (2008) Complex I Dysfunction Redirects Cellular and Mitochondrial Metabolism in Arabidopsis. Plant Physiol 148 (3):1324–1341. https://doi.org/10.1104/pp.108.125880

Garza-Lombó C, Pappa A, Panayiotidis MI, and Franco R (2020) Redox Homeostasis, Oxidative Stress and Mitophagy. Mitochondrion 51:105-117. https://doi.org/10.1016/j.mito.2020.01.002

Genovese G, Fromer M, Stahl EA, et al. (2016) Increased Burden of Ultra-Rare Protein-Altering Variants among 4,877 Individuals with Schizophrenia. Nat Neurosci 19 (11):1433–1441. https://doi.org/10.1038/nn.4402

Gerke J, Koenig AM, Conrad D, et al. (2018) Childhood Maltreatment as Risk Factor for Lifetime Depression: The Role of Different Types of Experiences and Sensitive Periods. Ment Health Prev 10:56–65. https://psycnet.apa.org/doi/10.1016/j.mhp.2018.03.002

Gholami M, Balajam NZ, Rakhsha S, et al. (2025) GLP-1 Receptor Agonists Efficacy in Managing Comorbidities Associated with Diabetes Mellitus: A Narrative Review. J Diabetes Metab Disord 24 (1):92. https://doi.org/10.1007/s40200-025-01604-w

Glancy B, Kim Y, Katti P, and Willingham TB (2020) The Functional Impact of Mitochondrial Structure across Subcellular Scales. Front Physiol 11:541040. https://doi.org/10.3389/fphys.2020.541040

Glausier JR, III JFE, and Lewis DA (2020) Diagnosis- and Cell Type-Specific Mitochondrial Functional Pathway Signatures in Schizophrenia and Bipolar Disorder. Am J Psychiatry 177 (12):1140–1150. https://doi.org/10.1176/appi.ajp.2020.19111210
Glausier JR, and Lewis DA (2013) Dendritic Spine Pathology in Schizophrenia. Neuroscience 251:90–107. https://doi.org/10.1016/j.neuroscience.2012.04.044
Gohil VM, Nilsson R, Belcher-Timme CA, et al. (2010) Mitochondrial and Nuclear Genomic Responses to Loss of *LRPPRC* Expression. J Biol Chem 285 (18):13742–13747. https://doi.org/10.1074/jbc.m109.098400
Gola H, Engler A, Morath J, et al. (2014) Reduced Peripheral Expression of the Glucocorticoid Receptor A Isoform in Individuals with Posttraumatic Stress Disorder: A Cumulative Effect of Trauma Burden. PLoS One 9 (1):e86333. https://doi.org/10.1371/journal.pone.0086333
Goldfarb M, De Hert M, Detraux J, et al. (2022) Severe Mental Illness and Cardiovascular Disease: *JACC* State-of-the-Art Review. J Am Coll Cardiol 80 (9):918–933. https://doi.org/10.1016/j.jacc.2022.06.017
Gomez-Casati D, Grisolía MJ, and Busi MV (2016) The Significance of Metabolomics in Human Health. In: Kumar D, Antonarakis S (eds) Medical and Health Genomics. Academic Press, Cambridge, MA, pp 89–100. doi:https://doi.org/10.1016/C2013-0-12872-3
Gonçalves VF, Cappi C, Hagen CM, et al. (2018) A Comprehensive Analysis of Nuclear-Encoded Mitochondrial Genes in Schizophrenia. Biol Psychiatry 83 (9):780–789. https://doi.org/10.1016/j.biopsych.2018.02.1175
Gong Y, Chai Y, Ding J-H, Sun X-L, and Hu G (2011) Chronic Mild Stress Damages Mitochondrial Ultrastructure and Function in Mouse Brain. Neurosci Lett 488 (1):76–80. https://doi.org/10.1016/j.neulet.2010.11.006
Grundy SM, Cleeman JI, Daniels SR, et al. (2005) Diagnosis and Management of the Metabolic Syndrome: An American Heart Association/National Heart, Lung, and Blood Institute Scientific Statement. Circulation 112 (17):2735–2752. https://doi.org/10.1161/circulationaha.105.169404
Grünewald A, Rygiel KA, Hepplewhite PD, et al. (2016) Mitochondrial DNA Depletion in Respiratory Chain-Deficient Parkinson Disease Neurons. Ann Neurol 79 (3):366–378. https://doi.org/10.1002/ana.24571
Gubert C, Stertz L, Pfaffenseller B, et al. (2013) Mitochondrial Activity and Oxidative Stress Markers in Peripheral Blood Mononuclear Cells of Patients with Bipolar Disorder, Schizophrenia, and Healthy Subjects. J Psychiatr Res 47 (10):1396–1402. https://doi.org/10.1016/j.jpsychires.2013.06.018
Gumpp AM, Behnke A, Bach AM, et al. (2021) Mitochondrial Bioenergetics in Leukocytes and Oxidative Stress in Blood Serum of Mild to Moderately Depressed Women. Mitochondrion 58:14–23. https://doi.org/10.1016/j.mito.2020.12.009
Gumpp AM, Behnke A, Ramo-Fernández L, et al. (2023) Investigating Mitochondrial Bioenergetics in Peripheral Blood Mononuclear Cells of Women with Childhood Maltreatment from Post-Parturition Period to One-Year Follow-Up. Psychol Med 53 (9):3793–3804. https://doi.org/10.1017/s0033291722000411
Guo C, Sun L, Chen X, and Zhang D (2013) Oxidative Stress, Mitochondrial Damage and Neurodegenerative Diseases. Neural Regen Res 8 (21):2003–2014. https://doi.org/10.3969/j.issn.1673-5374.2013.21.009
Hackenbrock CR (1981) Energy-Linked Condensed-Orthodox Ultrastructural Transformations in Mitochondria. Chemotherapy 27 (Suppl 2):21–26. https://doi.org/10.1159/000238042
Hall CN, Klein-Flügge MC, Howarth C, and Attwell D (2012) Oxidative Phosphorylation, Not Glycolysis, Powers Presynaptic and Postsynaptic Mechanisms Underlying Brain Information Processing. J Neurosci 32 (26):8940–8951. https://doi.org/10.1523/jneurosci.0026-12.2012

Hall J, Trent S, Thomas KL, O'Donovan MC, and Owen MJ (2015) Genetic Risk for Schizophrenia: Convergence on Synaptic Pathways Involved in Plasticity. Biol Psychiatry 77 (1):52–58. https://doi.org/10.1016/j.biopsych.2014.07.011

Harris JJ, Jolivet R, and Attwell D (2012) Synaptic Energy Use and Supply. Neuron 75 (5):762–777. https://doi.org/10.1016/j.neuron.2012.08.019

Havrankova J, Schmechel D, Roth J, and Brownstein M (1978) Identification of Insulin in Rat Brain. PNAS 75 (11):5737–5741. https://doi.org/10.1073/pnas.75.11.5737

Hazlett EA, Vaccaro DH, Haznedar MM, and Goldstein KE (2019) $^{F\text{-}18}$fluorodeoxyglucose Positron Emission Tomography Studies of the Schizophrenia Spectrum: The Legacy of Monte S. Buchsbaum, M.D. Psychiatry Res 271:535–540. https://doi.org/10.1016/j.psychres.2018.12.030

Henkel ND, Wu X, O'Donovan SM, et al. (2022) Schizophrenia: A Disorder of Broken Brain Bioenergetics. Mol Psychiatr 27 (5):2393–2404. https://doi.org/10.1038/s41380-022-01494-x

Herberth M, Koethe D, Cheng TMK, et al. (2011) Impaired Glycolytic Response in Peripheral Blood Mononuclear Cells of First-Onset Antipsychotic-Naive Schizophrenia Patients. Mol Psychiatr 16 (8):848–859. https://doi.org/10.1038/mp.2010.71

Hevner RF, and Wong-Riley MT (1991) Neuronal Expression of Nuclear and Mitochondrial Genes for Cytochrome Oxidase (CO) Subunits Analyzed by in Situ Hybridization: Comparison with CO Activity and Protein. J Neurosci 11 (7):1942–1958. https://doi.org/10.1523/jneurosci.11-07-01942.1991

Hoang U, Goldacre MJ, and Stewart R (2013) Avoidable Mortality in People with Schizophrenia or Bipolar Disorder in England. Acta Psychiatr Scand 127 (3):195–201. https://doi.org/10.1111/acps.12045

Howes OD, Bukala BR, and Beck K (2024) Schizophrenia: From Neurochemistry to Circuits, Symptoms and Treatments. Nat Rev Neurol 20 (1):22–35. https://doi.org/10.1038/s41582-023-00904-0

Huang Q, Monzel AS, Rausser S, et al. (2024) The Energetic Stress Marker GDF15 Is Induced by Acute Psychosocial Stress. https://doi.org/10.1101/2024.04.19.590241

Jay JA, Kumar V, Garrels E, et al. (2023) Management of Neuropsychiatric Disturbances in Huntington's Disease: A Literature Review and Case Report. Prim Care Companion CNS Disord 25 (1). https://doi.org/10.4088/PCC.22cr03265

Jenkins LM, Wang L, Rosen H, and Weintraub S (2022) A Transdiagnostic Review of Neuroimaging Studies of Apathy and Disinhibition in Dementia. Brain 145 (6):1886–1905. https://doi.org/10.1093/brain/awac133

Jiang J, Peng C, Sun L, et al. (2019) Leukocyte Proteomic Profiling in First-Episode Schizophrenia Patients: Does Oxidative Stress Play Central Roles in the Pathophysiology Network of Schizophrenia? Antioxid Redox Signal 31 (8):579–588. https://doi.org/10.1089/ars.2019.7805

Justs KA, Lu Z, Chouhan AK, et al. (2022) Presynaptic Mitochondrial Volume and Packing Density Scale with Presynaptic Power Demand. J Neurosci 42 (6):954–967. https://doi.org/10.1523/JNEUROSCI.1236-21.2021

Kann O (2016) The Interneuron Energy Hypothesis: Implications for Brain Disease. Neurobiol Dis 90:75–85. https://doi.org/10.1016/j.nbd.2015.08.005

Kann O, Papageorgiou IE, and Draguhn A (2014) Highly Energized Inhibitory Interneurons Are a Central Element for Information Processing in Cortical Networks. J Cereb Blood Flow Metab 34 (8):1270–1282. https://doi.org/10.1038/jcbfm.2014.104

Karabatsiakis A, Böck C, Salinas-Manrique J, et al. (2014) Mitochondrial Respiration in Peripheral Blood Mononuclear Cells Correlates with Depressive Subsymptoms and Severity of Major Depression. Transl Psychiatry 4 (6):e397. https://doi.org/10.1038/tp.2014.44

Karry R, Klein E, and Ben Shachar D (2004) Mitochondrial Complex I Subunits Expression Is Altered in Schizophrenia: A Postmortem Study. Biol Psychiatry 55 (7):676–684. https://doi.org/10.1016/j.biopsych.2003.12.012

Kasanin J (1926) The Blood Sugar Curve in Mental Disease: II. The Schizophrenic (Dementia Praecox) Groups. Arch Neurol Psychiatry 16 (4):414–419. https://doi.org/10.1001/ARCHNEURPSYC.1926.02200280022002

Khadimallah I, Jenni R, Cabungcal J-H, et al. (2022) Mitochondrial, Exosomal miR137-COX6a2 and Gamma Synchrony as Biomarkers of Parvalbumin Interneurons, Psychopathology, and Neurocognition in Schizophrenia. Mol Psychiatr 27 (2):1192–1204. https://doi.org/10.1038/s41380-021-01313-9

Khatri N, and Man H-Y (2013) Synaptic Activity and Bioenergy Homeostasis: Implications in Brain Trauma and Neurodegenerative Diseases. Front Neurol 4:199. https://doi.org/10.3389/fneur.2013.00199

Kim S-Y, Cohen BM, Chen X, et al. (2017) Redox Dysregulation in Schizophrenia Revealed by *in Vivo* NAD+/NADH Measurement. Schizophr Bull 43 (1):197–204. https://doi.org/10.1093/schbul/sbw129

Kirov G, Pocklington AJ, Holmans P, et al. (2012) *De Novo* CNV Analysis Implicates Specific Abnormalities of Postsynaptic Signalling Complexes in the Pathogenesis of Schizophrenia. Mol Psychiatr 17 (2):142–153. https://doi.org/10.1038/mp.2011.154

Klinedinst NJ, and Regenold WT (2015) A Mitochondrial Bioenergetic Basis of Depression. J Bioenerg Biomembr 47 (1-2):155–171. https://doi.org/10.1007/s10863-014-9584-6

Kolassa I-T, Ertl V, Eckart C, et al. (2010a) Association Study of Trauma Load and *SLC6A4* Promoter Polymorphism in Posttraumatic Stress Disorder: Evidence from Survivors of the Rwandan Genocide. J Clin Psychiatry 71 (5):1557. https://doi.org/10.4088/jcp.08m04787blu

Kolassa I-T, Ertl V, Eckart C, et al. (2010b) Spontaneous Remission from PTSD Depends on the Number of Traumatic Event Types Experienced. Psychol Trauma 2 (3):169. https://psycnet.apa.org/doi/10.1037/a0019362

Kooy FH (1919) Hyperglyaemia in Mental Disorders. Brain 42 (3):214–290. https://doi.org/10.1093/brain/42.3.214

Kotzaeroglou A, and Tsamesidis I (2022) The Role of Equilibrium between Free Radicals and Antioxidants in Depression and Bipolar Disorder. Medicines 9 (11):57. https://doi.org/10.3390/medicines9110057

Kraguljac NV, Reid M, White D, et al. (2012) Neurometabolites in Schizophrenia and Bipolar Disorder: A Systematic Review and Meta-Analysis. Psychiatry Res 203 (2–3):111–125. https://doi.org/10.1016/j.pscychresns.2012.02.003

Kuang H, Duong A, Jeong H, Zachos K, and Andreazza AC (2018) Lactate in Bipolar Disorder: A Systematic Review and Meta-Analysis. Psychiatry Clin Neurosci 72 (8):546–555. https://doi.org/10.1111/pcn.12671

Kula B, Antal B, Weistuch C, et al. (2024) D-β-Hydroxybutyrate Stabilizes the Hippocampal CA3-CA1 Circuit during Acute Insulin Resistance. PNAS Nexus 3 (5):196. https://doi.org/10.1093/pnasnexus/pgae196

Kumar P, Efstathopoulos P, Millischer V, et al. (2018) Mitochondrial DNA Copy Number Is Associated with Psychosis Severity and Anti-Psychotic Treatment. Sci Rep 8. https://doi.org/10.1038/s41598-018-31122-0

Lampe L, Zhang R, Beyer F, et al. (2019) Visceral Obesity Relates to Deep White Matter Hyperintensities via Inflammation. Ann Neurol 85 (2):194–203. https://doi.org/10.1002/ana.25396

Lawrence D, Hancock KJ, and Kisely S (2013) The Gap in Life Expectancy from Preventable Physical Illness in Psychiatric Patients in Western Australia: Retrospective Analysis of Population Based Registers. BMJ 346 (7909):f2539. https://doi.org/10.1136/bmj.f2539

Lax NZ, Grady J, Laude A, et al. (2016) Extensive Respiratory Chain Defects in Inhibitory Interneurones in Patients with Mitochondrial Disease. Neuropathol Appl Neurobiol 42 (2):180–193. https://doi.org/10.1111/nan.12238

Leboyer M, Godin O, Llorca PM, et al. (2022) Key Findings on Bipolar Disorders from the Longitudinal Fondamental Advanced Center of Expertise-Bipolar Disorder (FACE-BD) Cohort. J Affect Disord 307:149–156. https://doi.org/10.1016/j.jad.2022.03.053

Lee J, Giordano S, and Zhang J (2011) Autophagy, Mitochondria and Oxidative Stress: Cross-Talk and Redox Signalling. Biochem J 441 (2):523-540. https://doi.org/10.1042/BJ20111451

Lee SY, and Chung W-S (2024) Astrocytic Crosstalk with Brain and Immune Cells in Healthy and Diseased Conditions. Curr Opin Neurobiol 84:102840. https://doi.org/10.1016/j.conb.2024.102840

Lemche E, Killick R, Mitchell J, et al. (2024) Molecular Mechanisms Linking Type 2 Diabetes Mellitus and Late-Onset Alzheimer's Disease: A Systematic Review and Qualitative Meta-Analysis. Neurobiol Dis 196:106485. https://doi.org/10.1016/j.nbd.2024.106485

Leutner M, Dervic E, Bellach L, et al. (2023) Obesity as Pleiotropic Risk State for Metabolic and Mental Health Throughout Life. Transl Psychiatry 13 (1):175. https://doi.org/10.1038/s41398-023-02447-w

Li S, and Sheng Z-H (2022) Energy Matters: Presynaptic Metabolism and the Maintenance of Synaptic Transmission. Nat Rev Neurol 23 (1):4–22. https://doi.org/10.1038/s41583-021-00535-8

Liang HL, Ongwijitwat S, and Wong-Riley MTT (2006) Bigenomic Functional Regulation of All 13 Cytochrome C Oxidase Subunit Transcripts in Rat Neurons *in vitro* and *in Vivo*. Neuroscience 140 (1):177–190. https://doi.org/10.1016/j.neuroscience.2006.01.056

Ling E, Nemesh J, Goldman M, et al. (2024) A Concerted Neuron-Astrocyte Program Declines in Ageing and Schizophrenia. Nature 627 (8004):604–611. https://doi.org/10.1038/s41586-024-07109-5

Liu M-L, Zhang X-T, Du X-Y, et al. (2015) Severe Disturbance of Glucose Metabolism in Peripheral Blood Mononuclear Cells of Schizophrenia Patients: A Targeted Metabolomic Study. J Transl Med 13 (1):226. https://doi.org/10.1186/s12967-015-0540-y

Liu M-L, Zheng P, Liu Z, et al. (2014) Gc-MS Based Metabolomics Identification of Possible Novel Biomarkers for Schizophrenia in Peripheral Blood Mononuclear Cells. Mol Biosyst 10 (9):2398–2406. https://doi.org/10.1039/c4mb00157e

Lockhart SM, Saudek V, and O'Rahilly S (2020) GDF15: A Hormone Conveying Somatic Distress to the Brain. Endocr Rev 41 (4):bnaa007. https://doi.org/10.1210/endrev/bnaa007

Mansilla N, Racca S, Gras DE, Gonzalez DH, and Welchen E (2018) The Complexity of Mitochondrial Complex IV: An Update of Cytochrome *C* Oxidase Biogenesis in Plants. Int J Mol Sci 19 (3):662. https://doi.org/10.3390/ijms19030662

Martins-de-Souza D, Harris LW, Guest PC, and Bahn S (2011) The Role of Energy Metabolism Dysfunction and Oxidative Stress in Schizophrenia Revealed by Proteomics. Antioxid Redox Signal 15 (7):2067–2079. https://doi.org/10.1089/ars.2010.3459

Mendelsohn R, Garcia GC, Bartol TM, et al. (2022) Morphological Principles of Neuronal Mitochondria. J Comp Neurol 530 (6):886–902. https://doi.org/10.1002/cne.25254

Mendez MF (2021) Degenerative Dementias: Alterations of Emotions and Mood Disorders. In: Heilman KM, Nadeau SE (eds) Handbook of Clinical Neurology, vol 183. Handbook of Clinical Neurology. Elsevier, Amsterdam, pp 261–281. doi:https://doi.org/10.1016/b978-0-12-822290-4.00012-8

Milaneschi Y, Lamers F, Berk M, and Penninx BWJH (2020) Depression Heterogeneity and Its Biological Underpinnings: Toward Immunometabolic Depression. Biol Psychiatry 88 (5):369–380. https://doi.org/10.1016/j.biopsych.2020.01.014

Milaneschi Y, Lamers F, and Penninx BWJH (2021) Dissecting Depression Biological and Clinical Heterogeneity: The Importance of Symptom Assessment Resolution. JAMA Psychiatry 78 (3):341. https://doi.org/10.1001/jamapsychiatry.2020.4373

Minzenberg MJ, Laird AR, Thelen S, Carter CS, and Glahn DC (2009) Meta-Analysis of 41 Functional Neuroimaging Studies of Executive Function in Schizophrenia. Arch Gen Psychiatry 66 (8):811–822. https://doi.org/10.1001/archgenpsychiatry.2009.91

Mitelman SA, Bralet M-C, Mehmet Haznedar M, et al. (2018) Positron Emission Tomography Assessment of Cerebral Glucose Metabolic Rates in Autism Spectrum Disorder and Schizophrenia. Brain Imaging Behav 12 (2):532–546. https://doi.org/10.1007/s11682-017-9721-z

Montanari S, Jansen R, Schranner D, et al. (2025) Acylcarnitines Metabolism in Depression: Association with Diagnostic Status, Depression Severity and Symptom Profile in the Nesda Cohort. Transl Psychiatry 15 (1):65. https://doi.org/10.1038/s41398-025-03274-x

Morán M, Rivera H, Sánchez-Aragó M, et al. (2010) Mitochondrial Bioenergetics and Dynamics Interplay in Complex I-Deficient Fibroblasts. Biochim Biophys Acta 1802 (5):443–453. https://doi.org/10.1016/j.bbadis.2010.02.001

Morath J, Moreno-Villanueva M, Hamuni G, et al. (2014) Effects of Psychotherapy on DNA Strand Break Accumulation Originating from Traumatic Stress. Psychother Psychosom 83 (5):289–297. https://doi.org/10.1159/000362739

Neupane SP, Daray FM, Ballard ED, et al. (2023) Immune-Related Biomarkers and Suicidal Behaviors: A Meta-Analysis. Eur Neuropsychopharmacol 75:15–30. https://doi.org/10.1016/j.euroneuro.2023.05.009

Nitzan K, Benhamron S, Valitsky M, et al. (2019) Mitochondrial Transfer Ameliorates Cognitive Deficits, Neuronal Loss, and Gliosis in Alzheimer's Disease Mice. J Alzheimers Dis 72 (2):587–604. https://doi.org/10.3233/JAD-190853

Nold V, Sweatman C, Karabatsiakis A, et al. (2019) Activation of the Kynurenine Pathway and Mitochondrial Respiration to Face Allostatic Load in a Double-Hit Model of Stress. Psychoneuroendocrinology 107:148–159. https://doi.org/10.1016/j.psyneuen.2019.04.006

Nordentoft M, Wahlbeck K, Hällgren J, et al. (2013) Excess Mortality, Causes of Death and Life Expectancy in 270,770 Patients with Recent Onset of Mental Disorders in Denmark, Finland and Sweden. PLoS One 8 (1):e55176. https://doi.org/10.1371/journal.pone.0055176

O'Brien JT, Jalilvand SP, Suji NA, et al. (2024) Elevations in the Mitochondrial Matrix Protein Cyclophilin D Correlate with Reduced Parvalbumin Expression in the Prefrontal Cortex of Patients with Schizophrenia. Schizophr Bull 50 (5):1197–1207. https://doi.org/10.1093/schbul/sbae016

Oda E (2012) Metabolic Syndrome: Its History, Mechanisms, and Limitations. Acta Diabetol 49 (2):89–95. https://doi.org/10.1007/s00592-011-0309-6

Olsen L, Hansen T, Jakobsen KD, et al. (2008) The Estrogen Hypothesis of Schizophrenia Implicates Glucose Metabolism: Association Study in Three Independent Samples. BMC Med Genet 9 (1):39. https://doi.org/10.1186/1471-2350-9-39

Onozato M, Umino M, Shoji A, et al. (2020) Serum D- and L-Lactate, Pyruvate and Glucose Levels in Individuals with at-Risk Mental State and Correlations with Clinical Symptoms. Early Interv Psychiatry 14 (4):410–417. https://doi.org/10.1111/eip.12866

Owen MJ, Legge SE, Rees E, Walters JTR, and O'Donovan MC (2023) Genomic Findings in Schizophrenia and Their Implications. Mol Psychiatr 28 (9):3638–3647. https://doi.org/10.1038/s41380-023-02293-8

Pan A, Keum N, Okereke OI, et al. (2012) Bidirectional Association between Depression and Metabolic Syndrome: A Systematic Review and Meta-Analysis of Epidemiological Studies. Diabetes Care 35 (5):1171–1180. https://doi.org/10.2337/dc11-2055

Park H-J, Lee JD, Chun JW, et al. (2006) Cortical Surface-Based Analysis of 18F-FDG PET: Measured Metabolic Abnormalities in Schizophrenia Are Affected by Cortical Structural Abnormalities. Neuroimage 31 (4):1434–1444. https://doi.org/10.1016/j.neuroimage.2006.02.001

Pellerin L, and Magistretti PJ (1994) Glutamate Uptake into Astrocytes Stimulates Aerobic Glycolysis: A Mechanism Coupling Neuronal Activity to Glucose Utilization. PNAS 91 (22):10625–10629. https://doi.org/10.1073/pnas.91.22.10625

——— (1996) Excitatory Amino Acids Stimulate Aerobic Glycolysis in Astrocytes via an Activation of the Na^+/K^+ ATPase. Dev Neurosci 18 (5–6):336–342. https://doi.org/10.1159/000111426

Peng K, Yang L, Wang J, et al. (2017) The Interaction of Mitochondrial Biogenesis and Fission/Fusion Mediated by PGC-1α Regulates Rotenone-Induced Dopaminergic Neurotoxicity. Mol Neurobiol 54 (5):3783–3797. https://doi.org/10.1007/s12035-016-9944-9

Penninx BWJH, Milaneschi Y, Lamers F, and Vogelzangs N (2013) Understanding the Somatic Consequences of Depression: Biological Mechanisms and the Role of Depression Symptom Profile. BMC Med 11 (1):129. https://doi.org/10.1186/1741-7015-11-129

Perry BI, McIntosh G, Weich S, Singh S, and Rees K (2016) The Association between First-Episode Psychosis and Abnormal Glycaemic Control: Systematic Review and Meta-Analysis. Lancet Psychiatry 3 (11):1049–1058. https://doi.org/10.1016/s2215-0366(16)30262-0

Pillinger T, Beck K, Gobjila C, et al. (2017) Impaired Glucose Homeostasis in First-Episode Schizophrenia: A Systematic Review and Meta-Analysis. JAMA Psychiatry 74 (3):261–269. https://doi.org/10.1001/jamapsychiatry.2016.3803

Pontzer H (2018) Energy Constraint as a Novel Mechanism Linking Exercise and Health. Physiology 33 (6):384–393. https://doi.org/10.1152/physiol.00027.2018

Quintana A, Kruse SE, Kapur RP, Sanz E, and Palmiter RD (2010) Complex I Deficiency Due to Loss of Ndufs4 in the Brain Results in Progressive Encephalopathy Resembling Leigh Syndrome. PNAS 107 (24):10996–11001. https://doi.org/10.1073/pnas.1006214107

Raji CA, Meysami S, Hashemi S, et al. (2024) Visceral and Subcutaneous Abdominal Fat Predict Brain Volume Loss at Midlife in 10,001 Individuals. Aging Dis 15 (4):1831–1842. https://doi.org/d10.14336/AD.2023.0820

Raphael T, and Parsons JP (1921) Blood Sugar Studies in Dementia Praecox and Manic-Depressive Insanity. Arch Neurol Psychiatry 5 (6):687–709. https://doi.org/10.1001/arch-neurpsyc.1921.02180300052005

Reaven GM (2004) The Metabolic Syndrome or the Insulin Resistance Syndrome? Different Names, Different Concepts, and Different Goals. Endocrinol Metab Clin North Am 33 (2):283–303. https://doi.org/10.1016/j.ecl.2004.03.002

Reaven GM (2006) The Metabolic Syndrome: Is This Diagnosis Necessary? Am J Clin Nutr 83 (6):1237–1247. https://doi.org/10.1093/ajcn/83.6.1237

Redmann M, Darley-Usmar V, and Zhang J (2016) The Role of Autophagy, Mitophagy and Lysosomal Functions in Modulating Bioenergetics and Survival in the Context of Redox and Proteotoxic Damage: Implications for Neurodegenerative Diseases. Aging Dis 7 (2):150. https://doi.org/10.14336/AD.2015.0820

Reinecke F, Smeitink JAM, and van der Westhuizen FH (2009) OXPHOS Gene Expression and Control in Mitochondrial Disorders. Biochim Biophys Acta 1792 (12):1113–1121. https://doi.org/10.1016/j.bbadis.2009.04.003

Renshaw PF, Parow AM, Hirashima F, et al. (2001) Multinuclear Magnetic Resonance Spectroscopy Studies of Brain Purines in Major Depression. Am J Psychiatry 158 (12):2048–2055. https://doi.org/10.1176/appi.ajp.158.12.2048

Réus GZ, Manosso LM, Quevedo J, and Carvalho AF (2023) Major Depressive Disorder as a Neuro-Immune Disorder: Origin, Mechanisms, and Therapeutic Opportunities. Neurosci Biobehav Rev 155:105425. https://doi.org/10.1016/j.neubiorev.2023.105425

Richter-Dennerlein R, Oeljeklaus S, Lorenzi I, et al. (2016) Mitochondrial Protein Synthesis Adapts to Influx of Nuclear-Encoded Protein. Cell 167 (2):471–483.e410. https://doi.org/10.1016/j.cell.2016.09.003

Roberts RC (2017) Postmortem Studies on Mitochondria in Schizophrenia. Schizophr Res 187:17–25. https://doi.org/10.1016/j.schres.2017.01.056

Roberts RC, Barksdale KA, Roche JK, and Lahti AC (2015) Decreased Synaptic and Mitochondrial Density in the Postmortem Anterior Cingulate Cortex in Schizophrenia. Schizophr Res 168 (1–2):543–553. https://doi.org/10.1016/j.schres.2015.07.016

Robicsek O, Ene HM, Karry R, et al. (2018) Isolated Mitochondria Transfer Improves Neuronal Differentiation of Schizophrenia-Derived Induced Pluripotent Stem Cells and Rescues Deficits in a Rat Model of the Disorder. Schizophr Bull 44 (2):432–442. https://doi.org/10.1093/schbul/sbx077

Rosella L, Hurst M, Buajitti E, et al. (2025) A Population-Based Cohort Study of Mitochondrial Disease and Mental Health Conditions in Ontario, Canada. Orphanet J Rare Dis 20:177. https://doi.org/10.1186/s13023-025-03688-2

Rossi MJ, and Pekkurnaz G (2019) Powerhouse of the Mind: Mitochondrial Plasticity at the Synapse. Curr Opin Neurobiol 57:149–155. https://doi.org/10.1016/j.conb.2019.02.001

Rothman DL, De Feyter HM, Maciejewski PK, and Behar KL (2012) Is There *in Vivo* Evidence for Amino Acid Shuttles Carrying Ammonia from Neurons to Astrocytes? Neurochem Res 37 (11):2597–2612. https://doi.org/10.1007/s11064-012-0898-7

Rowland LM, Pradhan S, Korenic S, et al. (2016) Elevated Brain Lactate in Schizophrenia: A 7T Magnetic Resonance Spectroscopy Study. Transl Psychiatry 6 (11):e967. https://doi.org/10.1038/tp.2016.239

Ryan MCM, Collins P, and Thakore JH (2003) Impaired Fasting Glucose Tolerance in First-Episode, Drug-Naive Patients with Schizophrenia. Am J Psychiatry 160 (2):284–289. https://doi.org/10.1176/appi.ajp.160.2.284

Sahay S, Henkel ND, Vargas CF-A, McCullumsmith RE, and O'Donovan SM (2023) Activity of Protein Kinase a in the Frontal Cortex in Schizophrenia. Brain Sci 14 (1):13. https://doi.org/10.3390/brainsci14010013

Scaini G, Rezin GT, Carvalho AF, et al. (2016) Mitochondrial Dysfunction in Bipolar Disorder: Evidence, Pathophysiology and Translational Implications. Neurosci Biobehav Rev 68:694–713. https://doi.org/10.1016/j.neubiorev.2016.06.040

Scalettar BA, Abney JR, and Hackenbrock CR (1991) Dynamics, Structure, and Function Are Coupled in the Mitochondrial Matrix. PNAS 88 (18):8057–8061. https://doi.org/10.1073/pnas.88.18.8057

Schizophrenia Working Group of the Psychiatric Genomics Consortium (2014) Biological Insights from 108 Schizophrenia-Associated Genetic Loci. Nature 511 (7510):421–427. https://doi.org/10.1038/nature13595

Schneider F, Althaus A, Backes V, and Dodel R (2008) Psychiatric Symptoms in Parkinson's Disease. Eur Arch Psychiatry Clin Neurosci 258 Suppl 5 (S5):55–59. https://doi.org/10.1007/s00406-008-5012-4

Schoonover KE, Dienel SJ, and Lewis DA (2020) Prefrontal Cortical Alterations of Glutamate and GABA Neurotransmission in Schizophrenia: Insights for Rational Biomarker Development. Biomark Neuropsychiatry 3:100015. https://doi.org/10.1016/j.bionps.2020.100015

Sercel AJ, Sturm G, Gallagher D, et al. (2024) Hypermetabolism and Energetic Constraints in Mitochondrial Disorders. Nat Metab 6 (2):192–195. https://doi.org/10.1038/s42255-023-00968-8

Shan D, Mount D, Moore S, et al. (2014) Abnormal Partitioning of Hexokinase 1 Suggests Disruption of a Glutamate Transport Protein Complex in Schizophrenia. Schizophr Res 154 (1–3):1–13. https://doi.org/10.1016/j.schres.2014.01.028

Shaulson ED, Cohen AA, and Picard M (2024) The Brain–Body Energy Conservation Model of Aging. Nat Aging 4 (10):1354–1371. https://doi.org/10.1038/s43587-024-00716-x

Silva Ramos E, Larsson N-G, and Mourier A (2016) Bioenergetic Roles of Mitochondrial Fusion. Biochim Biophys Acta 1857 (8):1277–1283. https://doi.org/10.1016/j.bbabio.2016.04.002

Skilton MR, Moulin P, Terra J-L, and Bonnet F (2007) Associations between Anxiety, Depression, and the Metabolic Syndrome. Biol Psychiatry 62 (11):1251–1257. https://doi.org/10.1016/j.biopsych.2007.01.012

Somerville SM, Lahti AC, Conley RR, and Roberts RC (2011) Mitochondria in the Striatum of Subjects with Schizophrenia: Relationship to Treatment Response. Synapse 65 (3):215–224. https://doi.org/10.1002/syn.20838

Sørensen NV, and Benros ME (2023) The Immune System and Depression: From Epidemiological to Clinical Evidence. In: Savitz J, Yolken RH (eds) Microorganisms and Mental Health, vol 61. Current Topics in Behavioral Neurosciences. Springer, Cham, pp 15–34. doi:https://doi.org/10.1007/7854_2022_369

Stein A, Zhu C, Du F, and Öngür D (2023) Magnetic Resonance Spectroscopy Studies of Brain Energy Metabolism in Schizophrenia: Progression from Prodrome to Chronic Psychosis. Curr Psychiatry Rep 25 (11):659–669. https://doi.org/10.1007/s11920-023-01457-1

Steullet P, Cabungcal JH, Monin A, et al. (2016) Redox Dysregulation, Neuroinflammation, and NMDA Receptor Hypofunction: A "Central Hub" in Schizophrenia Pathophysiology? Schizophr Res 176 (1):41–51. https://doi.org/10.1016/j.schres.2014.06.021

Stone WS, Faraone SV, Su J, et al. (2004) Evidence for Linkage between Regulatory Enzymes in Glycolysis and Schizophrenia in a Multiplex Sample. Am J Med Genet B Neuropsychiatr Genet 127B (1):5–10. https://doi.org/10.1002/ajmg.b.20132

Stroud DA, Surgenor EE, Formosa LE, et al. (2016) Accessory Subunits Are Integral for Assembly and Function of Human Mitochondrial Complex I. Nature 538 (7623):123–126. https://doi.org/10.1038/nature19754

Sukumar N, Sabesan P, Anazodo U, and Palaniyappan L (2020) Neurovascular Uncoupling in Schizophrenia: A Bimodal Meta-Analysis of Brain Perfusion and Glucose Metabolism. Front Psychiatry 11:754. https://psycnet.apa.org/doi/10.3389/fpsyt.2020.00754

Sullivan CR, Koene RH, Hasselfeld K, et al. (2019a) Neuron-Specific Deficits of Bioenergetic Processes in the Dorsolateral Prefrontal Cortex in Schizophrenia. Mol Psychiatr 24 (9):1319–1328. https://doi.org/10.1038/s41380-018-0035-3

Sullivan CR, Mielnik CA, Funk A, et al. (2019b) Measurement of Lactate Levels in Postmortem Brain, iPSCs, and Animal Models of Schizophrenia. Sci Rep 9 (1):5087. https://doi.org/10.1038/s41598-019-41572-9

Sullivan CR, Mielnik CA, O'Donovan SM, et al. (2019c) Connectivity Analyses of Bioenergetic Changes in Schizophrenia: Identification of Novel Treatments. Mol Neurobiol 56 (6):4492–4517. https://doi.org/10.1007/s12035-018-1390-4

Sullivan CR, O'Donovan SM, McCullumsmith RE, and Ramsey A (2018) Defects in Bioenergetic Coupling in Schizophrenia. Biol Psychiatry 83 (9):739–750. https://doi.org/10.1016/j.biopsych.2017.10.014

Swerdlow NS, and Wilkins HM (2020) Mitophagy and the Brain. Int J Mol Sci 21 (24):9661. https://doi.org/10.3390/ijms21249661

Szatkiewicz JP, O'Dushlaine C, Chen G, et al. (2014) Copy Number Variation in Schizophrenia in Sweden. Mol Psychiatr 19 (7):762–773. https://doi.org/10.1038/mp.2014.40

Tabrizi SJ, Schobel S, Gantman EC, et al. (2022) A Biological Classification of Huntington's Disease: The Integrated Staging System. Lancet Neurol 21 (7):632–644. https://doi.org/10.1016/S1474-4422(22)00120-X

Tanner LB, Goglia AG, Wei MH, et al. (2018) Four Key Steps Control Glycolytic Flux in Mammalian Cells. Cell Syst 7 (1):49–62.e48. https://doi.org/10.1016/j.cels.2018.06.003

Thakore JH (2005) Metabolic Syndrome and Schizophrenia. British Journal of Psychiatry 186:455–456. https://doi.org/10.1192/bjp.186.6.455

Thakore JH, Mann JN, Vlahos I, Martin A, and Reznek R (2002) Increased Visceral Fat Distribution in Drug-Naive and Drug-Free Patients with Schizophrenia. Int J Obes Relat Metab Disord 26 (1):137–141. https://doi.org/10.1038/sj.ijo.0801840

Trubetskoy V, Pardinas AF, Qi T, et al. (2022) Mapping Genomic Loci Implicates Genes and Synaptic Biology in Schizophrenia. Nature 604 (7906):502–508. https://doi.org/10.1038/s41586-022-04434-5

Uittenbogaard M, Sen K, Whitehead M, et al. (2021) Genetic and Mitochondrial Metabolic Analyses of an Atypical Form of Leigh Syndrome. Front Cell Dev Biol 9:767407. https://doi.org/10.3389/fcell.2021.767407

Urlacher SS, Snodgrass JJ, Dugas LR, et al. (2019) Constraint and Trade-Offs Regulate Energy Expenditure during Childhood. Sci Adv 5 (12):eaax1065. https://doi.org/10.1126/sciadv.aax1065
Valiente-Pallejà A, Torrell H, Alonso Y, et al. (2020) Increased Blood Lactate Levels during Exercise and Mitochondrial DNA Alterations Converge on Mitochondrial Dysfunction in Schizophrenia. Schizophr Res 220:61–68. https://doi.org/10.1016/j.schres.2020.03.070
van Waveren C, and Moraes CT (2008) Transcriptional Co-Expression and Co-Regulation of Genes Coding for Components of the Oxidative Phosphorylation System. BMC Genomics 9:18. https://doi.org/10.1186/1471-2164-9-18
Vincent AE, Ng YS, White K, et al. (2016) The Spectrum of Mitochondrial Ultrastructural Defects in Mitochondrial Myopathy. Sci Rep 6:30610. https://doi.org/10.1038/srep30610
Viscomi C, Bottani E, Civiletto G, et al. (2011) *In Vivo* Correction of COX Deficiency by Activation of the AMPK/PGC-1α Axis. Cell Metab 14 (1):80–90. https://doi.org/10.1016/j.cmet.2011.04.011
Vogelzangs N, Beekman ATF, Boelhouwer IG, et al. (2011) Metabolic Depression: A Chronic Depressive Subtype? Findings from the Inchianti Study of Older Persons. J Clin Psychiatry 72 (5):598–604. https://doi.org/10.4088/JCP.10m06559
Volz HP, Rzanny R, Riehemann S, et al. (1998) ^{31}P Magnetic Resonance Spectroscopy in the Frontal Lobe of Major Depressed Patients. Eur Arch Psychiatry Clin Neurosci 248 (6):289–295. https://doi.org/10.1007/s004060050052
Vreijling SR, van Haeringen M, Milaneschi Y, et al. (2023) Sociodemographic, Lifestyle and Clinical Characteristics of Energy-Related Depression Symptoms: A Pooled Analysis of 13,965 Depressed Cases in 8 Dutch Cohorts. J Affect Disord 323:1–9. https://doi.org/10.1016/j.jad.2022.11.005
Walker ER, McGee RE, and Druss BG (2015) Mortality in Mental Disorders and Global Disease Burden Implications: A Systematic Review and Meta-Analysis: A Systematic Review and Meta-Analysis. JAMA Psychiatry 72 (4):334–341. https://doi.org/10.1001/jamapsychiatry.2014.2502
Wang JF, Shao L, Sun X, and Young LT (2009) Increased Oxidative Stress in the Anterior Cingulate Cortex of Subjects with Bipolar Disorder and Schizophrenia. Bipolar Disord 11 (5):523–529. https://doi.org/10.1111/j.1399-5618.2009.00717.x
Wang X, Memon AA, Palmér K, et al. (2022) The Association of Mitochondrial DNA Copy Number with Incident Mental Disorders in Women: A Population-Based Follow-up Study. J Affect Disord 308:111–115. https://doi.org/10.1016/j.jad.2022.04.064
Wang Y, Ni J, Gao C, et al. (2019) Mitochondrial Transplantation Attenuates Lipopolysaccharide- Induced Depression-Like Behaviors. Prog Neuropsychopharmacol Biol Psychiatry 93:240–249. https://doi.org/10.1016/j.pnpbp.2019.04.010
Weidling IW, and Swerdlow RH (2020) Mitochondria in Alzheimer's Disease and Their Potential Role in Alzheimer's Proteostasis. Exp Neurol 330:113321. https://doi.org/10.1016/j.expneurol.2020.113321
Weintraub D, and Mamikonyan E (2019) The Neuropsychiatry of Parkinson Disease: A Perfect Storm. Am J Geriatr Psychiatry 27 (9):998–1018. https://doi.org/10.1016/j.jagp.2019.03.002
Weske G, Berger M, Riemann D, and Voderholzer U (2001) Neurobiological Findings in a Patient with 48-Hour Rapid Cycling Bipolar Affective Disorder. A Case Report. Nervenarzt 72 (7):549–554. https://doi.org/10.1007/s001150170081
Whittaker RG, Turnbull DM, Whittington MA, and Cunningham MO (2011) Impaired Mitochondrial Function Abolishes Gamma Oscillations in the Hippocampus through an Effect on Fast-Spiking Interneurons. Brain 134 (7):e180. https://doi.org/10.1093/brain/awr018
Williams RS, Garcia-Moll M, Mellor J, Salmons S, and Harlan W (1987) Adaptation of Skeletal Muscle to Increased Contractile Activity. Expression Nuclear Genes Encoding Mitochondrial Proteins. J Biol Chem 262 (6):2764–2767. https://doi.org/10.1016/s0021-9258(18)61572-8

Wingo TS, Liu Y, Gerasimov ES, et al. (2022) Shared Mechanisms across the Major Psychiatric and Neurodegenerative Diseases. Nat Commun 13 (1):4314. https://doi.org/10.1038/s41467-022-31873-5

Wong-Riley MTT (2012) Bigenomic Regulation of Cytochrome *C* Oxidase in Neurons and the Tight Coupling between Neuronal Activity and Energy Metabolism. Adv Exp Med Biol 748:283–304. https://doi.org/10.1007/978-1-4614-3573-0_12

Xiang L, and Peng Y (2025) Impact of Glucagon-Like Peptide-1 Receptor Agonists on Mental Illness: Evidence from a Mendelian Randomization Study. Int J Mol Sci 26 (6):2741. https://doi.org/10.3390/ijms26062741

Yuksel C, Chen X, Chouinard V-A, et al. (2021) Abnormal Brain Bioenergetics in First-Episode Psychosis. Schizophr Bull Open 2 (1). https://doi.org/10.1093/schizbullopen/sgaa073

Yuksel C, Tegin C, O'Connor L, et al. (2015) Phosphorus Magnetic Resonance Spectroscopy Studies in Schizophrenia. J Psychiatr Res 68:157–166. https://doi.org/10.1016/j.jpsychires.2015.06.014

Zachos KA, Godin O, Choi J, et al. (2024) Diet Quality and Associations with Lactate and Metabolic Syndrome in Bipolar Disorder. J Affect Disord 364:167–177. https://doi.org/10.1016/j.jad.2024.05.167

Zakia H, Pradana K, and Iskandar S (2023) Risk Factors for Psychiatric Symptoms in Patients with Long Covid: A Systematic Review. PLoS One 18 (4):e0284075. https://doi.org/10.1371/journal.pone.0284075

Zhang J (2013) Autophagy and Mitophagy in Cellular Damage Control. Redox Biol 1 (1):19-23. https://doi.org/10.1016/j.redox.2012.11.008

Zhang L, Trushin S, Christensen TA, et al. (2016) Altered Brain Energetics Induces Mitochondrial Fission Arrest in Alzheimer's Disease. Sci Rep 6:18725. https://doi.org/10.1038/srep18725

Zhang Y, Wang J, Ye Y, et al. (2023) Peripheral Cytokine Levels across Psychiatric Disorders: A Systematic Review and Network Meta-Analysis. Prog Neuropsychopharmacol Biol Psychiatry 125:110740. https://doi.org/10.1016/j.pnpbp.2023.110740

Zhou J, and Seeley WW (2014) Network Dysfunction in Alzheimer's Disease and Frontotemporal Dementia: Implications for Psychiatry. Biol Psychiatry 75 (7):565–573. https://doi.org/10.1016/j.biopsych.2014.01.020

Zimmet P, and Alberti G (2005) The Metabolic Syndrome: Perhaps an Etiologic Mystery but Far from a Myth: Where Does the International Diabetes Federation Stand? 7(2)

Open Access This chapter is licensed under the terms of the Creative Commons Attribution-NonCommercial-NoDerivatives 4.0 International License (http://creativecommons.org/licenses/by-nc-nd/4.0/), which permits any noncommercial use, sharing, distribution and reproduction in any medium or format, as long as you give appropriate credit to the original author(s) and the source, provide a link to the Creative Commons license and indicate if you modified the licensed material. You do not have permission under this license to share adapted material derived from this chapter or parts of it.

The images or other third party material in this chapter are included in the chapter's Creative Commons license, unless indicated otherwise in a credit line to the material. If material is not included in the chapter's Creative Commons license and your intended use is not permitted by statutory regulation or exceeds the permitted use, you will need to obtain permission directly from the copyright holder.

Systemic Metabolic Aspects of Neuropsychiatric Disorders

9

Links between Inflammation and Metabolism in Periphery and Brain

Lessons from Epidemiological Research on Depression

Yuri Milaneschi

Abstract This review provides an overview of the latest research exploring the role of immune and metabolic pathways in depression. Findings from large-scale cohorts and biobanks, including studies using -omics technologies, consistently point toward an involvement of immuno-metabolic dysregulations in depression. Alterations in these biological pathways may explain the considerable comorbidity between depression and cardiometabolic conditions and represent a promising target for intervention. Specific depression clinical profiles are described, such as atypical neurovegetative symptoms, that are more strongly connected with immuno-metabolic alterations. These alterations are linked to brain circuits involved in energy homeostasis, potentially connecting peripheral immuno-metabolic dysregulations with specific behavioral symptom profiles. Further, a new generation of ongoing clinical trials is discussed that test treatment add-ons targeting immuno-metabolic pathways in subsets of depressed patients, selected based on specific bio-clinical profiles. Findings highlight the potential for a stratified approach to depression treatment based on underlying immuno-metabolic dysregulations.

Keywords Inflammation, depression, immuno-metabolic pathways, energy homeostasis

Yuri Milaneschi (✉)
Dept. of Psychiatry, Amsterdam UMC, Vrije Universiteit Amsterdam, Amsterdam 1081HJ, The Netherlands
E-Mail: y.milaneschi@amsterdamumc.nl

© Frankfurt Institute for Advanced Studies (FIAS) 2025

D. Öngür, J. M. Ford (eds.), *Metabolic Neuropsychiatry*, Strüngmann Forum Reports,
https://doi.org/10.1007/978-3-032-05630-6_9

1 Introduction

Biological pathways involved in inflammation and metabolism have been proposed to play a role in the partially shared pathophysiology of cardiometabolic (e.g., cardiovascular disease, diabetes) and neuropsychiatric conditions. There is now a major and growing literature highlighting the importance of these pathways in psychiatry. For example, a recent large-scale collaborative study (Tian et al. 2023) across biobanks in the United States, United Kingdom, and Australia derived composite scores from a wide array of biomarkers to index the health of various body systems and of brain in ~85,000 subjects with neuropsychiatric disorders (schizophrenia, bipolar disorder, depression, generalized anxiety disorder) and ~87,000 healthy controls. Scores indexing immune and metabolic health showed the greatest deviation, even larger than the brain health score, from normative reference ranges determined in healthy controls for all neuropsychiatric disorders. This type of evidence convincingly points toward a role in neuropsychiatric disorders for dysregulations in immune and metabolic (hereafter referred to as immuno-metabolic) pathways.

Among neuropsychiatric disorders, depression is the leading contributor to disability burden worldwide (Marx et al. 2023). Clinically relevant depression could be defined either through self-reported symptoms (e.g., applying established cut-offs to questionnaire scores) or through an interview-based psychiatric diagnosis of major depressive disorder (MDD). Diagnostic criteria for MDD encompass persistent depressed mood, loss of interest or pleasure in previously enjoyable activities (anhedonia), recurrent thoughts of death, and various physical and cognitive symptoms. The term "depression," if not otherwise specified, is used throughout this chapter in its broadest sense, including both clinically significant depressive symptoms assessed via self-report methods and the clinical diagnosis of MDD ascertained by psychiatric interviews.

The detrimental impact of depression on public health is due to sequelae that extend beyond mental health, increasing the risk of somatic diseases, such as cardiovascular disease and diabetes (Gold et al. 2020). Accumulating evidence indicates that alterations in immuno-metabolic pathways may represent the pathophysiological substrate connecting depression with somatic disease, and thus they are a promising target for intervention. Nevertheless, important research questions remain to be fully addressed. These include identifying specific clinical profiles within the heterogeneous disorder of depression that are linked to immuno-metabolic dysregulations, elucidating the neurobiological pathways connecting peripheral immuno-metabolic alterations with specific clinical manifestations, and determining whether this knowledge can be harnessed to improve treatment outcomes.

This chapter provides a focused overview of the latest research into the role of immuno-metabolic alterations in depression. It adopts the perspective of *Epidemiology*, surveying findings from large-scale cohorts and biobanks to describe reliable associations at population level. Recent findings are summarized on the association between depression and immuno-metabolic markers, including studies using -omics technologies, and different lines of evidence are described that identify clinical profiles more strongly connected with immuno-metabolic alterations. The potential role of brain circuits involved in energy homeostasis in mediating these associations is discussed, followed by a description of ongoing clinical trials testing targeted treatment add-ons in subsets of depressed patients selected based on specific bioclinical profiles. Discussion concludes by highlighting key issues to be addressed by future research in this field.

2 Immuno-Metabolic Pathways Regulating Energy Homeostasis

Biological pathways involved in immune function and metabolism interact in a signaling system conserved across species (Hotamisligil 2017). Immuno-metabolic signals are highly integrated and jointly orchestrate various downstream kinase pathways with multisystem effects. For instance, insulin and interleukin-6 (IL-6) related pathways co-regulate serum/glucocorticoid-regulated kinase 1 (SGK1) involved in both neural plasticity and cardiac remodeling (Bian et al. 2023).

Here I postulate that dysregulations in immuno-metabolic pathways involved in energy homeostasis constitute one module of the composite pathophysiology of depression. In this multisystem model (Figure 9.1), the brain acts as a broker allocating a finite energy budget. Immuno-metabolic pathways provide a reading of the potential imbalance in energy supply versus demand (metabolic stress) of various tissues and organs, leveraging various interoceptive messengers such as hormones (e.g., insulin, glucagon-like peptide-1 [GLP-1]), cytokines (e.g., IL-6, tumor necrosis factor [TNF]) and adipokines (e.g., leptin, ghrelin). Interoceptive signals reach brain hubs, like the hypothalamus and the insula, involved in the integration of a range of physiological responses, such as thermoregulation, food intake, energy expenditure, and sleep/wakefulness (Adamantidis and De Lecea 2023; Brüning and Fenselau 2023). Informed by interoceptive signals on the body metabolic status, the brain reshapes the organism priorities by adjusting the salience and reward value associated with different stimuli and activities. This readjustment is orchestrated by reward/motivational circuits (e.g., ventral striatum) linked to homeostatic hubs and sensitive to immuno-metabolic signals (Goldsmith et al. 2023; Kroemer et al. 2022; Treadway et al. 2019). The output of the system is congruent physiological and behavioral responses aimed at rebalancing energy supply and demand (e.g., reduced activity, increased sleep, food intake).

It has been postulated that engagement of immuno-metabolic pathways conveying metabolic stress signals may be a module of the heterogeneous pathophysiology of depression (Milaneschi et al. 2020). Dysregulations in immuno-metabolic pathways, or in their integration within the central homeostatic hub, may result from the cumulative effect of underlying polygenic vulnerabilities and environmental factors such as socioeconomic deprivation, stressful life events, and childhood trauma. These factors have been shown to impact immuno-metabolic health and other mechanisms of depression. In the context of depression, dysregulated immuno-metabolic signaling may trigger dysfunctional behavioral outputs such as reduced activity and increased food intake. These behaviors, in turn, may contribute to metabolic stress, establishing a vicious cycle in which immuno-metabolic dysregulations and depressive symptoms sustain each other, creating a detrimental downward spiral in health.

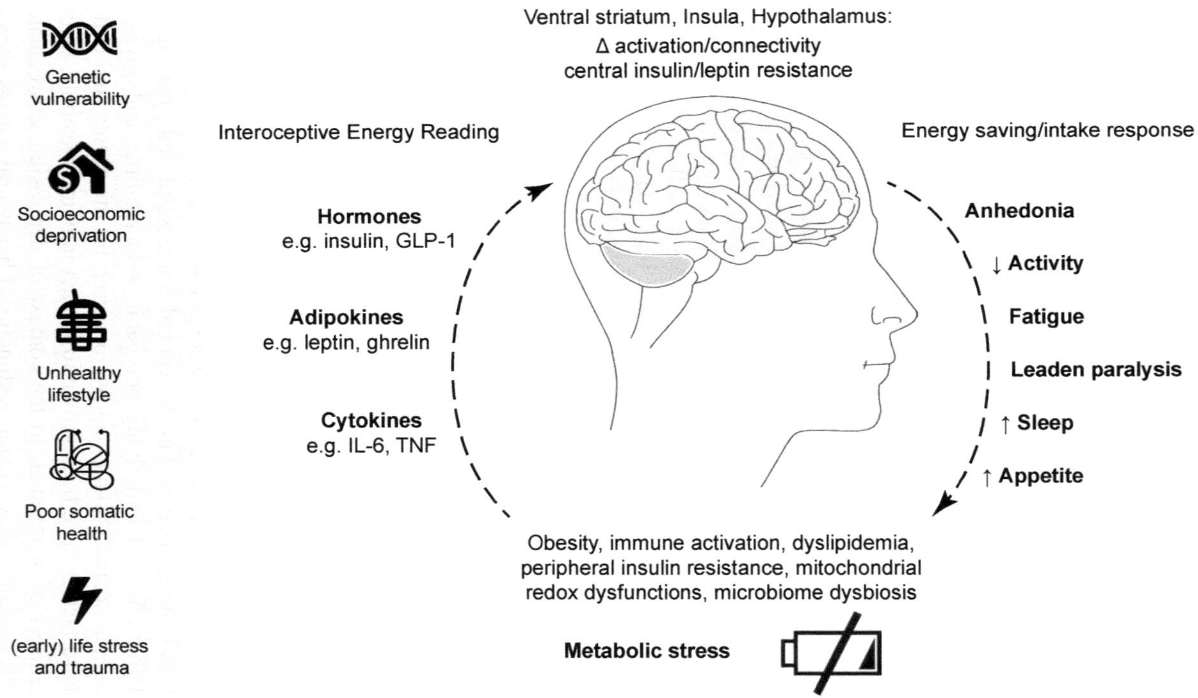

Figure 9.1 A schematic representation of the potential involvement of immuno-metabolic pathways regulating energy homeostasis in depression pathophysiology. Information on imbalance in energy supply and demand (metabolic stress) is conveyed through immuno-metabolic signals to the brain, which outputs physiological and behavioral energy-saving/intake responses. Disruptions of this homeostatic circuit may emerge from genetic predisposition and environmental challenges (e.g., trauma, stress, overnutrition). For instance, chronic inflammatory activation may lead to central insulin or leptin resistance: These interoceptive signals are then misread, pushing the system toward further enhancement of energy-saving/intake responses (e.g., increased appetite, reduced activity). These responses, in turn, further amplify the system's dysregulation (e.g., increasing inflammation) and set up a self-sustaining detrimental cycle.

3 Immuno-Metabolic Pathways in Depression

A large body of evidence, including large-scale meta-analyses (Milaneschi et al. 2019, 2020) that compared depressed persons versus healthy individuals, has revealed altered levels of peripheral markers of inflammation: C-reactive protein (CRP), IL-6, TNF, metabolism including adiposity (e.g., body mass index, BMI, and visceral fat, waist:hip circumference), metabolic syndrome and lipid profiles (e.g., high-density lipoprotein, HDL, and low-density lipoprotein, LDL), triglycerides, and insulin sensitivity. Of note, more recent meta-analyses have shown, in addition, higher cerebrospinal fluid (CSF) levels of IL-6 and TNF (Enache et al. 2019) as well as alterations in the immune cell repertoire (e.g., increased NK and T-cell counts) (Foley et al. 2022; Sørensen et al. 2022) in depressed subjects. The overall picture that emerges from this literature is that there are immuno-metabolic dysregulations in depressed subjects, including subclinical cases identified with self-report instruments and patients meeting psychiatric diagnostic criteria. These findings follow a dose-response pattern, with the most severe forms associated with stronger dysregulations. Overall, meta-analytic associations were of small effect and characterized by substantial heterogeneity. This could be because immuno-metabolic markers reflect only a portion of a broader pathophysiology and have poor specificity, since they can be altered under many different conditions. Alternatively, we propose that these findings may be partially explained by the bioclinical heterogeneity of depression (Milaneschi et al. 2020; Penninx et al. 2025), as they encompass very different symptom profiles that may represent the expression of different underlying pathophysiological processes (see Section 4). Furthermore, in several meta-analyses, estimates were substantially similar after adjustment for major sociodemographic, lifestyle, and health-related indicators (e.g., comorbid disease and medications), suggesting that the association between immuno-metabolic dysregulations and depression is not entirely explained by these distal factors and may rely on partially shared underlying biological pathways. Original studies pooled in the previous meta-analyses examined only one or just a few immuno-metabolic biomarkers. More recent cohort- and biobank-based research applied a broader approach that leveraged high-throughput -omics technologies to explore immuno-metabolic pathways in depression. Below, these newer findings from genomics and metabolomics are briefly summarized.

3.1 Genomic Evidence

A series of genome-wide association studies (GWAS), based on increasingly larger samples, have successfully started to unravel the complex genetic bases of depression (Howard et al. 2019; Levey et al. 2021; Wray et al. 2018). At the time of writing, the latest iteration from the Psychiatric Genomics Consortium (PGC) analyzed more than 680,000 cases with depression (based on MDD diagnosis or self-declared depression) and more than 4,000,000 healthy controls, and has identified 697 genetic variants spanning 635 genetic loci (Major Depressive Disorder Working Group of the Psychiatric Genomics Consortium 2025). This highly polygenic architecture has been consistently shown to be partially overlapping with that of immuno-metabolic traits. Different studies showed statistically significant genetic correlation estimates (*rg*) of small effect (*rg* ~0.1–0.2) between various measures of depression and traits, such as BMI, waist circumference, HDL and LDL cholesterol, fasting glucose, and CRP (Howard et al. 2019; Levey et al. 2021; Meng et al. 2024). This data suggests that a partially common genetic base influences both depression and immuno-metabolic traits.

Genomics further provided unique opportunities to investigate causality between traits applying new statistical tools, such as Mendelian randomization (MR), to summary-level data obtained from GWAS. MR is a technique that infers causality by leveraging genetic variants as proxy instruments, which are less likely to be related to confounders and not reversely affected by the phenotype. Applying MR, previous studies showed that genetically predicted levels of traits, such as adiposity indexes (e.g., BMI, waist circumference, body fat composition), triglycerides, and IL-6, were associated with increased odds of depression (Khandaker et al. 2020; Perry et al. 2021; Speed et al. 2019; Tyrrell et al. 2019). This suggests that higher levels of these markers, or the mechanism translating genetic variation to higher levels of such markers, are potentially involved causally in the development of depression. Meng et al. (2024) found evidence in the opposite causal direction: depression liability was associated with immuno-metabolic outcomes such as increased BMI, triglycerides, and CRP. In another recent MR study, Milaneschi et al. (2022) identified a potential causal role in depression onset for alterations in the metabolism of acylcarnitines, biogenic compounds involved in mitochondrial β-oxidation of fatty acids. A recent study by Montanari et al. (2025) examined whether the relationship between acylcarnitines and depression, predicted from previous genomic-based MR analyses, were expressed in the actual phenotypes that were measured in ~2,500 subjects from the NESDA (Netherlands Study of Depression and Anxiety) cohort. Findings confirmed the genomic predictions by showing, for instance, that low levels of acetylcarnitine (a downstream product of mitochondrial beta-oxidation of long-chain fatty acids) were linked to increased likelihood of MDD. These results align with recent molecular and cellular findings, highlighting the involvement of mitochondrial/bioenergetics and redox dysregulation in psychiatry (Dwir et al. 2023; Khadimallah et al. 2022; Kim et al. 2017). Increasingly, bioenergetic processes and mitochondrial dysfunction have been suggested to be a potential key pathophysiological mechanism linking depression to immuno-metabolic alterations (e.g., heightened inflammation and insulin resistance) as well as to cardiometabolic diseases. More recently, a large MR study (Bhattacharyya et al. 2025) combined data from two proteomic platforms that measured more than 3,000 proteins in ~35,000 subjects from UK Biobank and another ~35,000 subjects from deCODE Genetics in Iceland. Findings identified the potential causal role of multiple proteins largely shared across four phenotypes: MDD, bipolar disorder, schizophrenia, and cognitive task performance. Many of the identified proteins were related to immune functioning pathways, including interleukins, toll-like receptors, and complement factors. The strongest individual signal associated with increased risk of MDD and poorer cognition was IL23R, the receptor of cytokine IL23 involved in inflammatory responses in a variety of target organs. Other top associations for MDD included neuronal growth regulator 1 (NEGR1), expressed in one of the genomic loci most consistently associated with depression across all previous GWAS. NEGR1 modulates synaptic plasticity in brain areas crucial for mood and appetite regulation, such as the hypothalamus, where its expression is influenced by endocrine homeostatic signals (e.g., leptin, the hormone secreted by white adipose tissue that exerts a primary homeostatic function by suppressing nutritional intake and allowing energy expenditure). Studies combining genomics with another layer of -omics data consistently showed immune molecular signatures for psychiatric disorders. Lynall et al. (2022) integrated GWAS summary-level data from the cross-disorder analyses of PGC with different datasets, indexing epigenetic regulatory elements linked to histone modification. Their study showed that depression and schizophrenia genetic variants were enriched at epigenetically active sites in brain and T cells.

It is extremely important to note that the overwhelming majority of the discussed genomic analyses have been performed in samples of European ancestry, which is to date the most (over-)represented population in GWAS. As shown by Giannakopoulou et al. (2021) and Meng et al. (2024), genetic risk variants for depression have limited transferability across populations of different ancestries. Interestingly, Meng et al. (2024) and O'Loughlin et al. (2023) have also shown that the relationship between genetic risk for depression and BMI is reversed in a sample of East-Asian ancestry, with negative genetic correlation (rg –0.2) and a seemingly protective effect of genetically predicted BMI levels on depression risk in MR analyses. Thus, results from genetic risk factors should be carefully appraised and take into consideration important environmental and sociocultural differences. Ongoing multi-ancestry studies will be crucial to derive a more representative picture of the underlying genetic architecture of depression and its relationship with immunometabolic pathways.

3.2 Metabolomic Signature of Depression

Metabolomics enables a comprehensive analysis of a wide array of metabolites in a biological system, providing a snapshot of the dynamic changes that occur in an organism at the crossroads between genetic and environmental influences. Only recently has metabolomics been applied at scale using high-throughput platforms in epidemiological studies on depression. In the first study, which used the lipid-targeting Nightingale Health platform in 5,283 depressed cases and 10,145 healthy individuals, Bot et al. (2020) showed that depression is associated with a distinctive "immuno-metabolic" signature. This signature is characterized by higher levels of glycoprotein acetyls (a novel biomarker of systemic chronic inflammation indexing concentration and glycosylation of several acute phase proteins), isoleucine (a branched-chain amino acid implicated in the etiology of insulin resistance, diabetes, and obesity), triglycerides, very-low-density lipoproteins (VLDLs), and a lower level of high-density lipoproteins (HDLs). Associations were largely independent of age, gender, fasting status, use of lipid-modifying drugs, and smoking; they were also generally consistent across gender, age, and BMI strata. This biological signature overlaps with that identified by Ahola-Olli et al. (2019), using the same metabolomic platform in cardiometabolic conditions. In later studies that used the

Nightingale Health data measured in UK Biobank and various definitions of depression (from ~50,000 participants), Amin et al. (2023) and Julkunen et al. (2023) confirmed the previous metabolomic signature and identified altered levels of metabolites involved in mitochondrial functioning (e.g., alanine, citrate, pyruvate, and fatty acids). More recent studies have started to apply untargeted platforms such as Metabolon, which covers a larger proportion of the metabolome and encompasses a wider range of biochemical processes. A pooled analysis by van der Spek et al. (2023) of population-based cohorts (N ~18,000) showed that self-reported depressive symptoms measured with different questionnaires were associated with various lipids and other metabolites from amino acid, carbohydrates, cofactors, vitamins, and xenobiotic pathways. In a more recent study, Jansen et al. (2023) leveraged data from the NESDA cohort (N = 2,770) with extensive clinical phenotyping, including MDD psychiatric diagnosis and measurement of the Metabolon platform at two assessment waves. Findings identified a consistent metabolome-wide signature characterized by alterations in lipids involved in homeostatic processes, with downregulation of long-chain monounsaturated and saturated fatty acids and upregulation of lysophospholipids. Associations were substantially independent of sex, age, education, physical activity, smoking status, alcohol use, number of chronic diseases, BMI, and use of lipid-lowering medications. Of note, the classes of lipids identified in this study participate in important biological functions, including immunity, energy homeostasis, and brain development. Overall, accumulating evidence from large-scale epidemiological studies using -omics technologies supports the involvement of immuno-metabolic pathways in the pathophysiology of depression across multiple biological levels, from genetics to metabolomics

4 Immuno-Metabolic Pathways and Depression Heterogeneity

Depression is highly heterogenous. Patients within the same diagnostic category may endorse very different symptom profiles which, in turn, may represent the expression of different underlying pathophysiological processes. Thus, immuno-metabolic dysregulations may not be universally associated with all clinical manifestations of depression but rather more strongly correlated with specific clinical profiles.

Previous meta-analyses showed that subjects with MDD exhibit, on average, higher circulating inflammatory markers compared to healthy controls; nevertheless, elevated levels of inflammatory markers are reported only in a subset of patients with MDD. Pooling data from > 13,000 individuals with depression, a recent meta-analysis showed that low-grade inflammation—defined by circulating CRP > 3 mg/L (suggestive of high cardiovascular risk according to the Centers for Disease Control and Prevention and the American Heart Association)—was present in only 27% of the subjects (Osimo et al. 2019). This prevalence was independent from sample type (inpatients, outpatients, or population-based), antidepressant use, age, BMI, and ethnicity. Even after adopting a lower cutoff (CRP > 1 mg/L) to capture intermediate cardiovascular risk as well, evidence of inflammatory activation was found in only a subset (58%) of subjects. Studies have started to examine at scale the associations between immuno-metabolic biomarkers and individual symptoms of depression. Figure 9.2 reports the top nine associations with circulating CRP that were identified in two epidemiological studies: the first study examined ~150,000 UK Biobank participants and ~3,000 subjects from the NESDA cohort with ~7,000 repeated observations (Milaneschi et al. 2021a); the second analyzed >56,000 individuals from 15 population-based studies from the United Kingdom, the United States, Mexico, Costa Rica, and Taiwan (Frank et al. 2021). Despite the very different samples and instruments adopted in these two studies, findings converged substantially to identify a similar set of CRP-associated symptom domains, including neurovegetative dysregulations (appetite and sleep), lack of energy and fatigue, and motivational impairments (anhedonia).

In both studies, alteration in appetite (increase or decrease) was the symptom most strongly associated with CRP levels (Figure 9.2). Dysregulation of neurovegetative functions of appetite, weight, and sleep were shown to identify a more homogenous dimension across various clinical manifestations of depression (Milaneschi et al. 2020). Recently, van Loo et al. (2022) examined the factor structure of MDD symptoms in ~43,000 subjects with a lifetime diagnosis. Their analyses identified two distinct somatic dimensions of appetite/weight and sleep problems with distinctive patterns of associations with covariates, in particular a stronger association with BMI. This clustering of neurovegetative features identified at the phenotypic level

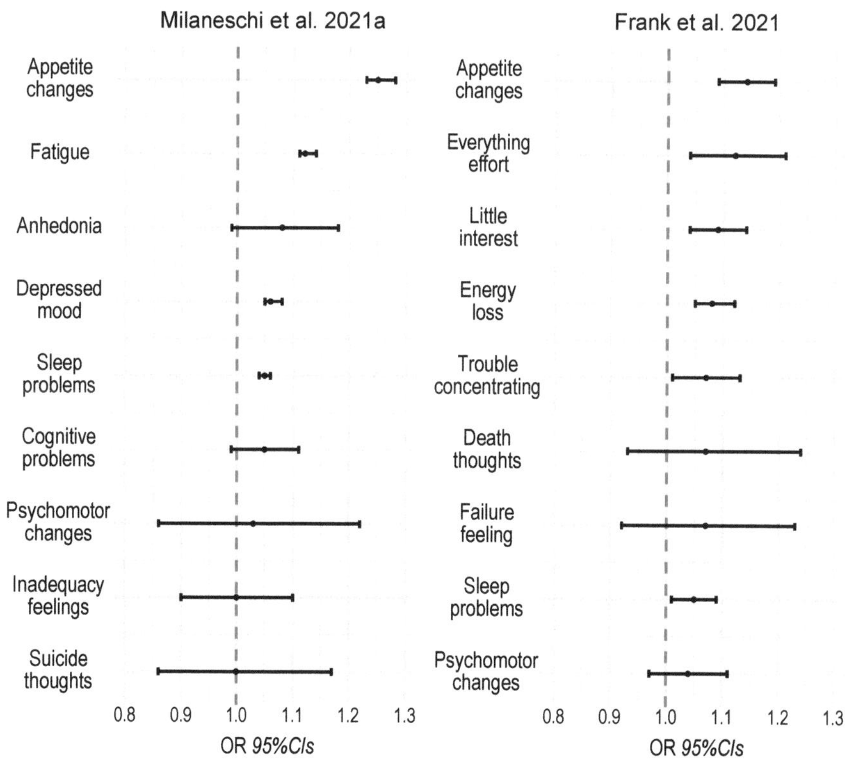

Figure 9.2 Associations between C-reactive protein levels and individual depressive symptoms found in two large-scale epidemiological studies: Milanschi et al. (2021a) and Frank et al. (2021). Associations were adjusted for sociodemographic, lifestyle, and health-related covariates.

map onto the underlying genetic level. An earlier study by Kendler et al. (2013) applied multivariate analyses to MDD symptoms, assessed in ~7,500 twins, and identified three main independent genetic factors that reflected cognitive/psychomotor, mood, and neurovegetative symptoms. The neurovegetative factor was characterized by high loadings from symptoms of appetite/weight, sleep alterations, and fatigue. A similar tripartite genetic structure was recently confirmed using genotype and MDD symptom data from ~100,000 participants in the UK Biobank (Huang et al. 2023). Application of genomic structural equation models (GenomicSEM) identified a cognitive/psychomotor, mood factor, and a neurovegetative factor characterized by symptoms of alterations in appetite/weight and sleep. Consistent findings by Adams et al. (2024) emerged from data of six large-scale cohorts and biobanks: PGC (N ~13,000), the Australian Genetics Depression Study (N ~21,000), the Generation Scotland: Scottish Family Health Study (N ~3,500), the Avon Longitudinal Study of Parents and Children (N ~14,000), the Estonian Biobank (N ~84,000), and the UK Biobank (N up to ~220,000). Based on GenomicSEM, analyses identified a best-fitting model that included a distinct factor for symptoms related to alterations in appetite/weight.

Neurovegetative symptoms of appetite, weight, and sleep included among diagnostic criteria for MDD may vary in opposite directions. During a depressive episode, patients may exhibit a loss of appetite and sleep, as is typically observed in the majority of cases in clinical settings. However, a subgroup of patients show instead "reversed" neurovegetative symptoms of increased appetite (and weight) and sleep. Such a "reversed" pattern, historically included in the atypical depression construct, emerged from a series psychopharmacological studies in the 1960s and was later integrated into the DSM Atypical Depression specifier. Overviews of the development of such construct and criticism of its validity have been published (Davidson and Thase 2007; Thase 2009). Large epidemiological studies often lack granular information on the specific direction of alteration in neurovegetative symptoms, because the majority of assessment instruments deployed conflate increased and decreased changes in single items (e.g., the items assess generic change in appetite rather than specific appetite increase or decrease). Consistent evidence from previous epidemiological studies that examined different immuno-metabolic parameters (including abdominal obesity, circulating levels of glucose, insulin, pro-inflammatory cytokines, and leptin) showed that strong associations emerged when control subjects were contrasted with patients who expressed symptoms of the atypical spectrum (Milaneschi et al. 2020). These associations reflect altered energy intake/expenditure balance, including the reversed neurovegetative symptoms of increased appetite, weight, and sleep together with fatigue and leaden paralysis (i.e., the feeling that limbs are weighed down). In contrast, associations with immuno-metabolic parameters were relatively weaker when all patients with an MDD diagnosis or other symptom profiles were considered. Consistent with this, a recent study on ~2,800 subjects from the NEDA cohort (de Kluiver et al. 2023) showed that the metabolomic signature (higher levels of glycoprotein acetyls, isoleucine, triglycerides, VLDLs, and lower lever of HDLs) of overall depression, previously identified in a large-scale meta-analysis using the Nightingale Health platform (Bot et al. 2020), was mainly driven by a specific symptom profile labeled "atypical, energy-related symptoms," including hyperphagia, hypersomnia, weight gain, fatigue, and leaden paralysis. Similarly, findings from a study based on NESDA data (Montanari et al. 2025) showed that the association between alterations in acylcarnitines involved in mitochondrial energy production and depression was specifically driven by this symptom profile. Highly convergent results have been obtained in studies using data-driven clustering methodologies. Previous research (Alshehri et al. 2023b), which involved 1,094 subjects with MDD from NESDA, clustered data from 149 metabolites of the Nightingale Health platform and 30 individual depressive symptoms using canonical correlation analyses. This method identified two main covariance dimensions, one of which showed relatively higher loadings for symptoms such as increased sleep, increased appetite, and low energy levels. In an independent replication sample of over 6,000 subjects from the Netherlands Epidemiology of Obesity (NEO) study, this dimension was associated with well-known cardiometabolic risk markers, including higher visceral adipose tissue, indices of insulin resistance, and lower HDL cholesterol.

The specificity of the connection between immuno-metabolic dysregulations and certain symptom profiles has also been detected at the genomic level. Findings from two independent studies from the PGC (N ~26,000; Milaneschi et al. 2017) and UK Biobank (N ~30,000; Badini et al. 2022) were consistent: only MDD cases that expressed, respectively, an increase in appetite and weight and an increase in sleep and weight during their index major depressive episode carried a higher number of genetic risk variants for immuno-metabolic traits like CRP, BMI, leptin triglycerides, and coronary artery disease. Such an increase in immuno-metabolic genetic risk was not observed in MDD cases that expressed an opposite pattern of alterations (decrease during index major depressive episode) of neurovegetative symptoms and was substantially diluted when all MDD cases were considered together. Overall, these findings indicate that depressed subjects who express reversed neurovegetative symptoms have a specific genetic predisposition to develop immuno-metabolic dysregulations.

Only a limited number of large-scale studies have systematically examined the association between markers of metabolic health and individual symptoms of depression. A recent meta-analysis pooled data from 15 population-based cohorts, comprising 57,532 individuals, and showed that BMI was more strongly associated with lack of energy and motivation (Frank et al. 2022). A portion of these specific associations was explained by elevated CRP and obesity-related diseases. In an earlier work based on ~6,600 subjects from the NEO study, Alshehri et al. (2019) examined the association between individual depressive symptoms and four overall and abdominal adiposity measures: BMI, total body fat, waist circumference, and visceral adipose tissue. Across these four measures, the most consistent and strongly associated symptoms were increased appetite, leaden paralysis, and lack of energy. In a follow-up study, Alshehri et al. (2023a) combined data from the NEO (N ~5,700) and NESDA (N ~2,200) cohorts and leveraged genomics to dissociate the effect of adiposity (body fat percentage) from that of metabolic dysregulations (e.g., dyslipidemia). They found that the previously identified symptoms of increased appetite, leaden paralysis, and lack of energy were linked solely to high genetic risk for increased body fat with metabolic dysregulations, not to high genetic risk of increased body fat without metabolic dysregulations. Thus, metabolic alterations may represent the underlying shared mechanism that connects adiposity to specific behavioral symptoms included among depression clinical manifestation. This conclusion is consistent with findings from other genomics studies (Badini et al. 2022; Milaneschi et al. 2017, 2021b) that leveraged large-scale GWAS datasets to show that genetic risk variants for mechanisms underlying metabolic alterations (e.g., BMI, leptin, tryglicerides) are specifically enriched in MDD patients exhibiting reversed neurovegetative symptoms. MR analyses further supported evidence of causality for this metabolic mechanism in the development of specific symptoms, such as hyperphagia, during a major depressive episode (Pistis et al. 2021).

Overall, two main research lines have emerged from the investigation of depression heterogeneity in relation to immuno-metabolic dysregulations. The first emerged within the established tradition of studies on sickness behavior that examined the impact of inflammation on depression, with a particular focus on clinical symptoms of motivational impairment, anhedonia, and anergia (Miller and Raison 2016). The second focused more on the set of "atypical, energy-related symptoms," including hyperphagia, hypersomnia, weight gain, fatigue, and leaden paralysis (Milaneschi et al. 2020). The specific association of this type of atypical-like depressive symptoms with inflammatory and metabolic dysregulations has been postulated to identify a dimension labeled "immuno-metabolic depression" (Milaneschi et al. 2020; Penninx et al. 2025), which maps the degree of expression of behavioral and biological processes overlapping with those in cardiometabolic phenotypes. Despite focusing on partially different clinical manifestations, these research lines suggest convergent mechanisms on neurobiological pathways related to energy homeostatic regulation.

5 Immuno-Metabolic Pathways and Brain Homeostatic Regulation

The specific clustering between behavioral symptoms characterized by altered energy input–output balance (e.g., fatigue, increased appetite, anhedonia) with markers of immuno-metabolic pathways points toward the engagement of neural circuits coordinating homeostatic regulation. At the center of these circuits are brain hubs, such as the hypothalamus and the insula, that are involved in the integration of peripheral, environmental, and neural signals, and which influence a range of physiological and behavioral responses, such as thermoregulation, food intake, energy expenditure, and sleep/wakefulness (Adamantidis and De Lecea 2023; Brüning and Fenselau 2023). Central homeostatic hubs are targeted by humoral signals from the interoceptive system communicating bodily physiological states to the brain. Neuroendocrine messengers like leptin, insulin, ghrelin, and GLP-1 provide a reading of the body energy status to the hypothalamus, whose neural circuits orchestrate physiological and behavioral responses (e.g., modulation of appetite) aimed at controlling energy balance and metabolism. Consistent with this function, alterations in leptin central signaling (due to impaired leptin transport across the blood-brain barrier, reduced function of leptin receptors, and defects in leptin signal transduction) shift energy balance from expenditure to accumulation by, for instance, disinhibiting feeding (van der Klaauw and Farooqi 2015). In humans, loss-of-function mutations in genes that encode leptin or melanocortin-4 receptors present in the hypothalamus result in rare forms of obesity characterized by severe hyperphagia (Farooqi et al. 2003; Farooqi et al. 2007). Similarly, disruption of insulin signaling in the hypothalamus and the insula may enhance physiological and behavioral responses (i.e., hyperphagia) aimed at increasing energy intake (Obici et al. 2002). An observational study by Simmons et al. (2020) showed that among 54 patients with a current MDD diagnosis, those reporting an increase in appetite during the current major depressive episode, compared to other patients and healthy individuals, had significantly different levels of a wide array of immuno-metabolic biomarkers, including higher leptin, insulin, insulin resistance, CRP, and IL-6, and lower ghrelin. Furthermore, the magnitude of their insulin resistance correlated positively with the insula response to food cues registered during functional magnetic resonance imaging (fMRI).

In addition, interoceptive pathways play a critical role in receiving peripheral immuno-inflammatory signals that communicate the body immune state, thus enabling the brain to adopt congruent energy-saving behavioral responses. Using an experimental medicine approach, previous studies in humans have shown that inflammatory challenges (e.g., administration of vaccines or lipopolysaccharides) produce behavioral symptoms, such as fatigue and sleepiness, at an intensity proportional to the degree of the elicited inflammatory response (Harrison et al. 2009; Lasselin et al. 2020). Interestingly, across various studies, increased activity in the insula has been consistently identified as the potential mediating neurocircuit (Savitz and Harrison 2018).

Brain homeostatic hubs that receive peripheral endocrine and immune interoceptive signals are tightly coupled to dopaminergic corticostriatal neurocircuitries that modulate reward sensitivity and hedonic sensing of various stimuli. Previous experimental neuroimaging studies have also shown an effect of inflammatory stimuli on motivational (e.g., anhedonia) and motor impairments in depression, mediated mainly by alterations in activation in the ventral striatum (Felger and Treadway 2017). More recently, neuroimaging studies that have examined resting-state functional connectivity (rsFC) have shown consistent associations between levels of inflammatory markers, dysconnectivity in motivational circuits, and symptoms such as anhedonia (Goldsmith et al. 2023).

The impact of immune activation on brain areas governing homeostatic processing is consistent with the concept of sickness behavior, an inflammation-triggered repertoire of illness-related behaviors (fatigue, malaise, sleep and appetite alterations, reduced motivation) shaped by evolution to divert energy from externally oriented activities to fight infection and promote wound healing. In this context, the impact of immune-related signals on reward/motivation neural pathways is interpreted as an adaptive behavioral mechanism aimed at saving energy. Consistent with this notion, Treadway et al. (2019) have shown that the impact of immunometabolic signaling on motivational outcomes such as anhedonia is mainly due to impairment in appetitive (energy effort and motivational vigor) rather than consummatory (perceived reward value) processes.

Intriguingly, immune-related mechanisms may interact and disrupt other interoceptive signaling pathways. For instance, inflammation favors the development of leptin and insulin central resistance by crippling the functionality of their receptors in the hypothalamus and insula (Cui et al. 2017), consequently enhancing homeostatic dysregulation via increased caloric intake (i.e., overeating) and reduced energy expenditure. For this reason, central hypothalamic inflammation is postulated to be a key mechanism for the development and maintenance of obesity-related metabolic diseases such as type 2 diabetes (Jais and Brüning 2017).

The complex interplay between inflammation, interoception, motivation, and behavioral symptoms has been elegantly captured by Cosgrove et al. (2020), who examined 64 subjects (31 unmedicated subjects with current MDD and 33 healthy controls) with measures of CRP and completed fMRI scans in which they rated the perceived pleasantness of various food stimuli. Compared to healthy controls, the MDD subjects showed a stronger positive correlation between circulating CRP and the coupling between activation in the striatum and food hedonic ratings, consistent with the notion of the influence of inflammation on reward circuitry in depression. Intriguingly, as compared to all other subjects, the MDD subjects who expressed appetite increase during their current major depressive episode showed a specific pattern of results: higher CRP was correlated to (a) stronger coupling between food hedonic rating and activation in the anterior insula, a region involved in salience processing, and (b) weaker coupling between food hedonic rating and activation of mid-posterior insula, regions primarily implicated in interoceptive reading of the body's homeostatic state. Cosgrove et al. interpreted this divergent pattern of results to indicate a potential role for neurobiological mechanisms to play in depressed subjects developing increased appetite during an acute episode, in whom a pro-inflammatory activation may impair the ability to integrate interoceptive signals effectively with salience evaluation of food stimuli. Another example of disconnection between interoceptive and motivational brain circuits in relation to a specific symptom has been shown in recent analyses (Kroemer et al. 2022) based on more than 800 subjects from the Marburg-Münster FOR 2107 Affective Disorder Cohort Study. Kroemer et al. examined the pattern of rsFC centered on the nucleus accumbens (NAcc), a reward processing area located in the ventral striatum. Results showed that among subjects with MDD, reduced NAcc-based rsFC to the insular ingestive cortex (involved in chemosensation necessary to control feeding behavior) was associated with increased appetite. These results suggest that the symptom (e.g., increased appetite) may emerge from a disconnect between the system that reads the current body energy availability (e.g., the insula receiving information from neuroendocrine peripheral messengers) and the system that evaluates the salience of the energy source (e.g., dopaminergic corticostriatal neurocircuitries).

In conclusion, transdiagnostic behavioral symptoms correlated with immuno-metabolic pathways, characterized by altered energy intake/output balance (e.g., fatigue, sleep alterations, increased appetite, anhedonia), are linked to brain centers that coordinate homeostasis (e.g., hypothalamus, insula) at the interface of interoceptive and motivational neurocircuitries. In this context, specific clinical symptoms may be postulated as emergent from disruptions in the processing and integration of signals in this complex system. Immuno-metabolic dysregulations have a key role in regulating and disrupting the activities of brain homeostatic neurocircuitries, reading body energy levels and allocating them to salient tasks. In this way, brain homeostatic neurocircuitries may represent the mediating substrate between peripheral immuno-metabolic dysregulations and specific behavioral transdiagnostic symptoms such as fatigue, increased appetite, and anhedonia.

6 A New Generation of Clinical Studies Targeting Immuno-Metabolic Pathways in Selected Patients

Immuno-metabolic dysregulations have been linked to poorer treatment response and a more chronic course of depression (Arteaga-Henríquez et al. 2019; Strawbridge et al. 2015; Vogelzangs et al. 2014). Emerging knowledge on the potential pathophysiological role of immuno-metabolic dysregulations for the development of specific clinical manifestations of depression stimulated the development of a new generation of clinical trials, which are testing treatment add-ons targeting immuno-metabolic pathways in subsets of depressed patients selected based on specific bioclinical profiles.

The majority of these new trials selected inflammation as a target for the add-on treatment in depression. This choice was based on promising results from meta-analyses in subjects with depression or somatic conditions, which showed a significant clinical effect on depressive symptoms of anti-inflammatory medications such as cytokine inhibitors, nonsteroidal anti-inflammatory drugs (NSAIDs), minocycline, and statins (Kappelmann et al. 2018; Köhler-Forsberg et al. 2019; Wittenberg et al. 2020). It is important to note that the databases pooled for the meta-analyses were composed of very heterogeneous sets of small-sample clinical studies with substantial risk of bias. More recent trials with larger samples did not confirm evidence of clinical efficacy for anti-inflammatory add-on in subjects with MDD.

The MinoTRD trial, including 173 MDD patients without sufficient response to antidepressants, did not show a significant statistical difference in depressive symptoms over a six-week follow-up between the group randomized to 200 mg/d minocycline add-on and the placebo add-on group (Hellmann-Regen et al. 2022). Similar negative results were also reported by Husain et al. (2020) for 266 subjects with bipolar depression randomized to treatment add-on with placebo, minocycline, and celecoxib (a first-line cox-2 inhibitor NSAID used in osteoarthritis and rheumatoid arthritis). Similarly, six weeks of vortioxetine augmented with 400 mg/d of celecoxib did not show superior clinical efficacy as compared to placebo add-on in 119 subjects with MDD from the PREDDICT study (Baune et al. 2021).

Interestingly, the PREDDICT trial was originally designed to stratify subjects into two pools, according to circulating CRP lower or higher than 3 mg/l (indicative of low-grade inflammation), and to perform randomization within each strata (Fourrier et al. 2018). This choice was consistent with previous empirical evidence derived from secondary reanalysis of an existing trial (Raison et al. 2013) and with the expectation that the potential clinical effect of anti-inflammatory medication could be magnified in the subset of MDD patients with signs of active inflammation. Unfortunately, the PREDDICT study was unable to perform the planned within-strata randomization due to difficulties in recruitment of subjects with high CRP. The originally planned approach would have overcome a major limitation of previous anti-inflammatory add-on trials in depression: the lack of selection of the relevant subject subgroup, those with clear engagement of the immune system.

More recently, studies have started to deploy selection criteria in an attempt to capture only subgroups of depressed subjects with active inflammatory processes. In one of the first studies by McIntyre et al. (2019), participants with bipolar depression were selected based on a wide array of biochemical and/or phenotypic indexes of inflammatory activation, including among others CRP≥5 mg/l, dyslipidemia, obesity, diabetes, rheumatologic disorders, migraine, and daily smoking. Treatment (N=28) with three infusions of infliximab (a monoclonal antibody neutralizing TNF pro-inflammatory signal) over 12 weeks did not produce significantly reduced depressive symptoms, as compared to placebo (N=30). Several limitations for the appraisal of the findings were highlighted by Berk et al. (2019), including the limited sample size and the broad heterogeneity of the selection criteria pragmatically adopted to ensure a feasible recruitment. In the MINDEP trial, Nettis et al. (2021) randomized MDD patients with CRP≥1 mg/l to add-on treatment with 200 mg/d minocycline (N=18) versus placebo (N=21) and found no significant differences when comparing four-week change in depressive symptoms across treatment groups. However, a secondary analyses that stratified subjects for CRP levels lower or higher than 3 mg/l showed the largest reduction in depressive symptoms in the group with high CRP receiving minocycline, as compared to patients with low CRP on minocycline and those on placebo.

Overall, previous clinical trials that aimed to target inflammation as an add-on strategy to improve depression outcomes were largely unsuccessful. Miller and Raison (2023) propose that one potential reason for these clinical failures is the lack of appropriate selection of depressed patients with bioclinical profiles, indicating a clear engagement of inflammatory pathways. Using substantially larger samples, a new generation of ongoing clinical trials is currently underway to test the possibility of whether subsets of depressed subjects can be selected, based on bioclinical profiles that are consistently associated with immuno-metabolic dysregulations, to match patients with targeted anti-inflammatory add-on therapies. The main features of these studies are summarized in Table 9.1. In all studies, patients with a current MDD diagnosis who are undergoing standard first-line antidepressant treatments are additionally screened for

- specific biological features, such as CRP≥3 mg/l (INSTA-MD, expected N=120) or BMI≥30 kg/m^2 (SIMCODE, expected N=160) (Otte et al. 2020) or
- a combination of biological parameters and clinical features, such as CRP≥3 mg/l, and (a) high endorsement of symptoms such as lack of pleasure, loss of energy, changes in sleep and appetite, concentration difficulty, tiredness or fatigue, and loss of interest in sex (INSIGHT, expected N=50) (Khandaker et al. 2018) or CRP≥1 mg/l and (b) high endorsement of atypical energy-related symptoms, such as increased appetite, sleep and weight, fatigue and leaden paralysis (INFLAMED, expected N=140) (Zwiep et al. 2023).

Follow-ups from four to twelve weeks test the efficacy of augmentation of standard treatment for depression with celecoxib (INFLAMED and INSTA-MD), minocycline (INSTA-MD), tocilizumab (INSIGHT), and simvastatin (SIMCODE).

Results from these new studies will provide key insights on the validity of this "stratified approach" to enhance the efficacy of anti-inflammatory add-on treatments for depression. Alternatively, if these clinical studies are unsuccessful, we should thoroughly reconsider the most efficient candidate disease mechanism to target. Inflammation is just one pathway involved in bioenergetic processes. Inflammatory

Table 9.1 Ongoing targeted add-on treatment trials for selected patients. CRP, C-reactive protein; BMI, body mass index.

Study	Country	Selection Criteria	Add-On Treatment	Total N (est.)	Registry ID
INSIGHT	UK	Somatic symptoms [a] and CRP ≥ 3 mg/l [b]	Tocilizumab (8 mg/kg)	100	ISRCTN16942542
SIMCODE	Germany	BMI ≥ 30 kg/m^2	Simvastatin (40 mg/d)	160	NCT04301271
INFLAMED	Netherlands	Atypical/energy-related symptoms [c] and CRP ≥ 31 mg/l	Celecoxib (400 mg/d)	140	NCT05415397
INSTA-MD	Belgium	CRP ≥ 3 mg/l [d]	Celecoxib (400 mg/d) Minocycline (200 mg/d)	240	NCT05644301

[a] Lack of pleasure, loss of energy, changes in sleep and appetite, concentration difficulty, tiredness or fatigue, and loss of interest in sex; instrument: Beck Depression Inventory (BDI-II)

[b] The study will enroll but not randomize to treatment subjects with CRP < 3 mg/l

[c] Atypical/energy-related symptoms: increased appetite, sleep and weight, fatigue and leaden paralysis; instrument: Inventory of Depressive Symptomatology Self-Report 30 items (IDS-SR_{30})

[d] The study will enroll and randomize to treatment subjects with CRP < 3 mg/l

markers like cytokines have broader functions beyond direct host defense; they are produced by multiple (nonimmune) cells, and their receptors are expressed in various tissues, including key energy-sensing structures within the brain (Medzhitov 2008; Uhlén et al. 2015). In this perspective, consistent with the model presented in Figure 9.1, cytokines act as messengers of metabolic stress in concert with other interoceptive signals (e.g., hormones and adipokines). Thus, developing and deploying interventions that target the sources of metabolic stress may represent a more effective treatment strategy and should be explored further through future research.

7 Research and Translational Perspectives

In this chapter, I have described different lines of evidence that identify relevant immuno-metabolic alterations in depression and their specific connection with clinical symptoms, characterized by altered energy intake/output balance (e.g., fatigue, increased appetite, anhedonia). Also discussed was how this specific connection may be mediated by brain neurocircuitries involved in energy homeostasis interfacing interoception and motivation, and how the knowledge of this specific connection is shaping the development of new clinical studies. Nevertheless, several issues remain to be addressed by future research.

It is important to adopt a transdiagnostic perspective that extends beyond depression. I propose that the energy homeostasis circuit depicted in Table 9.1 represents one of the multiple modules in which the complex and heterogeneous pathophysiology of depression could be decomposed. Alterations of this module may also be present in other psychiatric and somatic conditions that share similar clinical and biological profiles. The profile of symptoms identified is rather transdiagnostic and overlaps with other nosologic categories (e.g., atypical neurovegetative symptoms in bipolar disorder or anhedonia in psychosis) that have been previously shown to be characterized by alteration in markers of immuno-metabolic pathways. The examination of the transdiagnostic bioclinical clustering described here is consistent with the conceptual framework of the Research Domain Criteria (RDoC) (Cuthbert 2022), in which biology-based multilevel dimensions cut through diagnostic categories. Thus, it should extend not only to other psychiatric disorders but also to somatic conditions such as obesity, diabetes, cardiovascular disease, or neurodegenerative diseases.

The immuno-metabolic dysregulations discussed here, however, cover only a portion of the potential pathophysiological pathways involved in energy homeostasis. As described above, converging genetic (Milaneschi et al. 2022) and phenotypic (Montanari et al. 2025) evidence of alterations in depression involving the metabolism of acylcarnitines in mitochondrial fatty-acid oxidation suggest key involvement of altered bioenergetic mechanisms and mitochondrial dysfunctions, and deserve further investigation. This is consistent with recent molecular and cellular findings that support the involvement of mitochondrial/bioenergetics and redox dysregulation in psychiatry (Dwir et al. 2023; Khadimallah et al. 2022; Kim et al. 2017).

Furthermore, future research is needed to delineate causal connections in the complex network of associations between immuno-metabolic dysregulations, neural pathways, and specific symptoms. As described here, genetically informed techniques such as Mendelian randomization have been successfully applied for this purpose. Nevertheless, to properly disentangle causal mechanisms, findings need to be properly complemented and triangulated with evidence from experimental medicine approach and preclinical studies.

From a translational point of view, ongoing clinical trials that use bioclinical profiles linked to immuno-metabolic dysregulations aim at guiding targeted add-on treatments to enhance therapy response in depression. This approach is rooted in the theoretical expectation that patients stratified according to a bioclinical profile that indexes the activation of a specific pathway should be more likely to respond to therapy with an agent that engages a target on that pathway. Several key articulations of this "stratified medicine" approach deserve further inquiry. For instance, we have not yet established what constitutes the best panel of bioclinical features to be deployed in the selection of depressed patients with activated immuno-metabolic pathways. Valuable data from the new generation of trials described in Table 9.1 could be exploited in prediction-oriented analyses (e.g., machine learning) to select stratification features that will maximize the impact of different treatments. This line of research should also be directed toward the selection of the most parsimonious configuration of easily measurable bioclinical parameters that could be adopted in clinical contexts at scale.

Furthermore, future translational studies that adopt this "stratified medicine" approach must have valid and reliable repeated measures that trace the full pathway selected—from peripheral marker assays to neuroimaging signatures (e.g., circuit disconnection in rsFC)—as well as symptom assessment scales that reliably measure specific clinical profiles. This effort could drive the development of new instruments specifically tailored for the symptoms targeted. Measures that trace the full bio-neuro-behavioral pathway will be needed to sustain evidence of the actual engagement of the pathway and its potential change as a function of the treatment tested.

In addition, the issue of treatment selection that effectively targets immuno-metabolic pathways remains. Clinical studies discussed here adopted anti-inflammatory compounds to augment standard treatment for depression. Emerging empirical evidence, however, suggests that repurposing medications that act on metabolic processes in depressed patients may potentially be another viable option. For instance, biocomputational analyses in a recent multi-ancestry GWAS on major depression identified genes encoding targets of established drugs such as simvastatin and metformin, a first-line treatment for type 2 diabetes (Meng et al. 2024). Recently it has been shown that the effect of metformin on energy balance is exerted via the synthesis of the appetite-suppressing metabolite N-lactoyl phenylalanine in intestinal cells, a process triggered by the inhibition of metformin-targeted mitochondrial complex I assembly protein (Scott et al. 2024; Xiao et al. 2024). Animal models suggest associations between metformin administration and reduction in depressive- and anxiety-like behaviors (Zemdegs et al. 2019). Meta-analyses of human trials with available secondary measures of depressive symptoms, however, have not confirmed this potential clinical effect of metformin (Moulton et al. 2018; Nibber et al. 2022); nevertheless, previous trials were not designed to test metformin add-on in selected MDD patients. Furthermore, it remains to be determined whether newly developed compounds that act on metabolic processes have a role in psychiatry, such as GLP-1 agonist or receptor antagonists (mimicking GLP-1 interoceptive signaling on brain homeostatic hubs to regulate appetite) developed for diabetes. Intriguingly, a recent animal model study showed an anti-inflammatory effect of a GLP-1 agonist compound achieved directly through GLP-1 receptors and mediated by the brain (Wong et al. 2024). In addition, the systematic review by De Giorgi et al. (2025) provides an overview of available findings from studies using GLP-1 receptor antagonists (GLP-1RAs) in various neuropsychiatric disorders. Studies that include patients with mood disorders are limited in numbers and sample size, and have yielded conflicting results; for example, improvement in cognitive functions (Mansur et al. 2017), no effect on depressive symptoms (Cuomo et al. 2019), or even onset of relapse of depression in two case reports (Kohen and Lester 2008; Li et al. 2023). A small meta-analysis by Chen et al. (2024) of five trials that included subjects with obesity, diabetes, and Parkinson disease showed a significant reduction in secondary measures of depressive symptoms in subjects treated with GLP-1RAs. The same results were obtained from a largely overlapping meta-analysis that included subjects with obesity and diabetes, although only in analyses that excluded the largest study, which included nondiabetic subjects. Based on the knowledge currently available, no firm conclusion can be reached on the utility of GLP-1RAs in depression.

Finally, future translational studies aimed at targeting immuno-metabolic dysregulations in psychiatry may embrace the approach advocated by Miller and Raison (2023): Studies should focus on convergent mechanistic pathologies across diagnostic categories and adopt transdiagnostic clinical outcomes disentangled from specific neuropsychiatric diagnosis.

8 Conclusions

From this overview of research aimed at exploring the role of immuno-metabolic dysregulations in depression, there are three main takeaway messages: First, the identification of specific bioclinical profiles (e.g., atypical/energy-related symptoms) linked to immuno-metabolic alterations has important implications for understanding the heterogeneity of depression and developing targeted treatment approaches. Ongoing clinical trials that test treatment add-ons which target immuno-metabolic pathways in subsets of depressed patients, selected based on specific bioclinical profiles, represent a promising step toward a stratified approach to depression treatment.

Second, there are several key issues that remain to be addressed by future research. These include examining the generalizability of the findings to other neuropsychiatric disorders and somatic conditions, investigating the role of bioenergetic processes and mitochondrial dysfunction in the pathophysiology of depression, and further elucidating the causal connections between immuno-metabolic dysregulations, neural pathways, and specific depressive symptoms.

Third, translational efforts should focus on identifying the most parsimonious configuration of easily measurable bioclinical parameters for patient stratification, developing new instruments specifically tailored for assessing targeted symptoms and exploring the potential of repurposing medications that act on metabolic processes as novel treatment options for depression.

Acknowledgments Yuri Milaneschi is supported by Amsterdam UMC (Starter Grant Ronde 2) and Amsterdam Neuroscience (PoC funding 2024-2026).

References

Adamantidis AR, and De Lecea L (2023) Sleep and the Hypothalamus. Science 382 (6669):405–412. https://doi.org/10.1126/science.adh8285

Adams MJ, Thorp JG, Jermy BS, et al. (2024) Genome-Wide Meta-Analysis of Ascertainment and Symptom Structures of Major Depression in Case-Enriched and Community Cohorts. Psychol Med 54 (12):3459–3468. https://doi.org/10.1017/S0033291724001880

Ahola-Olli AV, Mustelin L, Kalimeri M, et al. (2019) Circulating Metabolites and the Risk of Type 2 Diabetes: A Prospective Study of 11,896 Young Adults from Four Finnish Cohorts. Diabetologia 62 (12):2298–2309. https://doi.org/10.1007/s00125-019-05001-w

Alshehri T, Boone S, de Mutsert R, et al. (2019) The Association between Overall and Abdominal Adiposity and Depressive Mood: A Cross-Sectional Analysis in 6459 Participants. Psychoneuroendocrinology 110:104429. https://doi.org/10.1016/j.psyneuen.2019.104429

Alshehri T, Mook-Kanamori DO, de Mutsert R, et al. (2023a) The Association between Adiposity and Atypical Energy-Related Symptoms of Depression: A Role for Metabolic Dysregulations. Brain Behav Immun 108:197–203. https://doi.org/10.1016/j.bbi.2022.12.005

Alshehri T, Mook-Kanamori DO, Willems van Dijk K, et al. (2023b) Metabolomics Dissection of Depression Heterogeneity and Related Cardiometabolic Risk. Psychol Med 53 (1):248–257. https://doi.org/10.1017/s0033291721001471

Amin N, Liu J, Bonnechere B, et al. (2023) Interplay of Metabolome and Gut Microbiome in Individuals with Major Depressive Disorder vs Control Individuals. JAMA Psychiatry 80 (6):597–609. https://doi.org/10.1001/jamapsychiatry.2023.0685

Arteaga-Henríquez G, Simon MS, Burger B, et al. (2019) Low-Grade Inflammation as a Predictor of Antidepressant and Anti-Inflammatory Therapy Response in MDD Patients: A Systematic Review of the Literature in Combination with an Analysis of Experimental Data Collected in the EU-MOODINFLAME Consortium. Front Psychiatry 10:458. https://doi.org/10.3389/fpsyt.2019.00458

Badini I, Coleman JRI, Hagenaars SP, et al. (2022) Depression with Atypical Neurovegetative Symptoms Shares Genetic Predisposition with Immuno-Metabolic Traits and Alcohol Consumption. Psychol Med 52 (4):726–736. https://doi.org/10.1017/s0033291720002342

Baune BT, Sampson E, Louise J, et al. (2021) No Evidence for Clinical Efficacy of Adjunctive Celecoxib with Vortioxetine in the Treatment of Depression: A 6-Week Double-Blind Placebo Controlled Randomized Trial. Eur Neuropsychopharmacol 53:34–46. https://doi.org/10.1016/j.euroneuro.2021.07.092

Berk M, Walker AJ, and Nierenberg AA (2019) Biomarker-Guided Anti-Inflammatory Therapies: From Promise to Reality Check. JAMA Psychiatry 76 (8):779–780. https://doi.org/10.1001/jamapsychiatry.2019.0673

Bhattacharyya U, John J, Lam M, et al. (2025) Circulating Blood-Based Proteins in Psychopathology and Cognition: A Mendelian Randomization Study. JAMA Psychiatry 95 (10):481–491. http://doi.org/10.1001/jamapsychiatry.2025.0033

Bian X, Xue H, Jing D, et al. (2023) Role of Serum/Glucocorticoid-Regulated Kinase 1 (SGK1) in Immune and Inflammatory Diseases. Inflammation 46 (5):1612–1625. https://doi.org/10.1007/s10753-023-01857-8

Bot M, Milaneschi Y, Al-Shehri T, et al. (2020) Metabolomics Profile in Depression: A Pooled Analysis of 230 Metabolic Markers in 5283 Cases with Depression and 10,145 Controls. Biol Psychiatry 87 (5):409–418. https://doi.org/10.1016/j.biopsych.2019.08.016

Brüning JC, and Fenselau H (2023) Integrative Neurocircuits That Control Metabolism and Food Intake. Science 381 (6665):eabl7398. https://doi.org/10.1126/science.abl7398

Chen X, Zhao P, Wang W, Guo L, and Pan Q (2024) The Antidepressant Effects of GLP-1 Receptor Agonists: A Systematic Review and Meta-Analysis. Am J Geriatr Psychiatry 32 (1):117–127. https://doi.org/10.1016/j.jagp.2023.08.010

Cosgrove KT, Burrows K, Avery JA, et al. (2020) Appetite Change Profiles in Depression Exhibit Differential Relationships between Systemic Inflammation and Activity in Reward and Interoceptive Neurocircuitry. Brain Behav Immun 83:163–171. https://doi.org/10.1016/j.bbi.2019.10.006

Cui H, López M, and Rahmouni K (2017) The Cellular and Molecular Bases of Leptin and Ghrelin Resistance in Obesity. Nat Rev Endocrinol 13 (6):338–351. https://doi.org/10.1038/nrendo.2016.222

Cuomo A, Bolognesi S, Goracci A, et al. (2019) Feasibility, Adherence and Efficacy of Liraglutide Treatment in a Sample of Individuals with Mood Disorders and Obesity. Front Psychiatry 9:784. https://doi.org/10.3389/fpsyt.2018.00784

Cuthbert BN (2022) Research Domain Criteria (RDoC): Progress and Potential. Curr Dir Psychol Sci 31 (2):107–114. https://doi.org/10.1177/09637214211051363

Davidson JRT, and Thase ME (2007) A History of the Concept of Atypical Depression. J Clin Psychiatry 68 (2):e03. https://doi.org/10.4088/jcp.0207e03

De Giorgi R, Ghenciulescu A, Dziwisz O, et al. (2025) An Analysis on the Role of Glucagon-Like Peptide-1 Receptor Agonists in Cognitive and Mental Health Disorders. Nat Ment Health 3 (3):354–373. https://doi.org/10.1038/s44220-025-00390-x

de Kluiver H, Jansen R, Penninx BWJH, et al. (2023) Metabolomics Signatures of Depression: The Role of Symptom Profiles. Transl Psychiatry 13 (1):198. https://doi.org/10.1038/s41398-023-02484-5

Dwir D, Khadimallah I, Xin L, et al. (2023) Redox and Immune Signaling in Schizophrenia: New Therapeutic Potential. Int J Neuropsychopharmacol 26 (5):309–321. https://doi.org/10.1093/ijnp/pyad012

Enache D, Pariante CM, and Mondelli V (2019) Markers of Central Inflammation in Major Depressive Disorder: A Systematic Review and Meta-Analysis of Studies Examining Cerebrospinal Fluid, Positron Emission Tomography and Post-Mortem Brain Tissue. Brain Behav Immun 81:24–40. https://doi.org/10.1016/j.bbi.2019.06.015

Farooqi IS, Keogh JM, Yeo GSH, et al. (2003) Clinical Spectrum of Obesity and Mutations in the Melanocortin 4 Receptor Gene. N Engl J Med 348 (12):1085–1095. https://doi.org/10.1056/nejmoa022050

Farooqi IS, Wangensteen T, Collins S, et al. (2007) Clinical and Molecular Genetic Spectrum of Congenital Deficiency of the Leptin Receptor. N Engl J Med 356 (3):237–247. https://doi.org/10.1056/nejmoa063988

Felger JC, and Treadway MT (2017) Inflammation Effects on Motivation and Motor Activity: Role of Dopamine. Neuropsychopharmacology 42 (1):216–241. https://doi.org/10.1038/npp.2016.143

Foley ÉM, Parkinson JT, Mitchell RE, Turner L, and Khandaker GM (2022) Peripheral Blood Cellular Immunophenotype in Depression: A Systematic Review and Meta-Analysis. Mol Psychiatr 28 (3):1004–1019. https://doi.org/10.1038/s41380-022-01919-7

Fourrier C, Sampson E, Mills NT, and Baune BT (2018) Anti-Inflammatory Treatment of Depression: Study Protocol for a Randomised Controlled Trial of Vortioxetine Augmented with Celecoxib or Placebo. Trials 19 (1):447. https://doi.org/10.1186/s13063-018-2829-7

Frank P, Jokela M, Batty GD, et al. (2021) Association between Systemic Inflammation and Individual Symptoms of Depression: A Pooled Analysis of 15 Population-Based Cohort Studies. Am J Psychiatry 178 (12):1107–1118. https://doi.org/10.1176/appi.ajp.2021.20121776

Frank P, Jokela M, Batty GD, et al. (2022) Overweight, Obesity, and Individual Symptoms of Depression: A Multicohort Study with Replication in UK Biobank. Brain Behav Immun 105:192–200. https://doi.org/10.1016/j.bbi.2022.07.009

Giannakopoulou O, Lin K, Meng X, et al. (2021) The Genetic Architecture of Depression in Individuals of East Asian Ancestry: A Genome-Wide Association Study. JAMA Psychiatry 78 (11):1258–1269. https://doi.org/10.1001/jamapsychiatry.2021.2099

Gold SM, Köhler-Forsberg O, Moss-Morris R, et al. (2020) Comorbid Depression in Medical Diseases. Nat Rev Dis Primers 6:1–22. https://doi.org/10.1038/s41572-020-0200-2

Goldsmith DR, Bekhbat M, Mehta ND, and Felger JC (2023) Inflammation-Related Functional and Structural Dysconnectivity as a Pathway to Psychopathology. Biol Psychiatry 93 (5):405–418. https://doi.org/10.1016/j.biopsych.2022.11.003

Harrison NA, Brydon L, Walker C, et al. (2009) Inflammation Causes Mood Changes through Alterations in Subgenual Cingulate Activity and Mesolimbic Connectivity. Biol Psychiatry 66 (5):407–414. https://doi.org/10.1016/j.biopsych.2009.03.015

Hellmann-Regen J, Clemens V, Grözinger M, et al. (2022) Effect of Minocycline on Depressive Symptoms in Patients with Treatment-Resistant Depression: A Randomized Clinical Trial. JAMA Netw Open 5 (9):e2230367. https://doi.org/10.1001/jamanetworkopen.2022.30367

Hotamisligil GS (2017) Inflammation, Metaflammation and Immunometabolic Disorders. Nature 542 (7640):177–185. https://doi.org/10.1038/nature21363

Howard DM, Adams MJ, Clarke T-K, et al. (2019) Genome-Wide Meta-Analysis of Depression Identifies 102 Independent Variants and Highlights the Importance of the Prefrontal Brain Regions. Nat Neurosci 22 (3):343–352. https://doi.org/10.1038/s41593-018-0326-7

Huang L, Tang S, Rietkerk J, et al. (2023) Polygenic Analyses Show Important Differences between Major Depressive Disorder Symptoms Measured Using Various Instruments. Biol Psychiatry 95 (12):1110–1121. https://doi.org/10.1016/j.biopsych.2023.11.021

Husain MI, Chaudhry IB, Khoso AB, et al. (2020) Minocycline and Celecoxib as Adjunctive Treatments for Bipolar Depression: A Multicentre, Factorial Design Randomised Controlled Trial. Lancet Psychiatry 7 (6):515–527. https://doi.org/10.1016/S2215-0366(20)30138-3

Jais A, and Brüning JC (2017) Hypothalamic Inflammation in Obesity and Metabolic Disease. J Clin Invest 127 (1):24–32. https://doi.org/10.1172/jci88878

Jansen R, Milaneschi Y, Schranner D, et al. (2023) The Metabolome-Wide Signature of Major Depressive Disorder. https://doi.org/10.21203/rs.3.rs-3127544/v1

Julkunen H, Cichońska A, Tiainen M, et al. (2023) Atlas of Plasma NMR Biomarkers for Health and Disease in 118,461 Individuals from the UK Biobank. Nat Commun 14 (1):604. https://doi.org/10.1038/s41467-023-36231-7

Kappelmann N, Lewis G, Dantzer R, Jones PB, and Khandaker GM (2018) Antidepressant Activity of Anti-Cytokine Treatment: A Systematic Review and Meta-Analysis of Clinical Trials of Chronic Inflammatory Conditions. Mol Psychiatr 23 (2):335–343. https://doi.org/10.1038/mp.2016.167

Kendler KS, Aggen SH, and Neale MC (2013) Evidence for Multiple Genetic Factors Underlying *DSM-IV* Criteria for Major Depression. JAMA Psychiatry 70 (6):599–607. https://doi.org/10.1001/jamapsychiatry.2013.751

Khadimallah I, Jenni R, Cabungcal J-H, et al. (2022) Mitochondrial, Exosomal Mir137-Cox6a2 and Gamma Synchrony as Biomarkers of Parvalbumin Interneurons, Psychopathology, and Neurocognition in Schizophrenia. Mol Psychiatr 27 (2):1192–1204. https://doi.org/10.1038/s41380-021-01313-9

Khandaker GM, Oltean BP, Kaser M, et al. (2018) Protocol for the Insight Study: A Randomised Controlled Trial of Single-Dose Tocilizumab in Patients with Depression and Low-Grade Inflammation. BMJ Open 8 (9):e025333. https://doi.org/10.1136/bmjopen-2018-025333

Khandaker GM, Zuber V, Rees JMB, et al. (2020) Shared Mechanisms between Coronary Heart Disease and Depression: Findings from a Large UK General Population-Based Cohort. Mol Psychiatr 25 (7):1477–1486. https://doi.org/10.1038/s41380-019-0395-3

Kim S-Y, Cohen BM, Chen X, et al. (2017) Redox Dysregulation in Schizophrenia Revealed by *in Vivo* NAD+/NADH Measurement. Schizophr Bull 43 (1):197–204. https://doi.org/10.1093/schbul/sbw129

Kohen I, and Lester P (2008) Exenatide-Induced Depression in a Geriatric Patient. Int J Geriatr Psychiatry 23 (4):443–444. https://doi.org/10.1002/gps.1937

Köhler-Forsberg O, N Lydholm C, Hjorthøj C, et al. (2019) Efficacy of Anti-Inflammatory Treatment on Major Depressive Disorder or Depressive Symptoms: Meta-Analysis of Clinical Trials. Acta Psychiatr Scand 139 (5):404–419. https://doi.org/10.1111/acps.13016

Kroemer NB, Opel N, Teckentrup V, et al. (2022) Functional Connectivity of the Nucleus Accumbens and Changes in Appetite in Patients with Depression. JAMA Psychiatry 79 (10):993–1003. https://doi.org/10.1001/jamapsychiatry.2022.2464

Lasselin J, Karshikoff B, Axelsson J, et al. (2020) Fatigue and Sleepiness Responses to Experimental Inflammation and Exploratory Analysis of the Effect of Baseline Inflammation in Healthy Humans. Brain Behav Immun 83:309–314. https://doi.org/10.1016/j.bbi.2019.10.020

Levey DF, Stein MB, Wendt FR, et al. (2021) Bi-Ancestral Depression GWAS in the Million Veteran Program and Meta-Analysis in >1.2 Million Individuals Highlight New Therapeutic Directions. Nat Neurosci 24 (7):954–963. https://doi.org/10.1038/s41593-021-00860-2

Li JR, Cao J, Wei J, and Geng W (2023) Case Report: Semaglutide-Associated Depression: A Report of Two Cases. Front Psychiatry 14:1238353. https://doi.org/10.3389/fpsyt.2023.1238353

Lynall M-E, Soskic B, Hayhurst J, et al. (2022) Genetic Variants Associated with Psychiatric Disorders Are Enriched at Epigenetically Active Sites in Lymphoid Cells. Nat Commun 13 (1):6102. https://doi.org/10.1038/s41467-022-33885-7

Major Depressive Disorder Working Group of the Psychiatric Genomics Consortium (2025) Trans-Ancestry Genome-Wide Study of Depression Identifies 697 Associations Implicating Cell Types and Pharmacotherapies. Cell 188 (3):640–652.e649. https://doi.org/10.1016/j.cell.2024.12.002

Mansur RB, Ahmed J, Cha DS, et al. (2017) Liraglutide Promotes Improvements in Objective Measures of Cognitive Dysfunction in Individuals with Mood Disorders: A Pilot, Open-Label Study. J Affect Disord 207:114–120. https://doi.org/10.1016/j.jad.2016.09.056

Marx W, Penninx BWJH, Solmi M, et al. (2023) Major Depressive Disorder. Nat Rev Dis Primers 9 (1):44. https://doi.org/10.1038/s41572-023-00454-1

McIntyre RS, Subramaniapillai M, Lee Y, et al. (2019) Efficacy of Adjunctive Infliximab vs Placebo in the Treatment of Adults with Bipolar I/II Depression: A Randomized Clinical Trial. JAMA Psychiatry 76 (8):783–790. https://doi.org/10.1001/jamapsychiatry.2019.0779

Medzhitov R (2008) Origin and Physiological Roles of Inflammation. Nature 454 (7203):428–435. https://doi.org/10.1038/nature07201

Meng X, Navoly G, Giannakopoulou O, et al. (2024) Multi-Ancestry Genome-Wide Association Study of Major Depression Aids Locus Discovery, Fine Mapping, Gene Prioritization and Causal Inference. Nat Genet 56 (2):222–233. https://doi.org/10.1038/s41588-023-01596-4

Milaneschi Y, Arnold M, Kastenmüller G, et al. (2022) Genomics-Based Identification of a Potential Causal Role for Acylcarnitine Metabolism in Depression. J Affect Disord 307:254–263. https://doi.org/10.1016/j.jad.2022.03.070

Milaneschi Y, Kappelmann N, Ye Z, et al. (2021a) Association of Inflammation with Depression and Anxiety: Evidence for Symptom-Specificity and Potential Causality from UK Biobank and NESDA Cohorts. Mol Psychiatr 26 (12):7393–7402. https://doi.org/10.1038/s41380-021-01188-w

Milaneschi Y, Lamers F, Berk M, and Penninx BWJH (2020) Depression Heterogeneity and Its Biological Underpinnings: Toward Immunometabolic Depression. Biol Psychiatry 88 (5):369–380. https://doi.org/10.1016/j.biopsych.2020.01.014

Milaneschi Y, Lamers F, and Penninx BWJH (2021b) Dissecting Depression Biological and Clinical Heterogeneity: The Importance of Symptom Assessment Resolution. JAMA Psychiatry 78 (3):341. https://doi.org/10.1001/jamapsychiatry.2020.4373

Milaneschi Y, Lamers F, Peyrot WJ, et al. (2017) Genetic Association of Major Depression with Atypical Features and Obesity-Related Immunometabolic Dysregulations. JAMA Psychiatry 74 (12):1214–1225. https://doi.org/10.1001/jamapsychiatry.2017.3016

Milaneschi Y, Simmons WK, van Rossum EFC, and Penninx BWJH (2019) Depression and Obesity: Evidence of Shared Biological Mechanisms. Mol Psychiatr 24 (1):18–33. https://doi.org/10.1038/s41380-018-0017-5

Miller AH, and Raison CL (2016) The Role of Inflammation in Depression: From Evolutionary Imperative to Modern Treatment Target. Nat Rev Immunol 16 (1):22–34. https://doi.org/10.1038/nri.2015.5

——— (2023) Burning Down the House: Reinventing Drug Discovery in Psychiatry for the Development of Targeted Therapies. Mol Psychiatr 28 (1):68–75. https://doi.org/10.1038/s41380-022-01887-y

Montanari S, Jansen R, Schranner D, et al. (2025) Acylcarnitines Metabolism in Depression: Association with Diagnostic Status, Depression Severity and Symptom Profile in the NESDA Cohort. Transl Psychiatry 15 (1):65. https://doi.org/10.1038/s41398-025-03274-x

Moulton CD, Hopkins CWP, Ismail K, and Stahl D (2018) Repositioning of Diabetes Treatments for Depressive Symptoms: A Systematic Review and Meta-Analysis of Clinical Trials. Psychoneuroendocrinology 94:91–103. https://doi.org/10.1016/j.psyneuen.2018.05.010

Nettis MA, Lombardo G, Hastings C, et al. (2021) Augmentation Therapy with Minocycline in Treatment-Resistant Depression Patients with Low-Grade Peripheral Inflammation: Results from a Double-Blind Randomised Clinical Trial. Neuropsychopharmacology 46 (5):939–948. https://doi.org/10.1038/s41386-020-00948-6

Nibber A, Singh H, Burnet P, Lennox B, and Minichino A (2022) Investigating the Pro-Cognitive and Anti-Depressant Efficacy of Metformin: A Systematic Review and Meta-Analysis of Randomised Controlled Trials. J Affect Disord 310:52–59. https://doi.org/10.1016/j.jad.2022.04.156

O'Loughlin J, Casanova F, Fairhurst-Hunter Z, et al. (2023) Mendelian Randomisation Study of Body Composition and Depression in People of East Asian Ancestry Highlights Potential Setting-Specific Causality. BMC Med 21 (1):37. https://doi.org/10.1186/s12916-023-02735-8

Obici S, Zhang BB, Karkanias G, and Rossetti L (2002) Hypothalamic Insulin Signaling Is Required for Inhibition of Glucose Production. Nat Med 8 (12):1376–1382. https://doi.org/10.1038/nm1202-798

Osimo EF, Baxter LJ, Lewis G, Jones PB, and Khandaker GM (2019) Prevalence of Low-Grade Inflammation in Depression: A Systematic Review and Meta-Analysis of CRP Levels. Psychol Med 49 (12):1958–1970. https://doi.org/10.1017/s0033291719001454

Otte C, Chae WR, Nowacki J, et al. (2020) Simvastatin Add-on to Escitalopram in Patients with Comorbid Obesity and Major Depression (SIMCODE): Study Protocol of a Multicentre, Randomised, Double-Blind, Placebo-Controlled Trial. BMJ Open 10 (12):e040119. https://doi.org/10.1136/bmjopen-2020-040119

Penninx B, Lamers F, Jansen R, et al. (2025) Immuno-Metabolic Depression: From Concept to Implementation. Lancet Reg Health Eur 48:101166. https://doi.org/10.1016/j.lanepe.2024.101166

Perry BI, Upthegrove R, Kappelmann N, et al. (2021) Associations of Immunological Proteins/Traits with Schizophrenia, Major Depression and Bipolar Disorder: A Bi-Directional Two-Sample Mendelian Randomization Study. Brain Behav Immun 97:176–185. https://doi.org/10.1016/j.bbi.2021.07.009

Pistis G, Milaneschi Y, Vandeleur CL, et al. (2021) Obesity and Atypical Depression Symptoms: Findings from Mendelian Randomization in Two European Cohorts. Transl Psychiatry 11 (1):96. https://doi.org/10.1038/s41398-021-01236-7

Raison CL, Rutherford RE, Woolwine BJ, et al. (2013) A Randomized Controlled Trial of the Tumor Necrosis Factor Antagonist Infliximab for Treatment-Resistant Depression: The Role of Baseline Inflammatory Biomarkers. JAMA Psychiatry 70 (1):31–41. https://doi.org/10.1001/2013.jamapsychiatry.4

Savitz J, and Harrison NA (2018) Interoception and Inflammation in Psychiatric Disorders. Biol Psychiatry Cogn Neurosci Neuroimaging 3 (6):514–524. https://doi.org/10.1016/j.bpsc.2017.12.011

Scott B, Day EA, O'Brien KL, et al. (2024) Metformin and Feeding Increase Levels of the Appetite-Suppressing Metabolite Lac-Phe in Humans. Nat Metab 6 (4):651–658. https://doi.org/10.1038/s42255-024-01018-7

Simmons WK, Burrows K, Avery JA, et al. (2020) Appetite Changes Reveal Depression Subgroups with Distinct Endocrine, Metabolic, and Immune States. Mol Psychiatr 25 (7):1457–1468. https://doi.org/10.1038/s41380-018-0093-6

Sørensen NV, Frandsen BH, Orlovska-Waast S, et al. (2022) Immune Cell Composition in Unipolar Depression: A Comprehensive Systematic Review and Meta-Analysis. Mol Psychiatr 28 (1):391–401. https://doi.org/10.1038/s41380-022-01905-z

Speed MS, Jefsen OH, Børglum AD, Speed D, and Østergaard SD (2019) Investigating the Association between Body Fat and Depression via Mendelian Randomization. Transl Psychiatry 9 (1):184. https://doi.org/10.1038/s41398-019-0516-4

Strawbridge R, Arnone D, Danese A, et al. (2015) Inflammation and Clinical Response to Treatment in Depression: A Meta-Analysis. Eur Neuropsychopharmacol 25 (10):1532–1543. https://doi.org/10.1016/j.euroneuro.2015.06.007

Thase ME (2009) Atypical Depression: Useful Concept, but It's Time to Revise the DSM-IV Criteria. Neuropsychopharmacology 34 (13):2633–2641. https://doi.org/10.1038/npp.2009.100

Tian YE, Di Biase MA, Mosley PE, et al. (2023) Evaluation of Brain-Body Health in Individuals with Common Neuropsychiatric Disorders. JAMA Psychiatry 80 (6):567–576. https://doi.org/10.1001/jamapsychiatry.2023.0791

Treadway MT, Cooper JA, and Miller AH (2019) Can't or Won't? Immunometabolic Constraints on Dopaminergic Drive. Trends Cogn Sci 23 (5):435–448. https://doi.org/10.1016/j.tics.2019.03.003

Tyrrell J, Mulugeta A, Wood AR, et al. (2019) Using Genetics to Understand the Causal Influence of Higher Bmi on Depression. Int J Epidemiol 48 (3):834–848. https://doi.org/10.1093/ije/dyy223

Uhlén M, Fagerberg L, Hallström BM, et al. (2015) Proteomics. Tissue-Based Map of the Human Proteome. Science 347 (6220):1260419. https://doi.org/10.1126/science.1260419

van der Klaauw AA, and Farooqi IS (2015) The Hunger Genes: Pathways to Obesity. Cell 161 (1):119–132. https://doi.org/10.1016/j.cell.2015.03.008

van der Spek A, Stewart ID, Kühnel B, et al. (2023) Circulating Metabolites Modulated by Diet Are Associated with Depression. Mol Psychiatr 28 (9):3874–3887. https://doi.org/10.1038/s41380-023-02180-2

van Loo HM, Aggen SH, and Kendler KS (2022) The Structure of the Symptoms of Major Depression: Factor Analysis of a Lifetime Worst Episode of Depressive Symptoms in a Large General Population Sample. J Affect Disord 307:115–124. https://doi.org/10.1016/j.jad.2022.03.064

Vogelzangs N, Beekman ATF, van Reedt Dortland AKB, et al. (2014) Inflammatory and Metabolic Dysregulation and the 2-Year Course of Depressive Disorders in Antidepressant Users. Neuropsychopharmacology 39 (7):1624–1634. https://doi.org/10.1038/npp.2014.9

Wittenberg GM, Stylianou A, Zhang Y, et al. (2020) Effects of Immunomodulatory Drugs on Depressive Symptoms: A Mega-Analysis of Randomized, Placebo-Controlled Clinical Trials in Inflammatory Disorders. Mol Psychiatr 25 (6):1275–1285. https://doi.org/10.1038/s41380-019-0471-8

Wong CK, McLean BA, Baggio LL, et al. (2024) Central Glucagon-Like Peptide 1 Receptor Activation Inhibits Toll-Like Receptor Agonist-Induced Inflammation. Cell Metab 36 (1):130–143.e135. https://doi.org/10.1016/j.cmet.2023.11.009

Wray NR, Ripke S, Mattheisen M, et al. (2018) Genome-Wide Association Analyses Identify 44 Risk Variants and Refine the Genetic Architecture of Major Depression. Nat Genet 50 (5):668–681. https://doi.org/10.1038/s41588-018-0090-3

Xiao S, Li VL, Lyu X, et al. (2024) Lac-Phe Mediates the Effects of Metformin on Food Intake and Body Weight. Nat Metab 6 (4):659–669. https://doi.org/10.1038/s42255-024-00999-9

Zemdegs J, Martin H, Pintana H, et al. (2019) Metformin Promotes Anxiolytic and Antidepressant-Like Responses in Insulin-Resistant Mice by Decreasing Circulating Branched-Chain Amino Acids. J Neurosci 39 (30):5935–5948. https://doi.org/10.1523/jneurosci.2904-18.2019

Zwiep JC, Bet PM, Rhebergen D, et al. (2023) Efficacy of Celecoxib Add-on Treatment for Immuno-Metabolic Depression: Protocol of the Inflamed Double-Blind Placebo-Controlled Randomized Controlled Trial. Brain Behav Immun Health 27:100585. https://doi.org/10.1016/j.bbih.2022.100585

Open Access This chapter is licensed under the terms of the Creative Commons Attribution-NonCommercial-NoDerivatives 4.0 International License (http://creativecommons.org/licenses/by-nc-nd/4.0/), which permits any noncommercial use, sharing, distribution and reproduction in any medium or format, as long as you give appropriate credit to the original author(s) and the source, provide a link to the Creative Commons license and indicate if you modified the licensed material. You do not have permission under this license to share adapted material derived from this chapter or parts of it.

The images or other third party material in this chapter are included in the chapter's Creative Commons license, unless indicated otherwise in a credit line to the material. If material is not included in the chapter's Creative Commons license and your intended use is not permitted by statutory regulation or exceeds the permitted use, you will need to obtain permission directly from the copyright holder.

10

Metabolism across the Lifespan

Implications for Body and Mind

Benjamin I. Perry, Bodyl A. Brand, Janna N. de Boer, Peter Swann, Toby Pillinger, Lena Pielke, Suchitra Varadarajan, Iain Campbell, and Iris-Tatjana Kolassa

Abstract Observations of comorbidity between disordered metabolism and neuropsychiatric illness have existed for over 100 years. From the first case reports and small cross-sectional studies of the early 20th century through to modern 21st century cohort studies comprising entire populations, evidence has accumulated confirming that metabolic manifestations are a key component of neuropsychiatric illness across the lifespan. Traditionally, observed comorbidity has been attributed to consequences of the illness (e.g., lifestyle factors), treatment (e.g., adverse metabolic effects), or health systems (e.g., health inequalities or diagnostic overshadowing). In recent decades, however, researchers have started to unearth putative pathophysiological links between metabolism and mind, thus heralding promise to transform our understanding of neuropsychiatric illnesses and uncover promising new avenues on how to treat them. Here, research is presented that links body and mind by metabolism at key life phases (pre- and postnatal periods through to older adulthood), with a consideration of the relevant sex differences. Promising strands of basic, epidemiological, and experimental research are presented alongside proposed directions for future research. This article underscores the life-long neuropsychiatric relevance of mitochondrial function, glucose-insulin homeostasis, sex hormones, and the immune system, and how these might be leveraged to develop novel therapies for neuropsychiatric illness.

Keywords Metabolic psychiatry, lifespan, childhood, adulthood, older adulthood, inflammation

Benjamin I. Perry (✉)
Institute for Mental Health, School of Psychology, University of Birmingham, Birmingham B15 2TT, UK
E-Mail: b.i.perry@bham.ac.uk

Affiliations for the coauthors are available in the List of Contributors.

© Frankfurt Institute for Advanced Studies (FIAS) 2025

D. Öngür, J. M. Ford (eds.), *Metabolic Neuropsychiatry*, Strüngmann Forum Reports,
https://doi.org/10.1007/978-3-032-05630-6_10

1 Introduction

In 1641, in *Meditations on First Philosophy*, the French philosopher Renee Descartes theorized the notion of dualism: "The difficulty is not merely that mind and body are different. It is that they are different in such a way that their interaction is impossible." Over the past few decades, the burgeoning field of metabolic psychiatry—a field that unites researchers from psychiatry, psychology, neuroscience, and metabolic medicine and links mind and body by metabolic processes—has started not only to reject mind-body dualism but to transform our pathophysiological understanding of mental disorders and unearth promising new avenues on how to treat them.

Clues to the importance of metabolism as a mediator between body and mind have long been found in observational epidemiological research. For example, in the years following Henry Maudsley's observations of type 2 diabetes (T2D) as "a disease which often shows itself in families in which insanity prevails" (Maudsley 1879), several small studies reported the first research evidence that disordered glucose-insulin metabolism was more likely to occur in individuals with mental disorders compared to those without (Kohen 2004). Those small early studies paved the way for modern whole-population analyses of millions of adults who were followed for decades (Momen et al. 2022), which found substantially higher rates of metabolic disorders among individuals diagnosed with mental disorders compared with the rest of the population.

Traditional attributions imply that the high burden of metabolic disruption in people with mental disorders follows wholly as a consequence of the mental disorder itself. For example, lifestyle factors that predispose to metabolic disorders (e.g., such as an unhealthy diet, lower levels of physical activity, disrupted sleep, and a higher prevalence of smoking) are common across the spectrum of mental disorders. Further, healthcare inequalities are prevalent and pervasive for people with mental disorders, such that they receive inadequate investigation and treatment for their physical health across healthcare settings.

In addition, psychotropic medications used to treat mental disorders can further exacerbate disturbances in appetite, sleep, and physical activity levels, with detectable metabolic impacts precipitating within weeks. Centrally, psychotropic modulation of dopaminergic, histamine-1, serotonin-2c, muscarinic, and adrenergic receptors contribute to changes in appetite and levels of sedation; interestingly, neurotransmitters such as dopamine and noradrenaline have peripheral action in metabolically relevant organs such as the pancreas and adipose tissue. In regard to the former, although the mechanisms remain incompletely understood, recent evidence suggests that pancreatic α- and β-cells both synthesize and signal catecholamines in order to assist the regulation of glucagon and insulin release (Aslanoglou et al. 2021). Regarding the latter, antipsychotics have direct effects on adipose tissue metabolism through mechanisms such as attenuating lipolysis, reducing adipocyte glucose uptake, and altering the expression of mitochondrial function genes and influencing inflammation (Chen et al. 2023).

While traditional attributions are important and warrant attention, a growing body of more recent observational research is beginning to suggest that they are unlikely to tell the full story. First, meta-analyses of observational studies confirm that metabolic dysfunction, particularly insulin resistance, is already detectable from at least the onset of mental disorders in young, psychotropic-naïve individuals (Perry et al. 2016), many years earlier than these metabolic disturbances are typically observed in the rest of the population. Second, evidence from large-scale genetic epidemiological studies of hundreds of thousands of individuals has pointed to the potential for overlapping, common genetic and environmental antecedents simultaneously predisposing to metabolic dysfunction and the onset of mental disorders (Perry et al. 2021a). Finally, metabolic dysfunction in adulthood is longitudinally associated with deterioration in cognitive function and a higher risk of neuropsychiatric disorders in older adulthood (Livingston et al. 2020).

Together, these traditional and novel bidirectional attributions result in metabolic comorbidity featuring as the predominant explanation for an average 15- to 20-year reduction in life expectancy for people with mental disorders, as summarized in Figure 10.1. In fact, recent analyses postulate that the physical manifestations are better predictors of mental disorders than those identifiable manifestations arising in the brain (Tian et al. 2023). Furthermore, the more complex, if not reversed direction of association between metabolic and mental disorders implied by more recent research suggests the potential for therapeutic modulation of metabolic pathways to treat not only the metabolic comorbidity, which is tightly bound to mental disorders, but the mental disorders themselves.

Here, we adopt a life-course approach to metabolism and discuss its implications for body and mind. We focus on emerging developments triangulated across basic, experimental, and epidemiological sciences at key life phases, from early life to late adulthood. Key implications at each life phase are highlighted, as are areas of promise requiring further exploration and research.

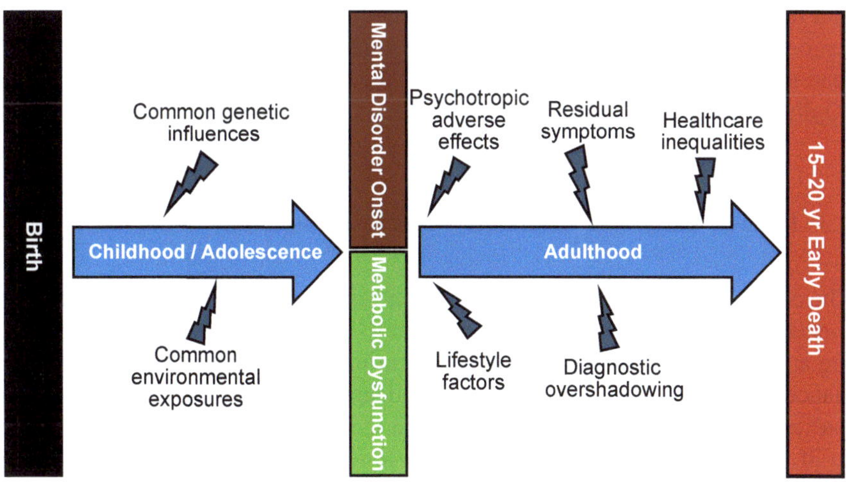

Figure 10.1 Causes of metabolic morbidity in mental disorders across the life course, contributing to preventable early mortality.

2 Conception to Early Life

2.1 Mitochondria as Core Regulators of Metabolism

Mitochondria are membrane-bound organelles found in the cells of most eukaryotes (e.g., animals and plants). Mitochondria generate cellular energy in the form of adenosine triphosphate (ATP) and are at the core of our metabolism: they are vital for healthy human oocyte, embryonic, and neonatal development. Mitochondria regulate cellular respiration and metabolism, stem cell differentiation, reactive oxygen species (ROS), cell division and signaling pathways, gene expression, apoptosis, steroid hormone synthesis, heat production, and membrane potential among other important processes. Dynamic regulation of mitochondria is essential for the coordination of key cellular events during embryogenesis, hence mitochondrial abnormalities and dysfunction have devastating consequences on fertility as well as embryonic and offspring health.

2.2 Environmental Influences on Developmental Imprinting and the Role of Mitochondria

Pregnancy and early childhood are sensitive periods during which noxious environmental influences (e.g., poor maternal diet, exposure to pollutants) can impact offspring health. Embryonic developmental imprinting occurs during gestation through gene imprinting (i.e., the partial or complete silencing of genes via DNA methylation and histone methylation), which in turn can be influenced by environmental or maternal physiological factors or through alterations in mitochondrial biology. Gestational exposure to noxious influences such as environmental pollutants and toxins, maternal obesity, poor maternal diet, maternal inflammation (e.g., in response to viral infection), and psychosocial stress affects the quality and bioenergetic capacity of mitochondrial DNA (mtDNA), and thereby the structural and functional properties of cells, tissues, and organs. On an organismic level, alterations in mtDNA quality, mitochondrial content, and bioenergetic capacity can lead to wide-ranging influences on health and a life-long susceptibility to disease (Gyllenhammer et al. 2020).

2.3 Infancy, Early-Life Traumatic Stress Load, and Metabolism

The consequences of exposure to early-life stressful life events (SLEs) on long-term mental health outcomes are well documented. Individuals with a history of SLEs are at an increased risk of a host of different mental disorders (e.g., anxiety, substance use, psychosis, eating disorders, and depression) that characteristically occur earlier, are chronic and more difficult to treat (Xiao et al. 2023). Similarly, exposure to SLEs increases the risk of early-onset age-related diseases such as T2D, autoimmune diseases, neurodegenerative diseases, cardiovascular disease, and cancer (Felitti et al. 2019).

Mitochondrial biology could play a key role in mediating the connection between early-life noxious stressors and later-life disease risk through effects on the immune system. Early-life SLEs can lead to permanent alterations in the metabolism of mitochondria, which are master regulators of ROS and inflammatory responses. Inflammation activates indoleamine 2, 3-dioxygenase (IDO1), an enzyme which metabolizes L-tryptophan via the kynurenine pathway ultimately producing nicotinamide adenine dinucleotide (NAD^+). NAD^+ is essential for DNA repair, but activation of the kynurenine pathway depletes L-tryptophan levels with impacts on the synthesis of serotonin, a neurotransmitter with implications in the pathophysiology of a number of neuropsychiatric disorders. Furthermore, high levels of DNA breakage, ROS, inflammation, and mitochondrial energy production have been observed to occur together in individuals who have experienced SLEs and individuals with mental disorders (Morath et al. 2014). DNA breakage is associated with a higher expression of the NAD^+ consuming DNA repair enzyme PARP1 (Behnke et al. 2022), which induces a metabolic shift from anaerobic glycolysis to mitochondria-dependent oxidative phosphorylation, which can carry effects in the periphery and CNS predisposing to the onset of wide-ranging neuropsychiatric disorders (Gimenez-Palomo et al. 2021).

Alterations in mtDNA copy number (mtDNAcn) and mitochondrial mutational load are additional putative mechanisms linking early-life stress with metabolism and the risk of mental disorders. Both have been observed in the placenta of children of traumatized mothers. In addition, maternal cumulative lifetime distress is linked to reduced placental mtDNAcn and higher mitochondrial mutational load in fetal placental tissue. Whether early-life or lifetime traumatic stress exposure, such as maternal or paternal history of SLEs, also effect germline mitochondrial quality and/or the generation of ROS and associated inflammation is a promising avenue for future research. Some evidence exists for inter- and transgenerational effects of early-life stress on the next generation via altered micro RNAs in the germline (Jawaid et al. 2021).

The intergenerational consequences of maternal traumatic experiences on metabolism are beginning to emerge. For example, higher maternal allostatic load (i.e., the aggregated score of biological factors such as body mass index [BMI], inflammation, and insulin resistance/T2D) may increase mitochondrial activity in offspring and be associated with deleterious effects on offspring behavior and higher risks of neuropsychiatric disorders in offspring adulthood (Nogueira Avelar et al. 2021).

2.4 Summary and Future Directions

Mitochondrial biopsychology is an emerging field in mental health research which holds the potential to revolutionize our pathophysiological understanding of mental disorders and how to treat them. Now required is a better understanding of how factors that alter mitochondrial energy production (e.g., infections, medication, environmental toxins, poor or pro-inflammatory diet, psychosocial stressors) during pregnancy affect the development of the embryo and its long-term repercussions on the development of mental and physical comorbidity. For example, it is currently unclear whether altered mitochondrial energy production affects vulnerability to specific or broad adverse mental and physical outcomes in which mental and physical diseases the vulnerability is increased, which environmental factors are most influential, and whether those factors may be susceptible to therapeutical targeting.

Similarly, it is unclear whether possible early-life noxious influences on mitochondria structure or function lead to systemic alterations that affect the whole organism or lead to specific changes in particularly sensitive organ systems. For example, organogenesis after gastrulation seems to be highly dependent on mitochondrial quality and integrity, and disturbances in mitochondrial functioning can even halt embryonic development. Yet, noxious stressors might also affect some indices of mitochondrial functioning more than others. Some mitochondrial indices might be more temporarily altered and could therefore be targeted by improving diet, gut health, or physical exercise, whereas others might be less easily reversed.

The need for DNA repair to maintain genomic integrity is equally important as maintaining a constant flow of cellular energy production via mitochondrial bioenergetics. Currently, no single measure to assess the structural and functional integrity of mitochondrial biology exists, because mitochondria are dynamic and are regulated in a complex manner not only via the crosstalk of mtDNA and nuclear DNA, but also in a system biological manner via gut-derived metabolites and dietary nutrients. Therefore, a panel of standardized biomarkers indexing distinct aspects of mitochondrial function and immunometabolic health will be essential to advance the field. Promising attempts for such measures have been proposed like the mitochondrial health index (Picard et al. 2018). Advanced research is required to develop more suitable and ready-to-use in practice indices to evaluate mitochondrial health and performance and to identify factors that help to improve or normalize immunometabolic alterations in mental health.

3 Puberty, Hormonal Influences, and Sex Differences

3.1 Sex Hormones in Males and Females

Biological differences between men and women are evident both within and outside of the CNS, many of which are the result of variations in sex hormones. The surge in sex hormones during puberty gives rise to two distinct biological systems, driven by testosterone in men and by progesterone and estrogen in women (Figure 10.2). In men, testosterone levels remain relatively stable throughout the lifespan, gradually declining once puberty has passed. In contrast, women undergo distinct life stages marked by fluctuations in sex hormones. After the onset of the menstrual cycle at puberty (i.e., menarche), estrogens and progesterone remain high during the reproductive years—with monthly fluctuations during each menstrual cycle—and decline after menopause. In addition, both sex hormones significantly rise throughout pregnancy and decrease following childbirth.

While both female sex hormones may have many relevant effects within and outside the CNS, the protective effects of estrogens are more well-established and acknowledged. Upon binding with genomic (ERα and ERβ) and nongenomic (GPER) receptors, estrogens influence the transcription of genes involved in energy metabolism and cell survival, and activate intracellular signaling pathways that reduce

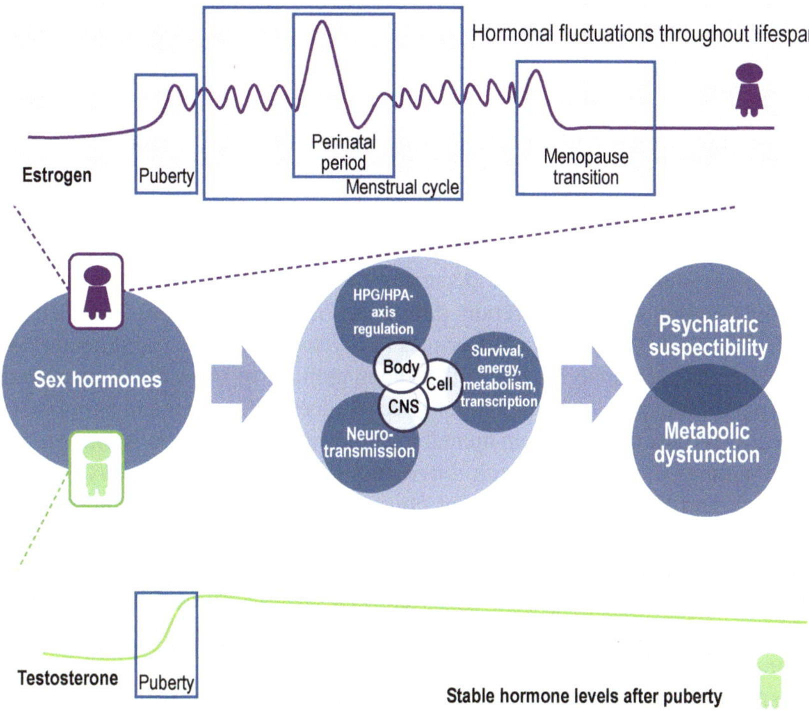

Figure 10.2 Sex hormone differences in males and females and associations with metabolic and mental disorders.

oxidative stress, apoptosis, and inflammation (Mauvais-Jarvis et al. 2013). In the CNS, estrogens strongly impact neuronal activity by modulating the signaling of neurotransmitters, including serotonin, dopamine, glutamate, and γ-aminobutyric acid (GABA). Moreover, estrogens enhance cell proliferation, synaptic sprouting, dendritic spine density, axon growth, and myelination. In women, both estrogen deficiency and excessive levels of androgens are associated with metabolic dysfunction as well as with higher risks for neuropsychiatric disorders including depression and psychosis (Brand et al. 2021).

3.2 Hormonal Disturbance and Polycystic Ovary Syndrome

The widespread consequences of dysregulated sex hormone levels in women is further exemplified by the endocrine disorder polycystic ovary syndrome (PCOS), which affects 8–13% of reproductive aged women and is characterized by dysregulated levels of estrogens and excessive androgen levels (hyperandrogenism) (Teede et al. 2018). PCOS is strongly linked to metabolic dysfunction, including insulin resistance, obesity, and glucose intolerance, and has also been associated with a variety of psychiatric problems, such as depression, psychosis, and anxiety. The exact cause of psychiatric vulnerability in PCOS remains uncertain, but it is hypothesized that, alongside dysregulated sex hormones, women with PCOS may have an altered sensitivity to allopregnanolone.

3.3 Broader Impacts of Hormonal Changes across the Lifespan

The connection between metabolic and mental health issues, along with their shared association with sex hormone levels, is not limited to endocrine disorders but also encompasses hormonal changes occurring throughout the lifespan (Figure 10.2). Several longitudinal studies report differences in pubertal timing between males and females that may relate to differences in psychopathology in adulthood (Hoyt et al. 2020), yet meta-analyses report that the magnitude of these differences could at least in part be explained by methodological choices (Ullsperger and Nikolas 2017). Nevertheless, sex differences in mental disorders are among the most intriguing and stable findings in psychiatry, with differences regarding prevalence, symptomatology, risk, and influencing factors, or course. Thus, it is well known that women have a higher lifetime prevalence of mood or anxiety disorders than men, and a later onset of schizophrenia and other psychotic disorders (Riecher-Rossler 2017). In addition, by early adulthood, there are differences between sexes in the character and progression of metabolic dysfunction, with early adiposity and dyslipidemia more common in females (Yoshida et al. 2022).

Earlier age of menarche has been associated with high BMI at pubertal age (Currie et al. 2012), possibly attributed to increased leptin production, which triggers the onset of puberty. Interestingly, both early menarche and BMI increase at pubertal age (Perry et al. 2021b) have been linked to increased susceptibility for depression later in life. While other factors may also influence BMI, menarche timing, and the development of depression (e.g., psychosocial, environmental, and genetic influences), BMI-associated early menarche could potentially disrupt the hypothalamic-pituitary-adrenal (HPA) axis by altering leptin signaling (Mauvais-Jarvis et al. 2013). Speculatively, this disruption may reduce resilience to depression and other stress-related disorders. Both early menarche and high BMI at pubertal age have also been associated with the development of PCOS, which underscores the complex interplay between metabolic and hormonal dysregulation in women. In addition, insulin sensitivity and thus glucose-insulin homeostasis change in females across different phases of the menstrual cycle, particularly in response to brain insulin action: clear differences between the follicular and luteal phases may contribute to risk of whole-body insulin resistance (Hummel et al. 2023).

Pregnancy is another important life period associated with negative long-term effects on both metabolic and mental health—effects attributed to extensive physiological changes that affect cardiac, immune, and metabolic functions as well as distinct endocrine signaling pathways that involve a variety of steroid and peptide hormones (Galea et al. 2018). Postnatally, some women are at increased risk of precipitant neuropsychiatric disorders, including affective and psychotic disorders (Davies 2017). While the underlying biological mechanisms remain unclear, the maternal body undergoes extreme physiological changes in the postpartum period, notably a massive drop in circulating estrogens upon expulsion of the placenta

(Jones et al. 2014). It has been suggested that abnormal sensitivity to this endocrinological disturbance may confer vulnerability to psychiatric disorders in some women, an idea supported by the fact that estrogen supplementation may be beneficial to some patients (Di Florio et al. 2014).

In men, metabolic and psychiatric health issues have been associated with androgen deficiency, instead of estrogen deficiency, reflecting the distinct effects of sex hormones in men and women. Additionally, the association between pubertal BMI increase and depression is less consistent in men than in women (Perry et al. 2021b). Increased leptin that results from increased BMI delays puberty in men, which indicates that leptin regulates puberty onset in a sexually dimorphic manner. Interestingly, persistently high levels of insulin throughout adolescence, rather than pubertal BMI increase, have been linked to the development of psychosis especially in men (Perry et al. 2021b). As estrogens regulate insulin sensitivity (Mauvais-Jarvis et al. 2013), women might be more resilient to insulin resistance and may therefore also be less prone to its consequences than men.

3.4 Menopause

Menopause is characterized by low levels of estrogens and progesterone and is associated with deteriorating metabolic health and increased vulnerability to various psychiatric symptoms (Brand et al. 2022). Estrogen-based hormone replacement therapy has been shown to improve both metabolic and mental health (Davis and Baber 2022). Moreover, postmenopausal women tend to respond less well to psychotropic medication, both in terms of efficacy and tolerability (Sommer et al. 2023). Furthermore, estrogen may amplify the effectiveness of psychotropics by modifying the binding to their drug target, resulting in decreased efficacy of psychotropics when estrogen levels decline. In terms of tolerability, estrogen's protective effects on insulin sensitivity, bone mineral density, and cardiovascular health make postmenopausal women more vulnerable to various side effects of psychotropics, including osteoporosis, cardiovascular disease, and metabolic syndrome (Brand et al. 2022; Mauvais-Jarvis et al. 2013).

3.5 Implications for Psychotropic Prescribing

Across different hormonal life phases, such as menopause, pregnancy, and possibly also the menstrual cycle, women may require drug-specific dose adjustments due to the influence of estrogens on the metabolism of many psychotropics. For example, estrogens inhibit CYP1A2/CYP2C19 activity but enhance CYP3A4 activity (Brand et al. 2022). After menopause, blood concentrations of drugs metabolized by CYP1A2/CYP2C19 (e.g., venlafaxine, citalopram, olanzapine, and clozapine) may decrease, whereas levels of drugs metabolized by CYP3A4 may increase (e.g., quetiapine, lurasidone, and cariprazine) (Brand et al. 2022).

3.6 Summary and Future Directions

Estrogen plays a crucial role in shaping sex differences across psychiatric and metabolic domains. Fluctuations in estrogen levels impact both the CNS and overall bodily functions, and have well-established protective effects. The link between sex hormones and metabolic function and mental health is illustrated by conditions like PCOS, as well as by natural hormonal transition phases such as puberty, pregnancy, and menopause. The decline in estrogens during menopause affects metabolic and mental health, highlighting the potential of hormone replacement therapy during this period. Female-specific hormonal transitions also affect pharmacotherapeutic treatment response and tolerability, possibly necessitating drug-specific dose adjustments and awareness of side effects to which women are more sensitive.

Future research should prioritize the understanding of the complex relationship between sex hormones, metabolic health, and mental disorders in a sex-specific manner. Taking a lifetime approach, which considers both metabolic and mental health across the lifespan, could provide insight into whether and how hormonal life phases are associated with psychiatric vulnerability. Importantly, associations between hormonal changes and psychiatric problems may differ between disorders (e.g., depression, anxiety, and psychotic disorders). Similarly, susceptibility to one psychiatric disorder might be linked to fluctuations in sex hormone levels, whereas another may be more strongly associated with consistently low sex hormone levels. Furthermore, we need to improve our understanding of how hormonal phases affect the effectiveness and tolerability of psychotropic medications and incorporate these potential effects into the optimization of current and future treatment strategies.

4 Adulthood

4.1 Glucose-Insulin Homeostasis in the Central Nervous System

Until the 1970s, neurons were considered to be insulin insensitive, and the action of insulin in the brain was considered negligible. Over the past fifty years, neuroscience research has gradually eroded this view and established the significant role of insulin receptors and insulin signaling in the CNS (Scherer et al. 2021). This research is highly relevant given the clear evidence for disruption to peripheral glucose-insulin homeostasis in young adults by the onset of mental disorders (Perry et al. 2016), implying a putative pathophysiological link.

Insulin crosses the blood-brain barrier via a saturable transport to bind to insulin receptors widely expressed throughout the brain; insulin signaling in the brain regulates systemic nutrient partitioning in animal models and humans (Scherer et al. 2021). In the brain, insulin rapidly binds to insulin and functions via signaling pathways, including the PI3K/Akt/glycogen synthase kinase-3β (GSK-3β) pathway, the Ras/Raf-1/extracellular signal-regulated kinase (ERK1 and ERK2, ERK1/2) pathway, and the mitogen-activated protein kinase pathway. These pathways have functions related to the regulation of cell apoptosis, antioxidant defense, synaptic plasticity, neuronal glucose metabolism, and regulation of the innate and adaptive immune systems, targeted to various intracellular locations including cytosol, mitochondria, and nucleus (Milstein and Ferris 2021).

The best-known brain insulin effect is the regulation of peripheral glucose transport and metabolism. This is mediated by glucose-sensing neurons in the arcuate nucleus of the hypothalamus, which respond to peripheral signals required for the regulation of eating behavior/body weight and energy homeostasis. Insulin is also thought to carry additional functions in the CNS other than glucose metabolism: neurite growth, regulation of monoamines (including dopamine, glutamate, and noradrenaline), myelination, neuronal survival, protein synthesis, synaptic plasticity, memory formation and storage, and neuroprotection. In fact, insulin administration has been shown to improve memory, learning, and attention in rats and healthy humans (after intranasal administration), without changes in peripheral glycemia (Duarte et al. 2012).

Yet, there remain some controversies and unanswered questions in the field. First, the role of brain insulin in suppressing hepatic glucose production remains controversial (Heni 2024). Second, the precise sources of brain insulin (e.g., whether the brain produces insulin *de novo*) remain debated (Duarte et al. 2012). Third, the involvement and interaction of insulin with other hormones and signaling factors, which undergo dynamic fluctuations in the brain (e.g., leptin, incretins), represents largely uncharted territory and requires further exploration (Heni 2024). This is of particular relevance given recent pharmacological advances in targeting postprandial signaling pathways. Finally, the lack of precise biomarkers for brain insulin requires addressing, given that current diagnostic procedures, such as fMRI combined with nasal administration of insulin, are costly and time-consuming. Developing easy-to-use, noninvasive tools such as biomarkers, digital tools, or combinations thereof could simplify diagnoses, enable accurate risk stratification, and facilitate monitoring of disease progression (Heni 2024).

4.2 Implications of Disrupted Glucose-Insulin Homeostasis

A primary driver of poor adult physical health in the modern age is an increasing incidence of metabolic disorders, catalyzed by changes in dietary habits and physical activity among other environmental factors. Metabolic dysfunction affects every organ of the body and leads to conditions, such as diabetic nephropathy, nonalcoholic fatty liver disease, atherosclerosis, and adiposity, that progress to T2D, cardiovascular disease, and ultimately death.

Insulin resistance, a form of metabolic dysfunction, represents the pathophysiological process of cells becoming sensitized to frequent and/or persistent stimulation by insulin to maintain stable blood glucose levels. This results in the need for upregulated insulin production by pancreatic beta cells. Heightened insulin production leads to hyperinsulinemia, which results in further reductions in cell insulin sensitivity until beta cells are unable to produce sufficient insulin to enable cells to adequately respond to hyperglycemia. Hyperglycemia in the presence of hyperinsulinemia characterizes the transition from insulin resistance to T2D. Hyperinsulinemia promotes inflammation through direct action on immune cells, which are known to express insulin receptors (Duarte et al. 2012). Visceral adiposity also contributes to insulin resistance, mediated by pro-inflammatory cytokines including interleukin-6 and tumor necrosis factor-alpha released by adipose cells, which disrupt normal insulin action in fat and muscle cells. Thus, insulin resistance both predisposes to and is predisposed by a pro-inflammatory state. The pro-inflammatory nature of insulin resistance is particularly relevant in the case of mental disorders. Several adult mental disorders including psychosis, depression, and bipolar disorder are thought to comprise pathophysiological immune abnormalities (Khandaker et al. 2017), and those immune abnormalities may provide a mechanistic link between adult mental disorders and insulin resistance. Furthermore, aberrant glucose-insulin homeostasis and inflammation both predispose to and follow disruptions in the HPA axis, which plays an important longitudinal role in the development of several mental disorders, including psychotic and affective disorders, and may be related to early-life stress (Murphy et al. 2022).

Acute increase in peripheral insulin levels leads to higher CSF insulin, whereas chronic peripheral hyperinsulinemia (as occurs in insulin resistance) downregulates insulin receptors at the blood-brain barrier, impairing insulin transport into the brain. Reduced insulin transport leads to an impairment of key CNS roles for insulin, including neurite growth, synaptic transmission of monoamines, myelination, neuronal survival, protein synthesis, and synaptic plasticity, memory formation and storage, and neuroprotection. Indeed, T2D-associated disruption between insulin activity and glucose metabolism results in decreased cerebral blood flow and oxidative glucose metabolism. Cerebral glucose hypometabolism is a feature shared by mental disorders, including schizophrenia, bipolar disorder, and depression.

Oxidative stress and mitochondrial dysfunction are involved in the long-term damaging effects of insulin-related metabolic dysfunction in the CNS. Although mechanisms are incompletely understood, it has been proposed that increased mitochondrial nitric oxide synthase (NOS) activity (and subsequent nitric oxide production) may inhibit mitochondrial complexes III and IV and ATP synthase, either by nitrosylation or protein thiol oxidation, leading to impairment of ATP production and cell death.

4.3 Glucose Restriction, Ketosis, and Medications Targeting Metabolic Function

Traditionally, nutrition has been used as adjunctive therapy for improving lipid profiles, blood glucose, insulin resistance, and T2D but has not been considered a metabolic therapy affecting the structure and function of the brain. There is, however, strong evidence that people with psychiatric disorders also prescribed common medications for metabolic health (e.g., metformin, statins) may have improved psychiatric outcomes (Hayes et al. 2019). There is also growing interest for the relevance of targeting metabolic dysfunction for the treatment of psychiatric disorders, although larger and better powered studies are required (Jones et al. 2021).

In ketosis, prolonged glucose restriction leads to an increased glucagon:insulin ratio in the periphery, leading to release of free fatty acids into the bloodstream. Free fatty acids are taken up into liver mitochondria, producing acetyl coenzyme A and then ketone bodies via ketogenesis. These ketone bodies, which provide 27% more free energy than glucose, then exit the liver where they are able to enter peripheral tissues and the CNS via monocarboxylic acid transporters. The extracellular changes that occur during ketosis lead to reductions in intracellular calcium and sodium concentrations (Norwitz et al. 2020).

During times of increased energetic demands or glucose deprivation, such as the insulin-resistant state, the brain has evolved to utilize ketones to preserve and augment critical central functions. Indeed, the switch to ketosis is accompanied by biological adaptations of neural networks in the brain that optimize their function. Ketones are anti-inflammatory. They decrease production of ROS, and upregulate mitochondrial biogenesis in the CNS, as well as regulate neurotransmitter metabolites, modulate GABA and glutamate levels, and provide a more efficient fuel for the brain than glucose.

Given the putative neuroprotective effects of ketogenesis, over the last decade there has been a burgeoning interest in the ketogenic diet as a potential treatment for adult mental disorders (Tillery et al. 2021). The ketogenic diet comprises a carbohydrate-restricted, high-fat diet with the aim of inducing lipolysis and the generation of ketone bodies, although the application and specifics of this diet vary. Nevertheless, there are numerous plausible biological mechanisms by which the ketogenic diet may treat mental disorders: GABA/glutamate modulation may have implications for psychotic disorders; immune-modulation may have implications for psychotic disorders and depression; the utilization of ketone bodies can ameliorate the glucose hypometabolism commonly associated with various mental disorders; and reductions in intracellular calcium and sodium are a common feature of mood-stabilizing medications used to treat symptoms of bipolar disorder (Norwitz et al. 2020). Recent systematic reviews comprising case reports, observational studies, and randomized controlled trials (RCTs) have found the ketogenic diet to be associated with symptomatic improvement across a range of neuropsychiatric disorders: Alzheimer disease (AD), anorexia nervosa, autism spectrum disorder, bipolar disorder, depression, narcolepsy, and psychotic disorders including schizophrenia (Tillery et al. 2021). Yet at present, there is a scarcity of adequately powered, robust RCTs of the ketogenic diet on psychiatric outcomes, and existing studies have been impacted by nonadherence and dropout.

4.4 Summary and Future Directions

Abnormal glucose-insulin homeostasis is a key feature of many major adult mental disorders, with growing evidence of bidirectionality. Far from being a consequence of commonly attributed lifestyle factors, health inequalities, or the adverse effects of treatment, a growing body of evidence indicates that disrupted peripheral glucose-insulin homeostasis is a marker of disease processes that extend into the CNS and herald pathophysiological relevance for the mental disorder itself. Research exploring the treatment effects of common metabolic treatments alongside the ketogenic diet as novel therapeutic targets for psychiatric disorders is beginning to emerge, but this field remains in its relative infancy.

Robust, adequately powered RCTs are now required to investigate the effects of metabolic medications as well as low carbohydrate and ketogenic dietary interventions on key mental disorder outcomes, including symptomatic amelioration, remission, and prevention of relapse. The comparison of different low carbohydrate diets with ketogenic diets is particularly relevant because, if efficacious, the less restrictive nature of low carbohydrate diets could increase generalizability and improve adherence. Clear guidance and frameworks must be developed to reduce the possibility that ketogenic diets and their maintenance/monitoring (e.g., calorie counting) might induce or accentuate traits of disordered eating. Evidence from RCTs could be improved with more precise means of measuring ketosis. For example, urinary ketone testing strips are cheap but unreliable, whereas blood ketone monitoring is reliable but impractical. Breath ketone analysis is a relatively recent development but may hold promise. Finally, RCTs must seek to explore longer-term outcomes and implications for safety, since there is some concern that a high-fat, low fiber diet may increase longer-term cardiovascular risk.

5 Older Adulthood

Later life presents unique challenges at the intersection of mind and metabolism. Aging is the primary risk factor for neurodegenerative disorders, for which the incidence is predicted to rise due to demographic changes. Delaying the onset of dementia at the population level will lead to great reductions in disability and associated healthcare costs (Zissimopoulos et al. 2018).

As current treatments have limited effectiveness, researchers are exploring alternative avenues to offset the emerging public health crisis. Metabolic dysregulation in midlife represents a key modifiable risk factor for dementia in later life (Livingston et al. 2020), and metabolic changes are intrinsic to the underlying neuropathology. Studies have implicated insulin resistance and degradation, as higher insulin levels divert resources from the processing of amyloid. Others have linked leptin resistance to higher plasma and CNS levels of pro-inflammatory cytokines, or chronic low-grade inflammation in T2D, with peripheral and central inflammation as initiators and drivers of neuropathology. Epidemiologically, there is clear evidence for an increased prevalence of mild cognitive impairment and dementia in individuals with mental health and metabolic disorders. In fact, some researchers have called for dementia to be termed "type 3 diabetes" due to its associated metabolic disruptions (de la Monte et al. 2018), and T2D is identified in over 45% of individuals with AD (Wang et al. 2012). Moreover, T2D itself is associated with neurocognitive deficits impacting working memory, verbal fluency, and attention (Munshi et al. 2006).

5.1 Metabolism in Later Life and Impacts on Brain Health

Later life is associated with changes in metabolism. Two major pathways, nutrient sensing and mitochondrial dysfunction, are considered hallmarks of biological aging (López-Otín et al. 2013). Impaired nutrient sensing impacts insulin-like growth factor-1 signaling, leading to altered anabolic metabolism and inflammation. Mitochondrial dysfunction, characterized by mutations in mtDNA and increased ROS, also contributes to chronic low-grade inflammation, via cytosolic DNA sensors and activation of the inflammasome.

AD pathology comprises amyloid beta plaques and tangles of hyperphosphorylated tau. Drug trials have focused on removing these protein aggregates and have shown improvements in cognitive and functional outcomes in clinical trials. Some, however, have questioned whether these benefits are clinically meaningful and advocate for a multifaceted approach to the treatment or prevention of AD, including pathways such as oxidative stress, lipid and metabolic dysregulation, and inflammation as crucial to progress (Korczyn and Grinberg 2024).

Vascular dementia is the second most common cause of dementia and is closely linked with cardiometabolic risk factors. It includes cognitive impairment following large infarcts, where obesity and hyperlipidemia are risk factors but also cerebral small vessel disease. The relationship between small vessel disease and lipid metabolism is more complex. Despite conflicting results, it appears that hypertriglyceridemia, rather than hypercholesterolemia, drives the association between lipids and white matter hyperintensities (Nägga et al. 2018). In T2D, the triad of insulin resistance, hyperglycemia, and free fatty acids are proposed to lead to oxidative stress and inflammation at the blood vessel epithelium. The current mainstay of treatment for the spectrum of vascular cognitive impairment is management of cardiometabolic risk factors.

5.2 Shared Mechanisms Underlying Metabolic and Neurodegenerative Disorders

Changes in metabolism are key features of neurodegenerative disease pathogenesis. Pathway analysis on large meta-analysis of genome-wide association studies in AD identified strong associations with lipid metabolism, including major regulators of cellular cholesterol, with close links to innate immunity (Kunkle et al. 2019). The most significant genetic risk factor in sporadic AD is the ε4 allele of the apolipoprotein E gene (*APOE4*). ApoE is highly expressed in microglia, the brain's resident macrophages. ApoE is crucial in neuronal cholesterol transport, and the *APOE4* allele is associated with changes in brain lipid regulation, culminating in dysregulated fatty acid metabolism in neurons, lipid accumulation in glia, and abnormalities in sphingolipid metabolism, as is the triggering receptor expressed on myeloid 2 (TREM2) cells. Rare variants in TREM2 are associated with significant increases in AD risk, driven by loss of (or reduction in) function.

TREM2 is a lipid binding receptor highly expressed by microglia and macrophages, and the primary variant associated with AD risk is unable to efficiently bind phospholipids, with reduced phagocytosis and increased production of pro-inflammatory cytokines (Colonna 2023). As a response to AD pathology, TREM2 is activated in microglia leading to a transcriptional profile linked with increased lipid metabolism and cholesterol efflux, interacting with ApoE. As microglia remove lipid debris from the brain, such as myelin, those with *APOE4* accumulate lipid deposits, which are pro-inflammatory.

5.3 Metabolism as a Discriminator of Neurodegenerative Disorder Subtypes

Changes in glucose metabolism, as measured by positron emission tomography (PET) imaging with fluorodeoxyglucose, can differentiate dementia subtypes. Regional patterns of hypometabolism are associated with AD, dementia with Lewy bodies, frontotemporal dementia, and other subtypes (Kato et al. 2016). The regional reductions in FDG-PET signal seen in neurodegeneration were proposed to be caused by synaptic loss. Understanding this signal is becoming more nuanced, as preclinical data shows astrocytic activation may be driving these changes in metabolic demand.

In those with AD, there are separate patterns of dysregulated metabolites peripherally and centrally, with different patterns seen in other dementias (Pan et al. 2024). These studies require replication in independent cohorts but show the potential of metabolomics in dementia diagnosis and prognostication. There is also a need to understand the complex interplay between the disease processes, and how behavioral changes associated with dementia can contribute to metabolic disorders. For instance, behavioral variant frontotemporal dementia is associated with eating abnormalities, which could contribute to the differences in lipid levels but also to higher levels of energy expenditure, which could be secondary to agitation.

5.4 Metabolic Interventions for Neurodegenerative Disorders

Nutritional, physical activity, and multidomain interventions have been trialed with limited success in dementia prevention. Trials of antihypertensives, aspirin, nutritional supplements, or drugs for T2D (pioglitazone or linagliptin) have also shown little benefit, and the evidence for population-level interventions is scarce (Walsh et al. 2022). There is, however, growing interest in clinical trials of treatments targeting metabolism and bioenergetic pathways (Cummings et al. 2023). For example, the crucial nature of ApoE4 in metabolism has led to increased interest in the potential relevance of metabolic treatments for ApoE4 carriers (Gibas 2017). The normalization of blood lipids may be associated with prevention of cognitive decline in carriers, and case reports show that clinically prescribed dietary ketogenesis may improve cognitive function, possibly via the modulation of cerebral metabolic flexibility by regulating nutrient-sensing pathways (Morrill and Gibas 2019).

5.5 Summary and Future Directions

There is clear evidence that changes to metabolism lead to an increased risk of neurodegenerative disorders. Dysregulated metabolism in the brain and periphery are putative pathways involved in neurodegeneration, potentially driven by chronic low-grade inflammation. To bring the field closer toward translating this growing evidence into tangible clinical benefits for patients, there are a number of key outstanding research areas that require advancement.

First, a greater understanding of the mechanisms by which T2D and obesity contribute to neuropathology requires more detailed longitudinal evidence, including delineation of potential mechanisms such as inflammation. Given the evidence that neurodegenerative processes begin many years before the onset of clinical symptoms, these studies must begin early enough in life to pick up the transition to preclinical and then clinical neurodegenerative pathology in order to identify the optimum window for intervention. Then, findings require translation into population-level interventions to lower dementia risk (Walsh et al. 2022).

Second, progress toward effective individual-level interventions depends on the development of disease-specific metabolic biomarkers that differentiate normal aging from clinically relevant neurodegeneration. For example, despite the growing and triangulated evidence for the implication of lipid metabolism in dementia across preclinical, clinical, and genetic studies, the failure to translate these findings into clinical benefit could be partly due to the lack of *in vivo* biomarkers. These biomarkers could aid stratification in clinical trials, direct resources to those most likely to benefit, or measure drug response. Blood metabolomics have shown promise for the development of biomarkers for AD, but challenges include limited replication between studies and overlap between neurodegenerative diseases, possibly due to methodological or disease heterogeneity. Future research should therefore prioritize the principles of collaboration between centers and data sharing. In addition, the majority of dementia biomarker research has focused on AD. To gain greater insight into the shared and distinct mechanisms of neurodegenerative diseases, the benefits of methodological and collaborative advances in AD research must be extended to other dementia subtypes; several discussed in this article are likely to comprise a metabolic pathophysiological component.

Finally, the development of new treatments targeting metabolic pathways in neurodegeneration could be accelerated with drug repurposing and platform trials. Drug repurposing benefits from the known safety profiles of licensed medications, while avoiding the high cost and time requirements of initial drug development. For example, licensed metabolic medications, such as the glucagon-like peptide-1 (GLP-1) agonist liraglutide, have been identified through the use of preclinical data as potentially holding promise for targeting neurodegeneration (Ballard et al. 2020). Platform trials, an advance of adaptive trial design, allow the simultaneous testing of multiple treatments. Studies continue to incorporate new potential options, drop those shown not to work, and can adopt new discoveries as a standard of care. This limits the extensive costs and time associated with continual trial set up and dismantling. Platform trials are already being employed for neurological disorders, such as multiple sclerosis, to maximize the chance of successfully identifying new treatments while learning from failures. They hold great potential to bring similar benefits to dementia research.

6 Concluding Remarks

In the early 20th century, the previously aligned fields of neurology and psychiatry deviated from one another through separate wards, training programs, journals and professional bodies. Given the Cartesian preconceptions on which such a system of care rests, it is perhaps not surprising that still today, people with neuropsychiatric disorders of all ages face the dilemma of taking treatments which optimize acute mental health outcomes at the expense of metabolic health and long-term physical health outcomes.

Neurology and psychiatry have taken diverging paths in grappling with the underlying biological reality of brain-based disorders, and yet they have intersected at key periods of treatment development. In the 1950s, the advent of the age of psychopharmacology promised to reunite neurobiology with psychiatric treatment. However, many of these new treatments continued to come with significant cardiometabolic and physical health consequences. Parallel to these developments, neuroscience has made significant advances in understanding the role of metabolic dysfunction in the brain, elucidating mechanisms of mitochondrial function, glucose-insulin homeostasis, sex hormones, and the immune system. Yet, these insights have yet to lead to new treatments for neuropsychiatric disorders.

Over the past few years, these independent research trajectories have shown promising indications of convergence toward the emerging field of metabolic psychiatry. This is a blossoming field not only within academia, but from a patient movement catalyzing renewed dialogue between separated disciplines (Campbell and Campbell 2019).

References

Aslanoglou D, Bertera S, Sanchez-Soto M, et al. (2021) Dopamine Regulates Pancreatic Glucagon and Insulin Secretion via Adrenergic and Dopaminergic Receptors. Transl Psychiatry 11 (1):59. https://doi.org/10.1038/s41398-020-01171-z

Ballard C, Aarsland D, Cummings J, et al. (2020) Drug Repositioning and Repurposing for Alzheimer Disease. Nat Rev Neurol 16 (12):661–673. https://doi.org/10.1038/s41582-020-0397-4

Behnke A, Mack M, Fieres J, et al. (2022) Expression of DNA Repair Genes and Its Relevance for DNA Repair in Peripheral Immune Cells of Patients with Posttraumatic Stress Disorder. Sci Rep 12 (1):18641. https://doi.org/10.1038/s41598-022-22001-w

Brand BA, de Boer JN, and Sommer IE (2021) Estrogens in Schizophrenia: Progress, Current Challenges and Opportunities. Curr Opin Psychiatry 34 (3):228–237. http://dx.doi.org/10.1097/YCO.0000000000000699

Brand BA, Haveman YRA, de Beer F, et al. (2022) Antipsychotic Medication for Women with Schizophrenia Spectrum Disorders. Psychol Med 52 (4):649–663. https://doi.org/10.1017/s0033291721004591

Campbell IH, and Campbell H (2019) Ketosis and Bipolar Disorder: Controlled Analytic Study of Online Reports. BJPsych Open 5 (4):e58. https://doi.org/10.1192/bjo.2019.49

Chen H, Cao T, Zhang B, and Cai H (2023) The Regulatory Effects of Second-Generation Antipsychotics on Lipid Metabolism: Potential Mechanisms Mediated by the Gut Microbiota and Therapeutic Implications. Front Pharmacol 14:1097284. https://doi.org/10.3389/fphar.2023.1097284

Colonna M (2023) The Biology of Trem Receptors. Nat Rev Immunol 23 (9):580–594. https://doi.org/10.1038/s41577-023-00837-1
Cummings J, Zhou Y, Lee G, et al. (2023) Alzheimer's Disease Drug Development Pipeline: 2023. Alzheimers Dement 9 (2):e12385. https://doi.org/10.1002/trc2.12385
Currie C, Ahluwalia N, Godeau E, et al. (2012) Is Obesity at Individual and National Level Associated with Lower Age at Menarche? Evidence from 34 Countries in the Health Behaviour in School-Aged Children Study. J Adolesc Health 50 (6):621–626. http://dx.doi.org/10.1016/j.jadohealth.2011.10.254
Davies W (2017) Understanding the Pathophysiology of Postpartum Psychosis: Challenges and New Approaches. World J Psychiatry 7 (2):77–88. https://doi.org/10.5498/wjp.v7.i2.77
Davis SR, and Baber RJ (2022) Treating Menopause: MHT and Beyond. Nat Rev Endocrinol 18 (8):490–502. https://doi.org/10.1038/s41574-022-00685-4
de la Monte SM, Tong M, and Wands JR (2018) The 20-Year Voyage Aboard the Journal of Alzheimer's Disease: Docking at 'Type 3 Diabetes', Environmental/Exposure Factors, Pathogenic Mechanisms, and Potential Treatments. J Alzheimers Dis 62 (3):1381–1390. https://doi.org/10.3233/jad-170829
Di Florio A, Jones L, Forty L, et al. (2014) Mood Disorders and Parity: A Clue to the Aetiology of the Postpartum Trigger. J Affect Disord 152–154 (100):334–339. https://doi.org/10.1016/j.jad.2013.09.034
Duarte AI, Moreira PI, and Oliveira CR (2012) Insulin in Central Nervous System: More Than Just a Peripheral Hormone. J Aging Res 2012:384017. https://doi.org/10.1155/2012/384017
Felitti VJ, Anda RF, Nordenberg D, et al. (2019) Reprint of: Relationship of Childhood Abuse and Household Dysfunction to Many of the Leading Causes of Death in Adults: The Adverse Childhood Experiences (ACE) Study. Am J Prev Med 56 (6):774–786. https://doi.org/10.1016/j.amepre.2019.04.001
Galea LAM, Qiu W, and Duarte-Guterman P (2018) Beyond Sex Differences: Short and Long-Term Implications of Motherhood on Women's Health. Curr Opin Physiol 6:82–88. https://doi.org/10.1016/j.cophys.2018.06.003
Gibas KJ (2017) The Starving Brain: Overfed Meets Undernourished in the Pathology of Mild Cognitive Impairment (MCI) and Alzheimer's Disease (AD). Neurochem Int 110:57–68. https://doi.org/10.1016/j.neuint.2017.09.004
Gimenez-Palomo A, Dodd S, Anmella G, et al. (2021) The Role of Mitochondria in Mood Disorders: From Physiology to Pathophysiology and to Treatment. Front Psychiatry 12:546801. https://doi.org/10.3389/fpsyt.2021.546801
Gyllenhammer LE, Entringer S, Buss C, and Wadhwa PD (2020) Developmental Programming of Mitochondrial Biology: A Conceptual Framework and Review. Proc Biol Sci 287 (1926):20192713. https://doi.org/10.1098/rspb.2019.2713
Hayes JF, Lundin A, Wicks S, et al. (2019) Association of Hydroxylmethyl Glutaryl Coenzyme a Reductase Inhibitors, L-Type Calcium Channel Antagonists, and Biguanides with Rates of Psychiatric Hospitalization and Self-Harm in Individuals with Serious Mental Illness. JAMA Psychiatry 76 (1):382–390. https://doi.org/10.1001/jamapsychiatry.2018.3907
Heni M (2024) The Insulin Resistant Brain: Impact on Whole-Body Metabolism and Body Fat Distribution. Diabetologia 67 (7):1181–1191. https://doi.org/10.1007/s00125-024-06104-9
Hoyt LT, Niu L, Pachucki MC, and Chaku N (2020) Timing of Puberty in Boys and Girls: Implications for Population Health. SSM Popul Health 10:100549. https://doi.org/10.1016/j.ssmph.2020.100549
Hummel J, Benkendorff C, Fritsche L, et al. (2023) Brain Insulin Action on Peripheral Insulin Sensitivity in Women Depends on Menstrual Cycle Phase. Nat Metab 5 (9):1475–1482. https://doi.org/10.1038/s42255-023-00869-w
Jawaid A, Jehle K-L, and Mansuy IM (2021) Impact of Parental Exposure on Offspring Health in Humans. Trends Genet 37 (4):373–388. https://doi.org/10.1016/j.tig.2020.10.006

Jones BDM, Farooqui S, Kloiber S, et al. (2021) Targeting Metabolic Dysfunction for the Treatment of Mood Disorders: Review of the Evidence. Life 11 (8):819. https://doi.org/10.3390/life11080819

Jones I, Chandra PS, Dazzan P, and Howard LM (2014) Bipolar Disorder, Affective Psychosis, and Schizophrenia in Pregnancy and the Post-Partum Period. Lancet 384 (9956):1789–1799. https://doi.org/10.1016/s0140-6736(14)61278-2

Kato T, Inui Y, Nakamura A, and Ito K (2016) Brain Fluorodeoxyglucose (FDG) PET in Dementia. Ageing Res Rev 30:73–84. https://doi.org/10.1016/j.arr.2016.02.003

Khandaker GM, Dantzer R, and Jones PB (2017) Immunopsychiatry: Important Facts. Psychol Med 47 (13):2229–2237. https://doi.org/10.1017/s0033291717000745

Kohen D (2004) Diabetes Mellitus and Schizophrenia: Historical Perspective. Br J Psychiatry Suppl 47:S64–S66. https://doi.org/10.1192/bjp.184.47.s64

Korczyn AD, and Grinberg LT (2024) Is Alzheimer Disease a Disease? Nat Rev Neurol 20 (4):245–251. https://doi.org/10.1038/s41582-024-00940-4

Kunkle BW, Grenier-Boley B, Sims R, et al. (2019) Genetic Meta-Analysis of Diagnosed Alzheimer's Disease Identifies New Risk Loci and Implicates Aβ, Tau, Immunity and Lipid Processing. Nat Genet 51 (3):414–430. https://doi.org/10.1038/s41588-019-0358-2

Livingston G, Huntley J, Sommerlad A, et al. (2020) Dementia Prevention, Intervention, and Care: 2020 Report of the Lancet Commission. Lancet 396 (10248):413–446. https://doi.org/10.1016/s0140-6736(20)30367-6

López-Otín C, Blasco MA, Partridge L, Serrano M, and Kroemer G (2013) The Hallmarks of Aging. Cell 153 (6):1194–1217. https://doi.org/10.1016/j.cell.2013.05.039

Maudsley H (1879) The Pathology of Mind. Macmillan, London

Mauvais-Jarvis F, Clegg DJ, and Hevener AL (2013) The Role of Estrogens in Control of Energy Balance and Glucose Homeostasis. Endocr Rev 34 (3):309–338. https://doi.org/10.1210/er.2012-1055

Milstein JL, and Ferris HA (2021) The Brain as an Insulin-Sensitive Metabolic Organ. Mol Metab 52:101234. https://doi.org/10.1016/j.molmet.2021.101234

Momen NC, Plana-Ripoll O, Agerbo E, et al. (2022) Mortality Associated with Mental Disorders and Comorbid General Medical Conditions. JAMA Psychiatry 79 (5):444–453. https://doi.org/10.1001/jamapsychiatry.2022.0347

Morath J, Moreno-Villanueva M, Hamuni G, et al. (2014) Effects of Psychotherapy on DNA Strand Break Accumulation Originating from Traumatic Stress. Psychother Psychosom 83 (5):289–297. https://doi.org/10.1159/000362739

Morrill SJ, and Gibas KJ (2019) Ketogenic Diet Rescues Cognition in ApoE4+ Patient with Mild Alzheimer's Disease: A Case Study. Diabetes Metab Syndr 13 (2):1187–1191. https://doi.org/10.1016/j.dsx.2019.01.035

Munshi M, Grande L, Hayes M, et al. (2006) Cognitive Dysfunction Is Associated with Poor Diabetes Control in Older Adults. Diabetes Care 29 (8):1794–1799. https://doi.org/10.2337/dc06-0506

Murphy F, Nasa A, Cullinane D, et al. (2022) Childhood Trauma, the HPA Axis and Psychiatric Illnesses: A Targeted Literature Synthesis. Front Psychiatry 13:748372. https://doi.org/10.3389/fpsyt.2022.748372

Nägga K, Gustavsson A-M, Stomrud E, et al. (2018) Increased Midlife Triglycerides Predict Brain Β-Amyloid and Tau Pathology 20 Years Later. Neurology 90 (1):e73–e81. https://doi.org/10.1212/WNL.0000000000004749

Nogueira Avelar ESR, Yu Y, Liew Z, et al. (2021) Associations of Maternal Diabetes During Pregnancy with Psychiatric Disorders in Offspring During the First 4 Decades of Life in a Population-Based Danish Birth Cohort. JAMA Netw Open 4 (10):e2128005. https://doi.org/10.1001/jamanetworkopen.2021.28005

Norwitz NG, Sethi S, and Palmer CM (2020) Ketogenic Diet as a Metabolic Treatment for Mental Illness. Curr Opin Endocrinol Diabetes Obes 27 (5):269–274. https://doi.org/10.1097/med.0000000000000564

Pan X, Donaghy PC, Roberts G, et al. (2024) Plasma Metabolites Distinguish Dementia with Lewy Bodies from Alzheimer's Disease: A Cross-Sectional Metabolomic Analysis. Front Aging Neurosci 15:1326780. https://doi.org/10.3389/fnagi.2023.1326780

Perry BI, Burgess S, Jones HJ, et al. (2021a) The Potential Shared Role of Inflammation in Insulin Resistance and Schizophrenia: A Bidirectional Two-Sample Mendelian Randomization Study. PLoS Med 18:e1003455. https://doi.org/10.1371/journal.pmed.1003455

Perry BI, McIntosh G, Weich S, Singh S, and Rees K (2016) The Association between First-Episode Psychosis and Abnormal Glycaemic Control: Systematic Review and Meta-Analysis. Lancet Psychiatry 3 (11):1049–1058. https://doi.org/10.1016/s2215-0366(16)30262-0

Perry BI, Stochl J, Upthegrove R, et al. (2021b) Longitudinal Trends in Childhood Insulin Levels and Body Mass Index and Associations with Risks of Psychosis and Depression in Young Adults. JAMA Psychiatry 78:416–425. https://doi.org/10.1001/jamapsychiatry.2020.4180

Picard M, Prather AA, Puterman E, et al. (2018) A Mitochondrial Health Index Sensitive to Mood and Caregiving Stress. Biol Psychiatry 84 (1):9–17. https://doi.org/10.1016/j.biopsych.2018.01.012

Riecher-Rossler A (2017) Sex and Gender Differences in Mental Disorders. Lancet Psychiatry 4 (1):8–9. https://doi.org/10.1016/s2215-0366(16)30348-0

Scherer T, Sakamoto K, and Buettner C (2021) Brain Insulin Signalling in Metabolic Homeostasis and Disease. Nat Rev Endocrinol 17 (8):468–483. https://doi.org/10.1038/s41574-021-00498-x

Sommer IE, Brand BA, Gangadin S, et al. (2023) Women with Schizophrenia-Spectrum Disorders after Menopause: A Vulnerable Group for Relapse. Schizophr Bull 49 (1):136–143. https://doi.org/10.1093/schbul/sbac139

Teede HJ, Misso ML, Costello MF, et al. (2018) Recommendations from the International Evidence-Based Guideline for the Assessment and Management of Polycystic Ovary Syndrome. Clin Endocrinol 89 (3):251–268. https://doi.org/10.1111/cen.13795

Tian YE, Di Biase MA, Mosley PE, et al. (2023) Evaluation of Brain-Body Health in Individuals with Common Neuropsychiatric Disorders. JAMA Psychiatry 80 (6):567–576. https://doi.org/10.1001/jamapsychiatry.2023.0791

Tillery EE, Ellis KD, Threatt TB, et al. (2021) The Use of the Ketogenic Diet in the Treatment of Psychiatric Disorders. Ment Health Clin 11 (3):211–219. https://doi.org/10.9740/mhc.2021.05.211

Ullsperger JM, and Nikolas MA (2017) A Meta-Analytic Review of the Association between Pubertal Timing and Psychopathology in Adolescence: Are There Sex Differences in Risk? Psychol Bull 143 (9):903–938. https://doi.org/10.1037/bul0000106

Walsh S, Merrick R, Richard E, Nurock S, and Brayne C (2022) Lecanemab for Alzheimer's Disease. BMJ 379:o3010. https://doi.org/10.1136/bmj.o3010

Wang KC, Woung LC, Tsai MT, et al. (2012) Risk of Alzheimer's Disease in Relation to Diabetes: A Population-Based Cohort Study. Neuroepidemiology 38 (4):237–244. https://doi.org/10.1159/000337428

Xiao Z, Murat Baldwin M, Wong SC, et al. (2023) The Impact of Childhood Psychological Maltreatment on Mental Health Outcomes in Adulthood: A Systematic Review and Meta-Analysis. Trauma Violence Abuse 24 (5):3049–3064. https://doi.org/10.1177/15248380221122816

Yoshida Y, Chen Z, Baudier RL, et al. (2022) Sex Differences in the Progression of Metabolic Risk Factors in Diabetes Development. JAMA Netw Open 5 (7):e2222070. https://doi.org/10.1001/jamanetworkopen.2022.22070

Zissimopoulos JM, Tysinger BC, St Clair PA, and Crimmins EM (2018) The Impact of Changes in Population Health and Mortality on Future Prevalence of Alzheimer's Disease and Other Dementias in the United States. J Gerontol B Psychol Sci Soc Sci 73 (Suppl_1):S38–S47. https://doi.org/10.1093/geronb/gbx147

Open Access This chapter is licensed under the terms of the Creative Commons Attribution-NonCommercial-NoDerivatives 4.0 International License (http://creativecommons.org/licenses/by-nc-nd/4.0/), which permits any noncommercial use, sharing, distribution and reproduction in any medium or format, as long as you give appropriate credit to the original author(s) and the source, provide a link to the Creative Commons license and indicate if you modified the licensed material. You do not have permission under this license to share adapted material derived from this chapter or parts of it.

The images or other third party material in this chapter are included in the chapter's Creative Commons license, unless indicated otherwise in a credit line to the material. If material is not included in the chapter's Creative Commons license and your intended use is not permitted by statutory regulation or exceeds the permitted use, you will need to obtain permission directly from the copyright holder.

11

Severe Mental Illness as a Multisystem Metabolic Disorder

From Brain to Body and Back Again

Toby Pillinger, Hannelore Ehrenreich, Zachary Freyberg, Margaret Hahn, Matthias Mack, Yuri Milaneschi, Benjamin I. Perry, and Rachel Upthegrove

Abstract Severe mental illnesses (SMIs), such as schizophrenia and bipolar disorder, are traditionally conceptualized as disorders of the central nervous system (CNS). Growing evidence, however, suggests that SMIs may be better understood as multisystem disorders, characterized by metabolic and immune dysregulation affecting both brain and body. In this chapter, the hypothesis that psychiatric symptoms may emerge because of systemic metabolic dysfunction is explored. Evidence is presented to show that metabolic abnormalities—including insulin resistance, dysglycemia, and immune activation—are present from onset of SMI, and that these abnormalities may contribute to core psychiatric symptoms such as amotivation, cognitive impairment, anhedonia, and neurovegetative alterations. The role of insulin and cytokine signaling pathways is discussed, with a focus on their reciprocal influence across the CNS and peripheral organs, and the potential intrinsic and extrinsic modulators of these pathways, including sex, stress, and nutrition, are outlined. A conceptual framework is proposed in which SMIs are understood as manifestations of dysregulated energy allocation and metabolic stress. Finally, a research strategy is outlined to investigate peripheral metabolic signatures as tools for stratification, prognosis, and treatment in SMIs, with the goal of advancing a precision medicine approach to psychiatric care.

Keywords Metabolic psychiatry; metabolic psychopathology, multisystem disorder; severe mental illness

---

Toby Pillinger (✉)
Institute of Psychiatry, Psychology and Neuroscience, 16 De Crespigny Park, Camberwell SE5 8AF, UK
E-Mail: toby.pillinger@kcl.ac.uk

Affiliations for the coauthors are available in the List of Contributors.

© Frankfurt Institute for Advanced Studies (FIAS) 2025

D. Öngür, J. M. Ford (eds.), *Metabolic Neuropsychiatry*, Strüngmann Forum Reports,
https://doi.org/10.1007/978-3-032-05630-6_11

Group photos (top left to bottom right) Toby Pillinger, Hannelore Ehrenreich, Margaret Hahn, Yuri Milaneschi, Group photo, Rachel Upthegrove, Matthias Mack, Zachary Freyberg, Yuri Milaneschi, Rachel Upthegrove, Group photo, Zachary Freyberg, Toby Pillinger, Group photo, Margaret Hahn, Matthias Mack, Hannelore Ehrenreich

1 Introduction

To date, psychiatry has focused its attention on the CNS in its attempt to define and better understand SMIs such as schizophrenia, bipolar disorder, and related conditions. It has used diverse methodological approaches, including those that examine pharmacology, genetics, metabolism, and immunity. However, focusing on the CNS in isolation has not successfully advanced treatments or improved outcomes for people with SMIs. Increasing evidence points to multi-organ pathology in SMIs, which involves metabolic dysfunction both inside and outside the brain (Pillinger et al. 2017, 2019; see also Chapter 6). Multisystem effects of SMIs are not surprising if at least some of the metabolic presentations of SMIs are due to genetics (Tomasik et al. 2019), as it is unlikely these would be limited to the CNS. Furthermore, there is evidence of a bidirectional physiological relationship between the CNS and peripheral metabolism (Roh et al. 2016), yet the mechanisms that explain specific symptoms related to metabolism across the brain and periphery are poorly defined. The reconceptualization of an SMI as a multisystem metabolic disorder raises the following issues:

1. It is unclear which systemic metabolic pathways are implicated in the emergence of psychiatric symptoms.
2. Intrinsic (e.g., sex hormones) and extrinsic (e.g., stress) modulators of metabolic pathways implicated in the pathoetiology of mental illness are poorly defined.
3. It is not known if variations in systemic metabolic signaling can add to the explanation of heterogeneity of symptomatology across SMIs.
4. The therapeutic potential of stratifying patients with SMIs based on a peripheral metabolic signature is unclear.

Here, we consider these issues, and the implications for both clinical practice and research, and propose a potential approach to address them with empirical evidence.

2 Systemic Metabolic Pathways Implicated in Severe Mental Illness

Though the brain constitutes only 2% of body weight, it accounts for 20% of total oxygen and caloric expenditures across the body (Raichle and Gusnard 2002). As one of the body's most metabolically active organs, the brain is deeply rooted within the larger framework of systemic metabolic regulation. Indeed, the CNS is exquisitely sensitive to changes in nutrient availability and is responsive to the metabolic activities of other organ systems. Hormones released in the periphery, such as glucagon-like peptide-1 (GLP-1), ghrelin, and leptin, signal in the hypothalamus to regulate satiety (Barakat et al. 2024). This signaling is bidirectional as brain circuits provide feedback to the periphery to modulate systemic energy metabolism (Wachsmuth et al. 2022). We must therefore view CNS function and dysfunction from a multisystem perspective, where the brain is in continuous communication with the periphery and vice versa. This interconnectedness between the CNS and the periphery may have profound implications on how we understand SMIs, previously conceptualized as primarily brain disorders, to metabolism. Critically, dysglycemia (e.g., glucose intolerance, insulin resistance) has increasingly been recognized not only as a hallmark of diabetes; it is present in psychotic spectrum illnesses, major depression, and bipolar disorder (Calkin 2019; Freyberg et al. 2017; Watson et al. 2021). While many psychiatric medications, including antipsychotic drugs, cause dysglycemia that ultimately culminates in type 2 diabetes mellitus, these metabolic manifestations are also evident in drug-naïve individuals during the first episodes of their illnesses (Freyberg et al. 2017; Lee et al. 2024a, b; Pillinger et al. 2017). This strongly suggests that metabolic symptoms are intrinsic to psychiatric illnesses and that these disorders can be viewed through a new lens of multi-organ dysfunction. Similarly, immune system dysfunction has increasingly been acknowledged as a feature of both SMIs and metabolic disease and may mediate some antipsychotic-induced metabolic side effects (Prestwood et al. 2021; Pillinger et al. 2020). Genetic studies have linked immune and inflammatory genes to schizophrenia. Likewise, both autoimmune illnesses and maternal infections are associated with SMIs (Debost et al. 2019; Eaton et al. 2006; Fabbri and Serretti 2017).

While recognizing the complexity and intersection of multiple biochemical pathways that contribute to metabolism, insulin and immune signaling are established as two key regulators of metabolism implicated in both the CNS and non-CNS effects of SMIs. There is also interplay between metabolic pathways and biochemical signaling mechanisms traditionally implicated in SMIs. For example, insulin and dopaminergic receptors play a reciprocal role in the dynamic regulation of metabolism via AKT-signaling pathways (Figure 11.1a) (Freyberg et al. 2010). An imbalance in these pathways may lead to metabolic disturbance in any organ where these receptor systems are co-located (e.g., the brain and pancreas). Although there is debate as to the source of CNS insulin (Gray et al. 2014), the importance of insulin signaling within the CNS is clear: it plays various roles, including the regulation of synaptic plasticity, memory formation, and control of peripheral metabolic homeostasis through feeding and glucose metabolism (Ferrario and Reagan 2018). Foremost, given insulin's importance as a key regulator of glucose availability and metabolism more generally, we posit that insulin signaling may be integral to ensure adequate energetic resources, not only in the periphery but also in the CNS. Nevertheless, existing data paint a complex metabolic picture of insulin's central versus peripheral action required to maintain metabolic balance. For example, insulin acts in the brain to boost the counterregulatory response to hypoglycemia by blunting CNS glucose sensing (Ren et al. 2019). Brain insulin signaling in the hypothalamus also impacts the balance between lipolysis and lipogenesis (Agrawal et al. 2021; Shin et al. 2017; Iwen et al. 2014). Perturbation of these physiological roles are implicated in SMIs; for example, alterations in CNS insulin signaling pathways that link the brain and periphery are observed in psychotic disorders (van Nimwegen et al. 2008). Impaired glucose homeostasis is observed in people with SMIs, even after accounting for lifestyle factors and medication (Pillinger et al. 2017). This is generally supported by genetic and epidemiological data (Liu et al. 2013; Perry et al. 2020). Insulin-induced suppression of hepatic glucose production by the hypothalamus is well described (Dash et al. 2015; Obici et al. 2002), and there is evidence that this process is impaired in antipsychotic-naïve patients with psychosis. Specifically, a study that performed hyperinsulinemic euglycemic clamps in antipsychotic-naïve

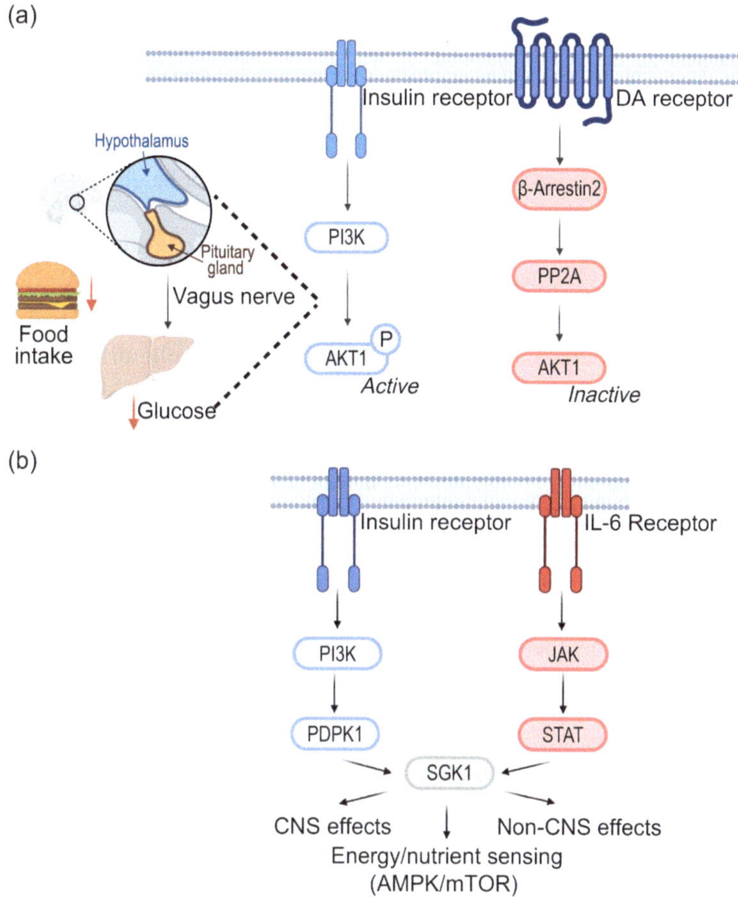

Figure 11.1 Dynamic interactions between insulin receptor and dopamine receptor signaling. (a) Upon insulin receptor activation, PI3 kinase (PI3K) is activated, which triggers phosphorylation of AKT1 kinase. Phospho-AKT1, the active form of the enzyme, acts on autonomic pathways. This includes central insulin-mediated actions via the vagus nerve that link metabolic signaling between the brain and the periphery to regulate systemic metabolism. This includes diminished food intake through CNS actions and suppression of hepatic glucose production via hypothalamic insulin signaling. Additionally, dopamine (DA) signaling through DA receptors may act in a counterregulatory manner. DA receptor activation leads to recruitment of a β-arrestin2-based complex, which includes the PP2A phosphatase, and dephosphorylates AKT1, resulting in the kinase's inactivation. Created in BioRender. Mack, M. (2025) https://BioRender.com/k0927cn. (b) Outside of the canonical AKT-signaling pathways, insulin and cytokine receptor signaling pathways converge on SGK1. SGK1 activation leads to effects in the CNS and on peripheral target organs. In both cases, microtubule formation is impacted, leading to effects on cell plasticity, energy, and nutrient sensing. Created in BioRender. Mack, M. (2025) https://BioRender.com/k0927cn.

patients with first-episode psychosis found that patients showed higher hepatic glucose production and hepatic insulin resistance, which could not be attributed to differences in adiposity or other confounders (van Nimwegen et al. 2008). Given the link between CNS insulin action and hepatic glucose production (Lewis et al. 2021), these findings suggest a primary metabolic signaling pathway abnormality in people with psychosis. This abnormality, however, remains to be examined in other SMIs.

Converging lines of evidence implicate immune dysregulation in SMIs (Goldsmith et al. 2016). However, key gaps in our knowledge are evident, including the following:

- Alterations in peripheral immunity may not directly reflect CNS immune activity and/or brain function (Engelhardt et al. 2017; Lalousis et al. 2023).
- Immune alterations, regardless of location, may not be observed in all patients with SMIs (Upthegrove and Goldsmith 2024).
- The functional role of immune parameters may differ within the CNS compared to the periphery (Gonzalez Caldito 2023).

Insulin resistance is both a cause and an effect of inflammation (Duarte et al. 2012). Immune cells are particularly sensitive to systemic metabolic disturbance since their energy stores rely on circulating blood glucose, and their functional integrity is essential in the control of downstream effects (Freyberg and Harvill 2017; Corsi-Zuelli et al. 2021). There is the potential for interactions between the canonical insulin receptor signal transduction network and cytokine receptor signaling (e.g., interleukin-6, IL6). For example, serum/glucocorticoid-regulated kinase 1 (SGK1) is regulated by both insulin- and IL6-related pathways, with the potential for alterations in both CNS (e.g., microtubule formation, neuronal plasticity, memory) and non-CNS (e.g., cardiovascular) function (Figure 11.1b) (Bian et al. 2023). Disruption in either canonical pathway could contribute to dysregulation of the other, leading to metabolic dysregulation and multisystem effects in SMIs.

3 What Are the Modulators of Systemic Metabolic Pathways in Severe Mental Illness?

A key intrinsic modulator of metabolic pathways implicated in SMIs is the biological sex (male/female) of an individual at birth, which accounts for well-established differences in prevalence and age of onset across SMIs. For instance, estrogens provide protection as they activate intracellular signaling pathways, which reduce oxidative stress, apoptosis, and inflammation (Straub 2007). Within the CNS, estrogens, in turn, can enhance cell proliferation and synaptic sprouting (Krolick et al. 2018). In the periphery, however, estrogen deficiency in females leads to glucose dysregulation (Yan et al. 2019).

While it is generally acknowledged that genetics significantly contributes to metabolic alterations in people with SMIs, the pathways are complex and do not necessarily reflect a single identifiable etiology. Some factors, such as adiposity, may represent both intrinsic and extrinsic modulatory pathways and involve complex interactions between inflammatory response and insulin/dopamine signaling that is sufficient to impact both CNS and metabolic function (Rasinska et al. 2022).

In terms of extrinsic modulators of metabolic pathways, stress of any form (e.g., maternal infection, obstetric complications, childhood trauma) during development is a key factor (Kivimaki et al. 2023). Furthermore, it is hypothesized that the timing of stress events may influence the psychiatric phenotype in the affected individual: stress events that occur earlier in development are more likely to cause neurodevelopmental disorders such as autism and schizophrenia whereas later stress events increase the chances of mood and anxiety disorders (Gee and Casey 2015; McLaughlin and Hatzenbuehler 2009). By extension, it is suggested that if SMIs represent multisystem dysfunction, then insults earlier in development may also increase the risk of co-morbidity (e.g., diabetes mellitus and cardiovascular disease). However, the specific mechanisms by which these stress modulators influence metabolic pathways leading to multisystem effects remain poorly defined. Various pathways may be implicated, including epigenetic effects of stress, the consequence of hypercortisolemia on intracellular metabolic signaling, and the direct effects of stress on metabolic pathways independent of hypothalamic-pituitary-adrenal (HPA) axis effects. Evidence to support the HPA axis independent effects of stress derives from a Mendelian randomization study showing that the association between childhood maltreatment and multimorbidity (depression and cardiometabolic disorders) is mediated by metabolic parameters (HbA1c, fasting insulin, and glucose) rather than cortisol (Baltramonaityte et al. 2025). The stress vulnerability model (Zubin and Spring 1977) may also be key to our understanding of extrinsic modulators of metabolic pathways and resultant development of SMIs. For example, during the COVID-19 pandemic, despite the widespread social isolation imposed by lockdowns, only a small proportion of individuals developed psychiatric disorders (Hastie et al. 2023). This suggests that environmental stressors alone are not sufficient to cause psychiatric illness in all people. Rather, it points to potential gene–environment interactions in combination with social factors, such as connectivity/loneliness on systemic health, metabolism, and emergence of psychiatric symptoms. Moreover, "risks factors" (e.g., lack of social connectivity) have a clear counterpoint (good social connectivity) and may indicate markers of resilience, with therapeutic implications.

Nutrition may also be a key extrinsic regulator of metabolic pathways implicated in SMIs. For example, three days of high fat/sucrose diet in rodents is sufficient to cause hypothalamic insulin resistance with resultant CNS effects (Clegg et al. 2011). Patients with SMI have a disproportionately poor diet as a result of many factors including cognitive dysfunction, negative symptoms, deprivation, and reduced access to non-high processed options (Firth et al. 2019). Furthermore, clinical data suggest that increased intake of ultra-processed foods alter mitochondrial metabolism (Coppola et al. 2023). Although an intervention such as a ketogenic diet would be logical in response to a glycolysis disorder, it is unclear if this is a tolerable or effective intervention in all patients with SMIs. Furthermore, there is uncertainty around the precise mechanism by which a ketogenic diet may exert its therapeutic effects in psychiatric disorders. For example, ketones themselves may be therapeutic, rather than the diet simply reducing intake of "toxic" glucose.

4 How Do Alterations in Metabolic Pathways Result in Symptomatic Heterogeneity?

Mental health research and treatment are based mainly around categorical diagnostic structures, such as depression and schizophrenia, as classified and defined in diagnostic manuals based on symptom checklists. Due to the polythetic structure of psychiatric diagnoses, two individuals may have the same diagnosis and no overlapping symptoms, or conversely a different diagnosis with multiple overlapping symptoms. Symptoms may also change over time both between and within individuals. It is plausible that the heterogeneity in clinical manifestations within and across diagnostic categories lies in variations in complex biological systems, particularly metabolic processes. Dysfunction in metabolic pathways may underlie psychiatric symptoms in ways that cut across traditional diagnostic boundaries. If common metabolic impairments are present in individuals with different psychiatric labels, they may present with shared clinical features, reflecting underlying biological overlap. As nature does not adhere to the constructs of psychiatric classification, it is entirely plausible that a "metabolic phenotype" may span multiple diagnostic categories. Conversely, heterogeneous alterations in specific metabolic pathways among individuals with the same diagnosis could drive the diversity of symptoms observed within that group. This biological variability may help explain why two patients with the same formal diagnosis respond differently to treatment or follow distinct clinical trajectories. A metabolic perspective thus offers a unifying, mechanistic lens through which to reinterpret psychiatric disorders, moving beyond rigid diagnostic silos toward more biologically informed and personalized models of care.

Psychiatric symptoms that accompany extreme metabolic and immune perturbations are well recognized, as in, for instance, delirium in the context of hypo/hyperglycemia or infection, hepatic encephalopathy, and depression associated with interferon-α. Furthermore, knockdown of insulin receptors in the hypothalamus in animals triggers depressive and anxiety-like behaviors (Grillo et al. 2011). However, the characteristic psychopathology that accompanies less extreme systemic metabolic dysregulation is less well defined. Of note, treatment of insulin resistance in mood disorders results in improvement in depressive and potentially cognitive symptoms (Jones et al. 2021; Calkin et al. 2022). The role of insulin signaling in cognition is well recognized (Kim and Feldman 2015), and comorbid metabolic dysfunction in depression predicts deficits across various cognitive domains (Maksyutynska et al. 2024). Furthermore, epidemiological studies in SMI have demonstrated that symptoms associated with metabolic/immune disturbance include appetite, sleep, energy, and motivational impairments (Milaneschi et al. 2020; Morales-Munoz et al. 2024).

In the hypothesized multisystem model, the brain may act as a broker allocating a finite energy budget, as highlighted by Rae et al. (Chapter 3). The interoceptive system provides the brain with a reading of the metabolic status of different organs and tissues via neuroendocrine messengers such as insulin, leptin, ghrelin, GLP-1, cytokines, and growth differentiation factor 15. These signals reach brain centers like the hypothalamus and the insula—hubs involved in the integration of a range of physiological responses, such as thermoregulation, food intake, energy expenditure, and sleep/wakefulness (Adamantidis and De Lecea 2023; Brüning and Fenselau 2023). Informed by interoceptive signals on the body's metabolic status, the brain modulates motivational and reward processing to output congruent energy-saving and energy-storing physiological and behavioral responses. Thus, it is hypothesized that transdiagnostic symptoms, such as impaired motivation, fatigue, sleep, and appetite alterations, may be a CNS response to systemic metabolic stress (Kim and

Lee 2021) or a consequence of dysregulation in the interoceptive pathways conveying metabolic stress signals or in their integration in central homeostatic-motivational hubs (Kroemer et al. 2022). Thus, the available evidence points toward a cluster of psychiatric symptoms associated with systemic metabolic disturbance, which we term "metabolic psychopathology." This includes cognitive deficits, amotivation, anhedonia, appetite and sleep disturbance. On this basis, it is hypothesized that a comprehensive metabolic psychiatric diagnostic battery should, in addition to various metabolic assessments, include a granular of assessment of metabolic psychopathology (e.g., using actigraphy and polysomnography for activity and sleep). Furthermore, to assess motivational processes, energy and effort expenditure could be measured, consistent with evidence showing that the impact of immuno-metabolic signaling on motivational outcomes is mainly due to impairment in appetitive processes (energy effort and motivational vigor) rather than consummatory processes (perceived reward value) (Treadway et al. 2019). Considering the potential for rapid fluctuations in both psychopathology and metabolism, symptom assessments should temporally align with the collection of biological samples. These efforts are required to better define the relationship between metabolic psychopathology and systemic biological alterations.

The question of causality between central (brain) and peripheral (metabolic) processes remains a key challenge in metabolic psychiatry. Rather than a unidirectional model, current evidence supports a bidirectional and dynamic interaction between brain and body. As we have discussed, peripheral metabolic signals—such as insulin and inflammatory cytokines—modulate central circuits involved in mood, cognition, and motivation. Conversely, psychiatric disorders can disrupt autonomic, endocrine, and immune function, thereby promoting insulin resistance, dyslipidemia, and visceral adiposity. While treating one domain may influence the other, neither acts in complete isolation. For example, improving peripheral insulin sensitivity through pharmacological or lifestyle interventions has shown promise in reducing depressive and cognitive symptoms, suggesting peripheral metabolic improvement can have central benefits (Calkin et al. 2022; De Giorgi et al. 2025; Firth et al. 2020). Similarly, effective treatment of mental illness can be associated with improvements in metabolic parameters (Gulley et al. 2022), although often confounded by medication effects (Pillinger et al. 2020). However, a third possibility is also plausible: that both psychiatric symptoms and metabolic dysfunction are downstream and independent manifestations of shared upstream drivers—such as mitochondrial inefficiency, inflammation, and/or early-life adversity (Pillinger et al. 2020).

5 Can We Stratify Psychiatric Presentations Based on a Peripheral Metabolic Signature, and What Are the Implications for Research?

There have been many attempts to stratify psychiatric conditions based on a given biological parameter (McIntyre et al. 2019; Watson et al. 2023; Nettis et al. 2021). These approaches have typically attempted to identify subgroups within a given SMI. If, as we have discussed, metabolic disturbances are associated with psychopathology that transcends traditional diagnostic constructs, then stratification based primarily on markers of systemic metabolic disturbance may advance this approach. Considering the putative role of insulin signaling in "metabolic psychopathology," insulin resistance is a reasonable first stratification step. However, there are key questions regarding the role of insulin resistance in the emergence of psychiatric symptoms. For example, despite the evidence described above, there is uncertainty as to whether insulin resistance truly represents a transdiagnostic or prognostic tool, and at which stage of any given SMI insulin resistance may be relevant; for example, insulin resistance is seen prior to the onset of psychosis, but the evidence in depression is weaker and may emerge later. Furthermore, the role of extrinsic and intrinsic moderators of insulin signaling in the developmental stages of psychiatric disease require further elucidation. Finally, the multisystem effects of metabolic interventions that target both CNS and non-CNS pathways are unclear.

We propose an ambitious interdisciplinary multisystem study to address these complex challenges, which require multiple approaches to validate insulin resistance as a treatment target in SMIs, using both computational and experimental methods. To avoid the repeated, effortful, and costly set up of multiple experiments, trials and observational biomarker validation using a collaborative, broad infrastructure approach is needed. An enriched, longitudinal cohort with an embedded novel trial design (e.g., platform, multiarm), able to host experimental medicine, early-stage and adaptive interventions across physical and mental health, could offer the field the best opportunity to address multiple issues (see Figure 11.2). In brief, the following is proposed: A cohort of young participants with insulin resistance (as defined, e.g., by age-specific HOMA-IR thresholds) and at risk for mental health disorder, but without a recognized SMI could be recruited. Participants would then undergo deep phenotyping, including genetic, cardiovascular, endocrine, and

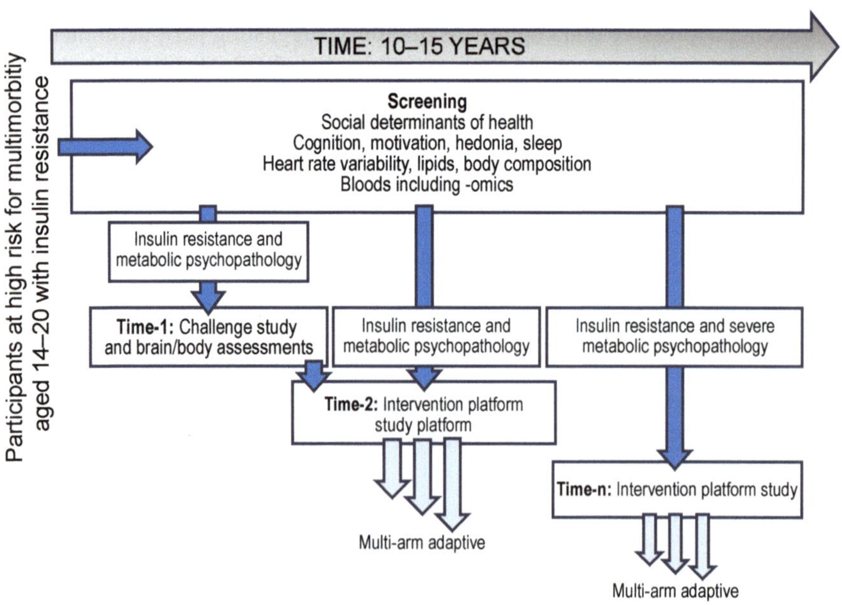

Figure 11.2 Proposed enriched longitudinal cohort with embedded platform trial.

metabolic assessments, alongside screening for psychopathology. Those participants presenting with identifiable early-stage "metabolic psychopathology" would then be invited to partake in preventive intervention trials, the first stage of which would involve psychological and physiological challenges to provide insight into pathways responsible for CNS and non-CNS alterations associated with systemic metabolic disturbance. The number of participants with insulin resistance and metabolic psychopathology would accumulate over time, and with increasing severity of illness, and would be recruited to subsequent stages of trials, which would see various interventions tested in a multiarm, multistage approach. As well as examining vulnerabilities to metabolic dysregulation and associated psychiatric disturbance, this project would also seek to define features of resilience for multisystem comorbidity, with implications for both population-level and personalized interventions.

6 Conclusions

Severe mental illness needs to be reconceptualized as a CNS manifestation of a multisystem disorder. Reemphasizing the mind as an integral part of the body offers a novel systemic approach to both the assessment and potential management of SMI. The following key issues, however, await resolution:

- Validate "metabolic psychopathology" as a transdiagnostic and/or prognostic marker in SMI
- Explore "metabolic psychopathology" across other multisystem disorders
- Explore the connection of microbiome with metabolic pathways related to SMI
- Explore sex differences in metabolic pathways related to SMI
- Explore novel metabolic interventions in SMI

With a view toward the development of novel therapeutics, we have put forth an experimental model aimed at supporting future efforts to better define causative metabolic signaling pathways and their modulators.

References

Adamantidis AR, and De Lecea L (2023) Sleep and the Hypothalamus. Science 382 (6669):405–412. https://doi.org/10.1126/science.adh8285

Agrawal R, Reno CM, Sharma S, et al. (2021) Insulin Action in the Brain Regulates Both Central and Peripheral Functions. Am J Physiol Endocrinol Metab 321 (1):E156–E163. https://doi.org/10.1152/ajpendo.00642.2020

Baltramonaityte V, Karhunen V, Felix JF, et al. (2025) Biological Pathways Underlying the Relationship between Childhood Maltreatment and Multimorbidity: A Two-Step, Multivariable Mendelian Randomisation Study. Brain Behav Immun 126:29–69. https://doi.org/10.1016/j.bbi.2025.01.024

Barakat GM, Ramadan W, Assi G, and Khoury NBE (2024) Satiety: A Gut-Brain-Relationship. J Physiol Sci 74 (1):11. https://doi.org/10.1186/s12576-024-00904-9

Bian X, Xue H, Jing D, et al. (2023) Role of Serum/Glucocorticoid-Regulated Kinase 1 (SGK1) in Immune and Inflammatory Diseases. Inflammation 46 (5):1612–1625. https://doi.org/10.1007/s10753-023-01857-8

Brüning JC, and Fenselau H (2023) Integrative Neurocircuits That Control Metabolism and Food Intake. Science 381 (6665):eabl7398. https://doi.org/10.1126/science.abl7398

Calkin CV (2019) Insulin Resistance Takes Center Stage: A New Paradigm in the Progression of Bipolar Disorder. Ann Med 51 (5–6):281–293. https://doi.org/10.1080/07853890.2019.1659511

Calkin CV, Chengappa KNR, Cairns K, et al. (2022) Treating Insulin Resistance with Metformin as a Strategy to Improve Clinical Outcomes in Treatment-Resistant Bipolar Depression (the Trio-Bd Study): A Randomized, Quadruple-Masked, Placebo-Controlled Clinical Trial. J Clin Psychiatry 83 (2):21m14022. https://doi.org/10.4088/jcp.21m14022

Clegg DJ, Gotoh K, Kemp C, et al. (2011) Consumption of a High-Fat Diet Induces Central Insulin Resistance Independent of Adiposity. Physiol Behav 103 (1):10–16. https://doi.org/10.1016/j.physbeh.2011.01.010

Coppola S, Paparo L, Trinchese G, et al. (2023) Increased Dietary Intake of Ultraprocessed Foods and Mitochondrial Metabolism Alterations in Pediatric Obesity. Sci Rep 13 (1):12609. https://doi.org/10.1038/s41598-023-39566-9

Corsi-Zuelli F, Deakin B, de Lima MHF, et al. (2021) T Regulatory Cells as a Potential Therapeutic Target in Psychosis? Current Challenges and Future Perspectives. Brain Behav Immun Health 17:100330. https://doi.org/10.1016/j.bbih.2021.100330

Dash S, Xiao C, Morgantini C, Koulajian K, and Lewis GF (2015) Intranasal Insulin Suppresses Endogenous Glucose Production in Humans Compared with Placebo in the Presence of Similar Venous Insulin Concentrations. Diabetes 64 (3):766–774. https://doi.org/10.2337/db14-0685

De Giorgi R, Ghenciulescu A, Dziwisz O, et al. (2025) An Analysis on the Role of Glucagon-Like Peptide-1 Receptor Agonists in Cognitive and Mental Health Disorders. Nat Ment Health 3 (3):354–373. https://doi.org/10.1038/s44220-025-00390-x

Debost JC, Larsen JT, Munk-Olsen T, et al. (2019) Childhood Infections and Schizophrenia: The Impact of Parental SES and Mental Illness, and Childhood Adversities. Brain Behav Immun 81:341–347. https://doi.org/10.1016/j.bbi.2019.06.031

Duarte AI, Moreira PI, and Oliveira CR (2012) Insulin in Central Nervous System: More Than Just a Peripheral Hormone. J Aging Res 2012:384017. https://doi.org/10.1155/2012/384017

Eaton WW, Byrne M, Ewald H, et al. (2006) Association of Schizophrenia and Autoimmune Diseases: Linkage of Danish National Registers. Am J Psychiatry 163 (3):521–528. https://doi.org/10.1176/appi.ajp.163.3.521

Engelhardt B, Vajkoczy P, and Weller RO (2017) The Movers and Shapers in Immune Privilege of the CNS. Nat Immunol 18 (2):123–131. https://doi.org/10.1038/ni.3666

Fabbri C, and Serretti A (2017) Role of 108 Schizophrenia-Associated Loci in Modulating Psychopathological Dimensions in Schizophrenia and Bipolar Disorder. Am J Med Genet B Neuropsychiatr Genet 174 (7):757–764. https://doi.org/10.1002/ajmg.b.32577

Ferrario CR, and Reagan LP (2018) Insulin-Mediated Synaptic Plasticity in the CNS: Anatomical, Functional and Temporal Contexts. Neuropharmacology 136 (Pt B):182–191. https://doi.org/10.1016/j.neuropharm.2017.12.001

Firth J, Siddiqi N, Koyanagi A, et al. (2019) The *Lancet Psychiatry* Commission: A Blueprint for Protecting Physical Health in People with Mental Illness. Lancet Psychiatry 6 (8):675–712. https://doi.org/10.1016/s2215-0366(19)30132-4

Firth J, Solmi M, Wootton RE, et al. (2020) A Meta-Review of "Lifestyle Psychiatry": The Role of Exercise, Smoking, Diet and Sleep in the Prevention and Treatment of Mental Disorders World Psychiatry 19 (3):360–380. https://doi.org/10.1002/wps.20773

Freyberg Z, Aslanoglou D, Shah R, and Ballon JS (2017) Intrinsic and Antipsychotic Drug-Induced Metabolic Dysfunction in Schizophrenia. Front Neurosci 11:432. https://doi.org/10.3389/fnins.2017.00432

Freyberg Z, Ferrando SJ, and Javitch JA (2010) Roles of the Akt/GSK-3 and Wnt Signaling Pathways in Schizophrenia and Antipsychotic Drug Action. Am J Psychiatry 167 (4):388–396. https://doi.org/10.1176/appi.ajp.2009.08121873

Freyberg Z, and Harvill ET (2017) Pathogen Manipulation of Host Metabolism: A Common Strategy for Immune Evasion. PLoS Pathog 13 (12):e1006669. https://doi.org/10.1371/journal.ppat.1006669

Gee DG, and Casey BJ (2015) The Impact of Developmental Timing for Stress and Recovery. Neurobiol Stress. Neurobiol Stress 1:184–194. https://doi.org/10.1016/j.ynstr.2015.02.001

Goldsmith DR, Rapaport MH, and Miller BJ (2016) A Meta-Analysis of Blood Cytokine Network Alterations in Psychiatric Patients: Comparisons between Schizophrenia, Bipolar Disorder and Depression. Mol Psychiatr 21 (12):1696–1709. https://doi.org/10.1038/mp.2016.3

Gonzalez Caldito N (2023) Role of Tumor Necrosis Factor-Alpha in the Central Nervous System: A Focus on Autoimmune Disorders. Front Immunol 14:1213448. https://doi.org/10.3389/fimmu.2023.1213448

Gray SM, Meijer RI, and Barrett EJ (2014) Insulin Regulates Brain Function, but How Does It Get There? Diabetes 63 (12):3992–3997. https://doi.org/10.2337/db14-0340

Grillo CA, Piroli GG, Kaigler KF, et al. (2011) Downregulation of Hypothalamic Insulin Receptor Expression Elicits Depressive-Like Behaviors in Rats. Behav Brain Res 222 (1):230–235. https://doi.org/10.1016/j.bbr.2011.03.052

Gulley LD, Shomaker LB, Kelly NR, et al. (2022) Examining Cognitive-Behavioral Therapy Change Mechanisms for Decreasing Depression, Weight, and Insulin Resistance in Adolescent Girls at Risk for Type 2 Diabetes J Psychosom Res 157:110781. https://doi.org/10.1016/j.jpsychores.2022.110781

Hastie CE, Lowe DJ, McAuley A, et al. (2023) True Prevalence of Long-Covid in a Nationwide, Population Cohort Study. Nat Commun 14 (1):7892. https://doi.org/10.1038/s41467-023-43661-w

Iwen KA, Scherer T, Heni M, et al. (2014) Intranasal Insulin Suppresses Systemic but Not Subcutaneous Lipolysis in Healthy Humans. J Clin Endocrinol Metab 99 (2):E246–E251. https://doi.org/10.1210/jc.2013-3169

Jones BDM, Farooqui S, Kloiber S, et al. (2021) Targeting Metabolic Dysfunction for the Treatment of Mood Disorders: Review of the Evidence. Life 11 (8):819. https://doi.org/10.3390/life11080819

Kim B, and Feldman EL (2015) Insulin Resistance as a Key Link for the Increased Risk of Cognitive Impairment in the Metabolic Syndrome. Exp Mol Med 47 (3):e149. https://doi.org/10.1038/emm.2015.3

Kim KH, and Lee MS (2021) GDF15 as a Central Mediator for Integrated Stress Response and a Promising Therapeutic Molecule for Metabolic Disorders and NASH. Biochim Biophys Acta Gen Subj 1865 (3):129834. https://doi.org/10.1016/j.bbagen.2020.129834

Kivimaki M, Bartolomucci A, and Kawachi I (2023) The Multiple Roles of Life Stress in Metabolic Disorders. Nat Rev Endocrinol 19 (1):10–27. https://doi.org/10.1038/s41574-022-00746-8

Kroemer NB, Opel N, Teckentrup V, et al. (2022) Functional Connectivity of the Nucleus Accumbens and Changes in Appetite in Patients with Depression. JAMA Psychiatry 79 (10):993–1003. https://doi.org/10.1001/jamapsychiatry.2022.2464

Krolick KN, Zhu Q, and Shi H (2018) Effects of Estrogens on Central Nervous System Neurotransmission: Implications for Sex Differences in Mental Disorders. Prog Mol Biol Transl Sci 160:105–171. https://doi.org/10.1016/bs.pmbts.2018.07.008

Lalousis PA, Schmaal L, Wood SJ, et al. (2023) Inflammatory Subgroups of Schizophrenia and Their Association with Brain Structure: A Semi-Supervised Machine Learning Examination of Heterogeneity. Brain, Behavior, and Immunity. Brain Behav Immun 113:166–175. https://doi.org/10.1016/j.bbi.2023.06.023

Lee J, Xue X, Au E, et al. (2024a) Glucose Dysregulation in Antipsychotic-Naive First-Episode Psychosis: In Silico Exploration of Gene Expression Signatures. Transl Psychiatry 14 (1):19. https://doi.org/10.1038/s41398-023-02716-8

Lee J, Xue X, Au E, et al. (2024b) Central Insulin Dysregulation in Antipsychotic-Naïve First-Episode Psychosis: In Silico Exploration of Gene Expression Signatures. Psychiatry Res 331:115636. https://doi.org/10.1016/j.psychres.2023.115636

Lewis GF, Carpentier AC, Pereira S, Hahn M, and Giacca A (2021) Direct and Indirect Control of Hepatic Glucose Production by Insulin. Cell Metab 33 (4):709–720. https://doi.org/10.1016/j.cmet.2021.03.007

Liu Y, Li Z, Zhang M, et al. (2013) Exploring the Pathogenetic Association between Schizophrenia and Type 2 Diabetes Mellitus Diseases Based on Pathway Analysis. BMC Med Genomics 6 (Suppl 1):S17. https://doi.org/10.1186/1755-8794-6-s1-s17

Maksyutynska K, Stogios N, Prasad F, et al. (2024) Neurocognitive Correlates of Metabolic Dysregulation in Individuals with Mood Disorders: A Systematic Review and Meta-Analysis. Psychol Med 54 (7):1245–1271. https://doi.org/10.1017/s0033291724000345

McIntyre RS, Subramaniapillai M, Lee Y, et al. (2019) Efficacy of Adjunctive Infliximab vs Placebo in the Treatment of Adults with Bipolar I/II Depression: A Randomized Clinical Trial. JAMA Psychiatry 76 (8):783–790. https://doi.org/10.1001/jamapsychiatry.2019.0779

McLaughlin KA, and Hatzenbuehler ML (2009) Mechanisms Linking Stressful Life Events and Mental Health Problems in a Prospective, Community-Based Sample of Adolescents. J Adolesc Health 44 (2):153–160. https://doi.org/10.1016/j.jadohealth.2008.06.019

Milaneschi Y, Lamers F, Berk M, and Penninx BWJH (2020) Depression Heterogeneity and Its Biological Underpinnings: Toward Immunometabolic Depression. Biol Psychiatry 88 (5):369–380. https://doi.org/10.1016/j.biopsych.2020.01.014

Morales-Munoz I, Marwaha S, Upthegrove R, and Cropley V (2024) Role of Inflammation in Short Sleep Duration across Childhood and Psychosis in Young Adulthood. JAMA Psychiatry 81 (8):825–833. https://doi.org/10.1001/jamapsychiatry.2024.0796

Nettis MA, Lombardo G, Hastings C, et al. (2021) Augmentation Therapy with Minocycline in Treatment-Resistant Depression Patients with Low-Grade Peripheral Inflammation: Results from a Double-Blind Randomised Clinical Trial. Neuropsychopharmacology 46 (5):939–948. https://doi.org/10.1038/s41386-020-00948-6

Obici S, Zhang BB, Karkanias G, and Rossetti L (2002) Hypothalamic Insulin Signaling Is Required for Inhibition of Glucose Production. Nat Med 8 (12):1376–1382. https://doi.org/10.1038/nm1202-798

Perry BI, Jones HJ, Richardson TG, et al. (2020) Common Mechanisms for Type 2 Diabetes and Psychosis: Findings from a Prospective Birth Cohort. Schizophr Res 223:227–235. https://doi.org/10.1016/j.schres.2020.08.006

Pillinger T, Beck K, Gobjila C, et al. (2017) Impaired Glucose Homeostasis in First-Episode Schizophrenia: A Systematic Review and Meta-Analysis. JAMA Psychiatry 74 (3):261–269. https://doi.org/10.1001/jamapsychiatry.2016.3803

Pillinger T, D'Ambrosio E, McCutcheon R, and Howes OD (2019) Is Psychosis a Multisystem Disorder? A Meta-Review of Central Nervous System, Immune, Cardiometabolic, and Endocrine Alterations in First-Episode Psychosis and Perspective on Potential Models. Mol Psychiatr 24 (6):776–794. https://doi.org/10.1038/s41380-018-0058-9

Pillinger T, McCutcheon RA, Vano L, et al. (2020) Comparative Effects of 18 Antipsychotics on Metabolic Function in Patients with Schizophrenia, Predictors of Metabolic Dysregulation, and Association with Psychopathology: A Systematic Review and Network Meta-Analysis. Lancet Psychiatry 7 (1):64–77. https://doi.org/10.1016/s2215-0366(19)30416-x

Prestwood TR, Asgariroozbehani R, Wu S, et al. (2021) Roles of Inflammation in Intrinsic Pathophysiology and Antipsychotic Drug-Induced Metabolic Disturbances of Schizophrenia. Behav Brain Res 402:113101. https://doi.org/10.1016/j.bbr.2020.113101

Raichle ME, and Gusnard DA (2002) Appraising the Brain's Energy Budget. PNAS 99 (16):10237–10239. https://doi.org/10.1073/pnas.172399499

Rasinska J, Klein C, Stahn L, et al. (2022) Transposon-Mediated Glial Cell Line-Derived Neurotrophic Factor Overexpression in Human Adipose Tissue-Derived Mesenchymal Stromal Cells: A Potential Approach for Neuroregenerative Medicine? J Tissue Eng Regen Med 16 (6):515–529. https://doi.org/10.1002/term.3296

Ren H, Vieira-de-Abreu A, Yan S, et al. (2019) Altered Central Nutrient Sensing in Male Mice Lacking Insulin Receptors in Glut4-Expressing Neurons. Endocrinology 160 (9):2038–2048. https://doi.org/10.1210/en.2019-00341

Roh E, Song DK, and Kim MS (2016) Emerging Role of the Brain in the Homeostatic Regulation of Energy and Glucose Metabolism. Exp Mol Med 48 (3):e216. https://doi.org/10.1038/emm.2016.4

Shin AC, Filatova N, Lindtner C, et al. (2017) Insulin Receptor Signaling in POMC, but Not AgRP, Neurons Controls Adipose Tissue Insulin Action. Diabetes 66 (6):1560–1571. https://doi.org/10.2337/db16-1238

Straub RH (2007) The Complex Role of Estrogens in Inflammation. Endocr Rev 28 (5):521–574. https://doi.org/10.1210/er.2007-0001

Tomasik J, Lago SG, Vázquez-Bourgon J, et al. (2019) Association of Insulin Resistance with Schizophrenia Polygenic Risk Score and Response to Antipsychotic Treatment. JAMA Psychiatry 76 (8):864–867. https://doi.org/10.1001/jamapsychiatry.2019.0304

Treadway MT, Cooper JA, and Miller AH (2019) Can't or Won't? Immunometabolic Constraints on Dopaminergic Drive. Trends Cogn Sci 23 (5):435–448. https://doi.org/10.1016/j.tics.2019.03.003

Upthegrove R, and Goldsmith DR (2024) The Psychoneuroimmunology of Psychosis: Peeking Past the Clouds of Heterogeneity. Brain, Behavior, and Immunity. Brain Behav Immun 119:709–712. https://doi.org/10.1016/j.bbi.2024.04.037

van Nimwegen LJ, Storosum JG, Blumer RM, et al. (2008) Hepatic Insulin Resistance in Antipsychotic Naive Schizophrenic Patients: Stable Isotope Studies of Glucose Metabolism. J Clin Endocrinol Metab 93 (2):572–577. https://doi.org/10.1210/jc.2007-1167

Wachsmuth HR, Weninger SN, and Duca FA (2022) Role of the Gut-Brain Axis in Energy and Glucose Metabolism. Exp Mol Med 54 (4):377–392. https://doi.org/10.1038/s12276-021-00677-w

Watson AJ, Giordano A, Suckling J, et al. (2023) Cognitive Function in Early-Phase Schizophrenia-Spectrum Disorder: IQ Subtypes, Brain Volume and Immune Markers. Psychol Med 53 (7):2842–2851. https://doi.org/10.1017/s0033291721004815

Watson KT, Simard JF, Henderson VW, et al. (2021) Incident Major Depressive Disorder Predicted by Three Measures of Insulin Resistance: A Dutch Cohort Study. Am J Psychiatry 178 (10):914–920. https://doi.org/10.1176/appi.ajp.2021.20101479

Yan H, Yang W, Zhou F, et al. (2019) Estrogen Improves Insulin Sensitivity and Suppresses Gluconeogenesis via the Transcription Factor Foxo1. Diabetes 68 (2):291–304. https://doi.org/10.2337/db18-0638

Zubin J, and Spring B (1977) Vulnerability: A New View of Schizophrenia. J Abnorm Psychol 86 (2):103–126. https://doi.org/10.1037//0021-843x.86.2.103

Open Access This chapter is licensed under the terms of the Creative Commons Attribution-NonCommercial-NoDerivatives 4.0 International License (http://creativecommons.org/licenses/by-nc-nd/4.0/), which permits any noncommercial use, sharing, distribution and reproduction in any medium or format, as long as you give appropriate credit to the original author(s) and the source, provide a link to the Creative Commons license and indicate if you modified the licensed material. You do not have permission under this license to share adapted material derived from this chapter or parts of it.

The images or other third party material in this chapter are included in the chapter's Creative Commons license, unless indicated otherwise in a credit line to the material. If material is not included in the chapter's Creative Commons license and your intended use is not permitted by statutory regulation or exceeds the permitted use, you will need to obtain permission directly from the copyright holder.

Metabolism-Based Therapies

12

Best Practices for Randomized Controlled Trials in Metabolic Neuropsychiatry Studies

Melanie M. Wall and Martina Pavlicova

Abstract Randomized controlled trials (RCTs) are a well-established scientific method for building evidence to support the efficacy of an intervention. Proposed metabolic interventions for neuropsychiatric problems range from ketogenic diets to probiotics and omega-3 supplements, among many others. Here, a comprehensive guide is presented for researchers and clinicians to use as they navigate the intricacies of planning, conducting, analyzing, and reporting on clinical trials in metabolic therapies for neuropsychiatric disorders. In-depth topics include the fundamental role of and ways to achieve randomization, trial design decisions, choice of control group and blinding, specification of outcomes including mechanisms of action, power calculations and sample size, as well as the intention to treat and account for lack of adherence. Issues related to ethical oversight, safety monitoring, standardized protocols, preregistration, and adherence to reporting standards (e.g., CONSORT) are presented as integral elements of a clinical trial's integrity, ensuring reliable and actionable outcomes. The chapter advocates for thoughtfully designed and analyzed RCTs that prioritize scientific rigor alongside clinical relevance.

Keywords Randomized controlled trial, methodology, metabolism, diet

---

Melanie M. Wall (✉)
Dept. of Biostatistics, Mailman School of Public Health, and Dept. of Psychiatry, Vagelos College of Physicians and Surgeons, Columbia University Irving Medical Center; New York State Psychiatric Institute, New York, NY 10032, USA
Email: mmwall@columbia.edu

Martina Pavlicova
Dept. of Biostatistics, Mailman School of Public Health, Columbia University Irving Medical Center, New York, NY 10032, USA

© Frankfurt Institute for Advanced Studies (FIAS) 2025

D. Öngür, J. M. Ford (eds.), *Metabolic Neuropsychiatry*, Strüngmann Forum Reports,
https://doi.org/10.1007/978-3-032-05630-6_12

1 Introduction

The current landscape of clinical trials that examine metabolic interventions for psychiatric disorders and cognitive function is a dynamic and evolving field. Increasingly, researchers are interested in the metabolic underpinnings of mental health, exploring how interventions such as diet, exercise, and metabolic-targeted drugs can improve psychiatric and cognitive outcomes. Clinical trials are investigating the efficacy of ketogenic diets, probiotics, and supplements such as omega-3 fatty acids in managing various conditions (e.g., depression, schizophrenia, and cognitive decline). Relatedly, metabolic pharmacologic therapies have recently been approved for treatment of obesity and diabetes (Chakhtoura et al. 2023; Tchang et al. 2000). There is also a growing focus on personalized medicine, which aims to tailor metabolic interventions based on individual patient profiles, including genetics and metabolism. Despite the promising nature of these approaches, the field must confront challenges, such as variability in study designs, small sample sizes, and the need for longer-term follow-up, before we fully understand the efficacy and safety of these interventions.

Here, we delve into various aspects that are important to consider when planning, conducting, and statistically analyzing results from clinical trials that test metabolic interventions in the field of neuropsychiatry. Discussion underscores the necessity of a well-structured framework to assess both the efficacy and safety of novel treatments as well as the methods to assess the targeted metabolic pathways that influence neuropsychiatric disorders. We will give a primer on why randomized controlled trials (RCTs) are considered the gold standard for confirming evidence of the efficacy of an intervention. Thereafter, we dive deeper into issues regarding trial phases and design, choice of primary outcome measures and their timing, randomization, blinding, sample size, and data analytic issues. Our goal is to provide a guide for researchers and clinicians, aiming to conduct and understand clinical trials in metabolic-based therapies for neuropsychiatric disorders.

2 Overview of Clinical Trials of Metabolic-Based Therapies for Neuropsychiatric Outcomes

Using the repository ClinicalTrials.gov, which offers information on active and completed clinical trials conducted around the world, a search revealed an increasing number of clinical trials aimed at investigating metabolic interventions for psychiatric and neurological disorders. Since 2010, over 1,400 trials have investigated dietary inventions and over 4,000 have addressed exercise interventions to treat psychiatric or neurological conditions. Sample studies include the following:

- Gluten- and casein-free dietary interventions for autism (Whiteley et al. 2010)
- Increased dietary protein for neurobiologic improvement and cognitive improvement in people with attention-deficit/hyperactivity disorder (Huberts-Bosch et al. 2024; ICH GCP 2024)

- Ketogenic diets to improve metabolic and psychiatric health in people with bipolar disorder and schizophrenia (Stanford Medicine Health Care 2025)
- Dietary supplementation with molecular hydrogen and pyrroloquinoline quinone to improve mitochondrial function and cognition in the elderly (Ames 2018)
- Aerobic exercise to improve inflammation, brain-derived neurotrophic factor, and cognition (McGurk et al. 2021) in severe mental illness

A common feature of many clinical trials investigating metabolic-based interventions for neuropsychiatric conditions is that the outcomes include both the biological targets (e.g., mitochondrial function, neurobiology, inflammation) and the primary psychological or cognitive symptoms of the patient (Figure 12.1). That is, it is common for these trials to have two goals. The first, and often primary, goal is to answer whether the metabolism-based therapy improves patient psychiatric symptoms and/or cognitive function (Figure 12.1a). Ultimately, determination of the treatment

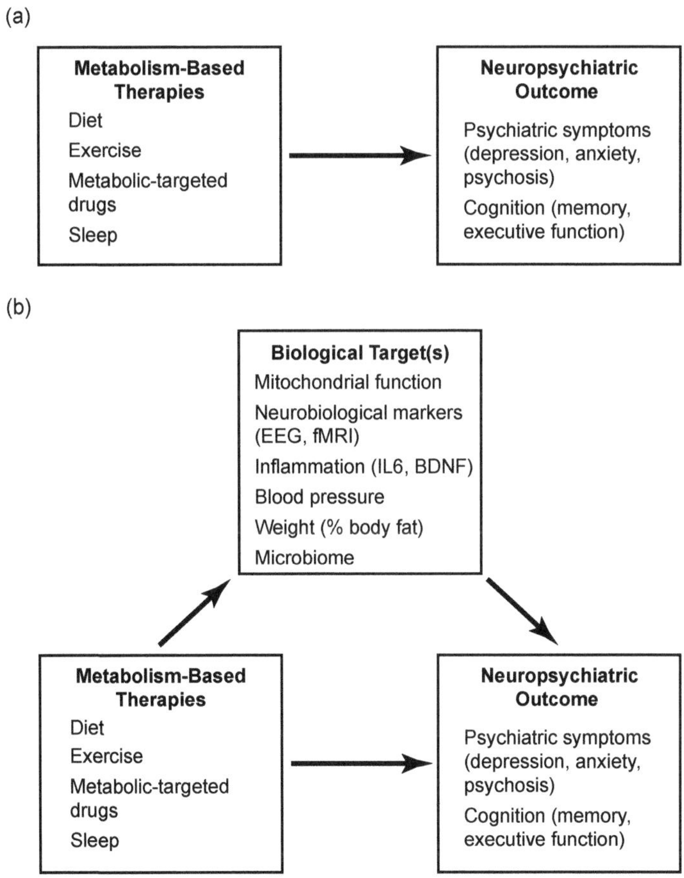

Figure 12.1 (a) Primary aim of the randomized controlled clinical trial: Does the metabolism-based therapy improve patient psychiatric symptoms or cognition? (b) Mechanism of action goal: Does the metabolism-based therapy affect the biological targets, which in turn improve patient symptoms?

effect on this primary outcome (i.e., that the therapy works and how well it works on average) is the main finding that matters for the livelihood and quality of life for the people who will take the treatment. The second goal, however, is also important, which is to understand and test the biological mechanism of action of the therapy (Figure 12.1b) (i.e., why it works).

The randomized intervention used in a clinical trial can also function as a probe for the metabolic mechanisms and ultimately expand our understanding of them. Here the goal is to further the scientific knowledge to understand the biological explanations for why the therapies work. In addition, this focus on the mechanisms through biological targets might potentially enable future treatments to be tailored or adapted to specific patient profiles that respond differentially. Of note, this focus in clinical trials on the mechanism of action was an explicit policy introduced in 2013 by the National Institute of Mental Health (Insel 2013).

3 Why Are Randomized Clinical Trials the Gold Standard for Assessing Treatment Effects?

RCTs follow rigorous design and methodology which help to ensure that the results are reliable and valid. The fundamental reason that RCTs are considered the gold standard for assessing the scientific evidence for whether a therapy works comes from the act of randomizing subjects to the therapy of interest or to a comparator ("control") group. Randomization ensures that differences between groups in the outcomes are due to the treatment they were given, rather than to other background factors. Theoretically, randomization ensures balance between the background characteristics between the groups, and thus provides the best means to establish causality between the intervention and the outcomes by minimizing the possibility of confounders. That is, unlike in the real world where people and their clinical providers choose what type of interventions they try, and those choices are confounded by prognostic factors (e.g., people who are more concerned or more severe may tend to choose certain therapies), in an RCT, the choice of whether a person receives an intervention or is in the control group is determined by a coin flip. Because this is the only factor that influences which treatment someone is given in an RCT, there are no confounders and hence, causal effects can be estimated directly. In an RCT, each randomized group can serve as the counterfactual for the other group and hence, the average causal treatment effect can be estimated simply by the difference between the groups. The fundamental role of randomization as a means to estimate causal effects has been well-formalized in the literature; for a good review, see Eichler et al. (2016).

Despite the elegant theory underlying why randomization provides a means for establishing causality in randomized trials, in practice, many things can go wrong during the processes of conducting the clinical trial, which can introduce biases and confounding. The following types of biases can occur:

- Selection bias, which happens when those who choose to enroll in trials are not representative of the broader population who would use the therapy (e.g., enrolling more highly functioning people with schizophrenia with strong social networks).

- Detection bias, which can happen from unblinded experimenters/raters or subjects who may report outcomes differentially based on their expectations of improvement from being in a trial rather than actual experience of improvement (e.g., a subject enrolled in a ketogenic diet reports improved symptoms because they believe their dramatic change in diet must be healthy so they discount or downplay continued symptoms).
- Attrition bias that stems from subjects not adhering to protocols or dropping out completely; for example, "lifestyle" interventions, such as nutrition or exercise changes, are common in metabolic neuropsychiatry and notoriously difficult to sustain adherence.
- Publication bias, which occurs when positive rather than null studies are more likely to be published.

To minimize such biases and uphold the status of RCTs as the gold standard for evidence, the scientific community has agreed upon several guidelines for conducting RCTs, which we review below (Evans 2010b; Meinert 2011).

3.1 Standardized Protocols

Clinical trials follow standardized protocols that detail how the study is conducted, including how participants are chosen, how the treatment is administered, and how outcomes are measured. When studying psychiatric disorders, to enroll enough subjects with particular disorders, it is common to use multisite studies where, for example, enrollment and administration is conducted at multiple outpatient psychiatric clinics. For studies conducted at multiple sites, standardization ensures consistency across the study and enhances the reliability of the findings. The Consolidated Standards of Reporting Clinical Trials (CONSORT) guides the reporting of essential components of randomized trial methods and findings in published reports (Schulz et al. 2010). It has been adopted and is required by most health journals for publication (Shamseer et al. 2016). One prominent recommendation of CONSORT is the inclusion of a flow diagram (see Schulz et al. 2010) that describes how subjects were involved—from the beginning (screening for eligibility) until the final analytic sample—as well as the delineation of the specific choices made within the trial (see CONSORT 2010 checklist of information in Schulz et al. 2010). Both help to assess the internal and external validity of the study.

3.2 Preregistration

Preregistration for clinical trials involves registering the study's protocol with a publicly accessible database (e.g., clinicaltrials.gov); this must be done before enrolling participants and starting the trial. The preregistration process includes detailing the study's objectives, methodology, design, population, and planned analyses. The aim here is to enhance transparency and credibility in research by providing a public record of the planned research before it begins. The importance of preregistration lies in its ability to prevent selective reporting and publication bias, ensuring that all results—positive, negative, or inconclusive—are available for public scrutiny. This transparency helps researchers avoid practices like "p-hacking" or "data dredging," where they might selectively report outcomes that were not originally designated as primary endpoints. By outlining the research plan in advance, preregistration commits researchers to follow the specified methods and analyses, thus contributing to the integrity and reproducibility of scientific research.

3.3 Ethical and Safety Oversight

Clinical trials are subject to rigorous ethical oversight to protect participants' rights and well-being. Ethical standards require that trials have scientific validity, provide potential benefits that outweigh risks, and ensure informed consent from participants. *Data Safety and Monitoring Boards* enable continuous monitoring of data, essential for identifying any adverse effects or ethical concerns. Interim analyses, conducted at predetermined points during the trial, can help in making decisions about continuing, modifying, or stopping the trial early for efficacy or safety reasons.

4 Designing and Conducting RCTs: Decisions That Need to Be Made

4.1 Phases of Clinical Trials

Depending upon the specific type of metabolic intervention (e.g., a new experimental drug not previously used in humans vs. a strict gluten-free dietary nutritional intervention) and the extent of research that has already been conducted, the phase of the clinical trial to be conducted will vary. The following standard definitions of clinical trial phases are used to delineate specific objectives and characteristics to ensure the safety and efficacy of new drugs before they become widely available to the public.

- *Preclinical* phase involves experimenting with the chemical/drug in nonhumans to understand its biological mechanisms. For example, rodents are typically used when exploring metabolic interventions, whereas nonhuman primates are used to investigate more complex human-like responses, especially in drug development, therapeutic, or neurobehavioral interventions (Mukherjee et al. 2022; Soufizadeh et al. 2024).
- *Phase I*, the initial phase of clinical testing, involves a small group of healthy human volunteers (10–50 participants) and focuses primarily on assessing the safety, tolerability, pharmacokinetics, and pharmacodynamics of the drug or intervention. The main goal is to determine the drug's most frequent and serious side effects and how it is metabolized and excreted by the body. One example of a phase I clinical trial evaluated the safety, tolerability, pharmacokinetics, and pharmacodynamics of TG103 injections in overweight or obese participants (Lin et al. 2024). The authors concluded

that all three doses of once-weekly TG103 were well tolerated with an acceptable safety profile and demonstrated preliminary 12-week weight loss without lifestyle intervention. Aztra Zeneca is currently conducting a phase I randomized, single-blind controlled study to evaluate the safety, tolerability, pharmacokinetics, and pharmacodynamics of AZD6234 in overweight or obese participants (Astrazeneca 2023).

- *Phase II* trials are larger (typically involving 50–200 participants) and aim to assess the preliminary efficacy of the drug, while continuing to evaluate its safety. This phase often includes patients who have the condition that the drug is intended to treat. Researchers try to ascertain the optimal dose and scheduling for the drug and further investigate its therapeutic effects. An example of phase II study is the MIRNA study, which investigated the efficacy and safety of a DGAT2 inhibitor in adults with nonalcoholic steatohepatitis (NASH) and fibrosis (Amin et al. 2022). Another is the phase II clinical trial by Boehringer Ingelheim and Zealand Pharma, which investigated the efficacy of survodutide (BI 456906), a novel glucagon/GLP-1 receptor dual agonist, in patients with obesity or overweight. The study demonstrated significant weight loss and also reported a favorable safety profile, with gastrointestinal issues as the primary reason for discontinuation, particularly during rapid dose escalation, which could potentially be mitigated with slower titration (Le Roux et al. 2023).
- *Phase III* trials expand in scale, involving several hundred to several thousand participants. Here the aim is to provide a more thorough understanding of the intervention's effectiveness, benefits, and the range of possible adverse reactions. Phase III trials are pivotal in determining whether a drug will be approved by regulatory authorities. They compare the new drug, often against a placebo or standard treatment, to evaluate its overall risk:benefit ratio. For instance, a clinical trial SURMOUNT-3 that was conducted by Eli Lilly, evaluated the efficacy of tirzepatide for weight loss in adults with obesity or overweight but without type 2 diabetes (Wadden et al. 2023). After an intensive 12-week lifestyle intervention program, participants who achieved at least 5% weight loss were randomized to receive tirzepatide or a placebo. At the 72-week follow-up, participants in the tirzepatide group achieved a 21.1% mean weight loss, compared to placebo participants who gained 3.3% bodyweight on average. Another example is an announcement by Boehringer Ingelheim and Zealand Pharma of three Phase III trials (SYNCHRONIZE-1, SYNCHRONIZE-2, and SYNCHRONIZE-CVOT) (Boehringer Ingelheim 2023) to further investigate the efficacy and safety of survodutide (BI 456906), a dual agonist of the glucagon and GLP-1 receptors, for treating obesity and overweight. The trials build on successful phase II data, mentioned above. The Phase III studies will focus on individuals with or without type 2 diabetes and cardiovascular risk factors.
- *Phase IV* trials, also known as post-marketing surveillance, are conducted after the drug has been marketed to monitor the long-term effectiveness and impact of the drug in a larger, more diverse population. These trials are essential for detecting rare or long-term adverse effects and can lead to further refinement of the drug's usage recommendations or, in some cases,

withdrawal from the market. An example of a phase IV clinical trial is the post-marketing surveillance study conducted on ocrelizumab for multiple sclerosis. In this trial, real-world data is collected to assess long-term safety and efficacy across diverse populations, often revealing additional information that was not captured in earlier phases (Genentech 2023).

Each phase is crucial for understanding the different aspects of the drug or intervention's safety and efficacy, contributing to the body of evidence that will be used by regulatory authorities (if it is an experimental drug) or by the general audience review before choosing to recommend a new medication or therapy.

4.2 Trial Design: A Case for the Parallel Arm Design

There are numerous types of experimental designs that can be utilized in clinical trials, such as parallel, factorial, crossover, adaptive, SMART, non-inferiority (for a well-annotated guide on the various types of designs and their strengths, see Brody 2016; Evans 2010a; Nair 2019). The most basic and by far most commonly used RCT design, however, is the two-arm parallel trial. The term "parallel" refers to the fact that both groups (i.e., those randomized to experimental intervention and those randomized to control) proceed through the study simultaneously, following the same timeline. The outcomes of each group are then compared at the end of the study period. This fundamental design in clinical research allows for a clear comparison of the effects of an intervention against a control, providing robust evidence of the intervention's efficacy. Because the design is straightforward and intuitive, it makes it easier to interpret the direct effects of the treatment. In addition, there is some time efficiency in this type of design since both groups are studied over the same period; thus, the trial can be completed in less time than when groups are studied sequentially. A commonly cited disadvantage of the parallel arm trial is that it theoretically requires a larger number of participants to detect differences between the groups than a within person (e.g., crossover) design might need. However, complications that can come with other types of designs—including crossover designs, (Krogh et al. 2019) which need careful consideration of wash-out periods especially in the context of metabolic interventions (e.g., diet) where there can be long-term carry over effects—lead most investigators to choose the simple parallel arm trial.

4.3 Choosing the Primary Outcome Measure

Determining the primary outcome in a metabolic intervention clinical trial is a critical step that should be guided by the trial's objectives, the condition under study, and the intervention's expected effects. There are several key considerations and steps a researcher should take to determine the primary outcome. First, the *clinical relevance* of the outcome measure needs to be considered. That is, the primary outcome should be clinically relevant, addressing a question of real importance to patient care or health outcomes. It should reflect a meaningful benefit to patients, such as improved survival, symptom relief, or enhanced quality of life. Moreover, the outcome should satisfy regulatory requirements, especially if the trial aims for new drug approval or changes to clinical practice.

A second consideration is the *feasibility and psychometric properties* of the outcome. By feasibility we simply mean it can be collected within the study's time frame and budget. Outcomes that require overly complex or expensive methods may not be practical. A measure having good psychometric properties means that it is both reliable and valid, ensuring that it accurately and consistently measures what it is intended to measure. It should also be sensitive enough to detect changes or differences that result from the intervention.

Another very important consideration is the *timing of the measurements*. The timing should align with the expected time frame of the intervention's effects, both immediate as well as long term. It is common to take regular outcome measurements during the active time of the intervention, at the end of intervention, and several weeks or months after finishing the intervention. For example, in a clinical trial which studies a 4-month long ketogenic diet plan, measures may be taken at baseline (prior to starting), at the end of month 1, 2, 3, and 4 (when the active intervention ends), as well as 6 months later (i.e., 10 months after start) to capture both the trajectory of effects while on active intervention and longer-term sustainability effects. Neuropsychiatric conditions often have a chronic course, and long-term effects of interventions may differ from short-term outcomes. Designing trials that allow for long-term follow-up assessments can also provide valuable information on the short- and long-term safety of the treatments, including when they are stopped.

Finally, researchers should consider *consultation with stakeholders*, including health-care professionals and people with lived experience, to gain an understanding of what outcomes are most important and relevant to them. The determination of the primary outcome is often an iterative process, incorporating new information and feedback as the trial design evolves. The primary outcome needs to be justified and defended in the study protocol and to regulatory bodies or ethical review committees. The primary outcome is the cornerstone for designing the study, analyzing the results, and interpreting the trial's success or failure.

Very often, researchers are interested in more than just one outcome; there may be *multiple secondary outcomes*. As shown in Figure 12.1, in metabolic interventions there are often secondary outcomes focused on the mechanism of action leading to the primary outcome. Insofar as those mechanisms relate to hypotheses that will be tested, those additional outcomes should be given careful consideration similar to the primary outcomes. One particular point to consider is the timing of measurement for potential mediators (i.e., mechanisms of action) such that change in them can be captured before change in the primary outcome has occurred (Davis 2020).

4.4 Devising the Control Groups in RCTs

The use of an appropriate control group or control group treatment, such as a placebo or an active comparator, is vital for interpreting the true effect of the intervention. In metabolic intervention studies, it is crucial to ensure that the control group receives a treatment that mimics the intervention in every aspect except for the active component that is being tested.

In dietary studies that test the impact of certain supplements (e.g., omega-3 supplements on cognitive functions), subjects in the placebo group would receive similar gelatin capsules that are filled with an inert oil mimicking the taste and appearance of the active treatment capsules. In studies that assess effects of certain diets (e.g., high protein diet or keto diet), a comparator group is often asked to continue their regular diet without any modification. In this case, the comparator group is the as-usual diet group, which helps to assess the impact of a specific dietary intervention against the standard eating habits. In behavioral studies, typical comparator groups are behavior-as-usual groups in which subjects are instructed not to change their behavior and receive no specific instructions regarding their diet or exercise. Common are also "treatment-as-usual" comparator groups, especially in the case when subjects require a certain amount of care or treatment. One example could involve examining the effect of a ketogenic diet on epilepsy, in which subjects randomized into the control group receive standard care for epilepsy without any specific dietary advice, whereas in the active treatment group, subjects are instructed and expertly guided to follow a ketogenic diet in addition to their typical treatment of epilepsy. When performing dietary intervention studies, it is important to consider the social component of dietary consulting and, if possible, to include some type of consulting or interaction with study staff in the control group to mimic that which is being received by the active intervention arm. Furthermore, asking the control group to make some type of lifestyle adjustment (e.g., a standardized diet and exercise regimen different from those in the intervention arm) to accompany nutritional counseling, to directly control the active treatment arm, would be a way to improve the precision of the findings to the specific nutritional intake rather than the lifestyle change or social interactions (Sherman 2020).

Generally, the specific selection of the control group or control intervention can either dampen or accentuate the observed treatment efficacy. One reason for such dampening could be a placebo effect. Even when the control group receives a placebo or inactive treatment, the expectations of the benefit alone or just the sole participation in the clinical trials and subsequent attention of the investigators and the staff can lead to measurable changes in biological markers or outcomes. Placebo-controlled designs are thus necessary to assess the accuracy and true efficacy of the active component that can be attributed to just the treatment and not to participants' expectations and their handling during the trial.

4.5 Avoiding Bias through "Blinding"

When possible, clinical trials should use blinding (or masking) so that participants, clinicians, and those analyzing the results do not know who is receiving the experimental treatment and the comparator treatment. The purpose of blinding is to prevent occurrence of biases in treatment administration and monitoring, as well as potentially in the interpretation of outcomes. Single or double blinding refers to how many parties are unaware of the treatment assignment. In single-blind studies, only one party, usually the study subject, is unaware of which treatment they receive, whereas in double-blind studies, both the study subject and researcher are unaware. In neuropsychiatric trials, where outcomes are often based on self-reported symptoms, the implementation of blinding or double blinding is even more important because of the placebo effect, which can substantially influence outcomes. The placebo effect is a phenomenon where a subject perceives improvement in their symptoms after receiving the intervention, even though it should have no therapeutic effect (i.e., they were in the control group and received an inert pill capsule). Such improvements in health are believed to be primarily due to the belief in or expectation of treatments helping them (Laferton et al. 2017). Blinding helps to offset the placebo effect by minimizing the expectations and biases of participants and evaluators from influencing the outcomes of the study. While the expectancy for improvement cannot be eliminated in subjects who enter trials hoping to improve, through a combination of blinding and randomization, those expectations will be balanced across randomized groups and controlled for in the same way that background characteristics are in an RCT. Blinding the investigators and evaluators also ensures that their expectations, behavior, or attitudes do not inadvertently affect the participants' responses to the treatment or the interpretation of the outcomes. For instance, a researcher who knows who is receiving the actual treatment might unconsciously convey positive expectations to the participant or might interpret ambiguous signs and symptoms more favorably, thus biasing the results.

Double blinding is, however, not always possible. For instance, in RCTs that explore the differences between diet interventions, such as ketogenic versus common diets, the blinding of participants is not possible. Still, the blinding of the evaluators should still be in effect to prevent observer biases.

Ideally, clinical trials should test the successfulness of the blinding by asking subjects and evaluators to guess the treatment allocation (Colagiuri and Boakes 2010). These checks should be carefully scheduled and executed to ensure that they do not inadvertently encourage participants or the evaluators to speculate about the participants' treatment assignments (Sackett 2007). Overall, such checks are important to assess the validity of the blinding process across the course of the trial. If the participants or the evaluators are correctly guessing the treatment allocation at rates that are significantly higher or lower than random chance would indicate, potential flaws in the blinding process should be investigated and reported.

Determining when to unblind subjects and researchers is also important. Generally, subjects and evaluators should only be unblinded at the conclusion of the whole study, after all data has been collected and preliminary analyses performed. This means that unblinding should not occur at the end of each subject's treatment period. For instance, if evaluators become aware of a subject's random assignment at the end of that subject's participation, it could help them to assess treatment overall efficacy and influence their future assessments, which would potentially introduce bias into their evaluations of future subjects within the trial. However, sometimes unblinding before the conclusion of the study may be necessary, such as when a participant experiences a severe adverse event (knowledge of their treatment allocation can be crucial for their medical management) or when interim analysis is necessary. Predefined protocols should be put in place to ensure that unblinding occurs in a controlled manner, with only specific people being unblinded. This safeguards the study's validity while addressing ethical and safety considerations.

4.6 Flipping the Coin: Simple, Stratified, and Blocked Randomization

As described earlier, the goal of randomization is to mitigate the influence of potential baseline confounding variables (i.e., background characteristics of the subjects) by evenly distributing those variables across the experimental intervention and control groups. Randomization aims to ensure that any observed differences seen in the primary and secondary outcomes can be attributed to the intervention rather than to underlying differences between the randomized groups. The term "balance" is often used to describe the goal of randomization; that is, we want the randomized groups to be balanced, in terms of their background characteristics, so that we are making an apples-to-apples comparison when testing the intervention. However, simple randomization (i.e., flipping a coin, using a random number generator, for each subject that enrolls to determine which treatment they get) does not *guarantee* perfect balance. The potential for imbalance from simple randomization is especially true in smaller sample sizes and to counteract this, *stratified randomization* should be considered.

Stratified randomization ensures that the treatment groups are similarly balanced with respect to certain predetermined baseline characteristics. When these are known or may influence the outcome of interest (e.g., age, disease severity, genetic markers, or site/clinic), baseline characteristics are useful to create distinct strata or subgroups. Participants within each stratum are then randomized between the different intervention and control groups. This stratification ensures that the randomized groups are balanced across the selected baseline characteristics which reduces confounding bias. However, stratified randomization should not be done with more than one or two baseline characteristics as it can quickly become complex and potentially unmanageable for subject enrollment if recordkeeping and checking of several characteristics is necessary prior to randomization. Moreover, with many strata it may lead to small sample sizes within each one.

To prevent potential imbalance of random assignment of subjects across the enrollment period, it is common to use *(random) block randomization*. Here, the enrollment period/sample is divided into smaller blocks of either equal or random size, and the participants are randomly assigned within each block to the different study group in a predetermined ratio. Blocks of random sizes are preferred to blocks of fixed sizes because random block size prevents simple guessing of random assignments within the block. Overall, block randomization minimizes the risk or imbalance across treatment groups, especially in the early stages of enrollment, and can be used in combination with stratified randomization.

4.7 How Many People Do We Need to Study? What Is Statistical Power and What Affects It?

Phase III clinical trials aim to confirm the primary hypothesis that the experimental intervention is having the intended effect on the primary and secondary outcomes. As shown in Figure 12.1, the goal of the RCT is to confirm whether the metabolic intervention improves the identified neuropsychiatric outcomes. Clinical trials are designed to have sufficient statistical power to detect meaningful differences in outcomes between treatment groups. This is achieved through careful sample size calculations with consideration to power. A detailed power assessment and sample size justification are essential for all study proposals to determine under what conditions and assumptions the given study can reasonably estimate the true effect of the intervention.

Statistical power (or just "power") is the ability of the clinical trial to detect whether the intervention has the expected effect when there is indeed a true effect to be detected. It is common to require the study to be designed to have at least 80% or 90% power to detect an effect in the primary outcome while using a predetermined level of statistical significance of 5%. In lay terms, 80% power means that if a study of an effective intervention is repeated 10 times, the intervention would, on average, be found to be statistically significant in eight cases; in the remaining two, the study would wrongly conclude that the intervention was not significant. Designing a study with higher power increases the likelihood of detecting a true effect and enhances the study's value to funders and researchers by maximizing the chances that the investment will yield a clear and actionable result.

The primary driver of the power of a clinical trial is the *sample size*, and every study needs to provide a power justification for how many subjects will need to be allocated to each randomized group. Larger sample sizes enhance the power of detecting treatment effects by increasing the precision of the estimates of those effects (i.e., this creates narrower confidence intervals for testing differences). When planning the sample size, it is necessary to consider the *expected true effect size* of the intervention as this also directly affects the power calculations. Sample size and expected true effect size of the intervention both go into the calculation of power (along with features of the design itself). Effect size is a quantitative measure of the magnitude of the experimental effect, such as the mean difference in improvements between the intervention and control groups, commonly standardized to eliminate the influence of the measurement scale. Cohen's d is a typically used measure of effect size in clinical trials, calculated as the mean difference in the primary outcome measure at the endpoint time between the treatment groups (or between treatment and control groups) scaled by the pooled standard deviation of the measure (typically collected at baseline). Common rules of thumb qualifying the magnitude of the effect size are: 0.2 small, 0.5 medium, and 0.8 large effects (Cohen 1988). For dichotomous outcomes, treatment effects are typically presented as differences in proportions; for example, the difference in percentage of depressed patients that remit after active treatment (40%) versus the percentage who remit after control (20%) yields 40 – 20 = 20% difference. Alternatively, the treatment effect can be presented as a relative risk (RR): 2 (40%/20%) times more subjects remitting in the active treatment. However, it is important to understand that for dichotomous outcomes, the effect size is directly influenced by the magnitude of the proportions themselves, not just their differences or ratio. For example, in Table 12.1, in a trial with equal sample size in both arms, the sample size needed to detect a RR = 2 with 80% power when the expected success in treatment and control groups are 40% and 20% is n = 164. The needed sample size more than doubles to n = 398 if the expected proportions are lower at 20% and 10% (still a RR = 2). Translation of proportions and their difference to a standard Cohen's d effect size scale can be done in standard power analysis software, for example, using the ES.h function in R from the pwr package (CRAN et al. 2020). Indeed, if the expected outcome proportions are 40% and 20%, this is approximately equal to a Cohen's d of 0.44, whereas if the outcomes are 20% and 10%, this equals a Cohen's d of only 0.28.

At the planning stage, a best guess of the true expected effect size of the intervention is needed as it is used to determine the required sample size to achieve the desired power. A good practice is to determine and use the smallest clinically relevant effect size, grounded in research, theory, and experiences for the power calculation. For example, in a four-month-long pilot study that investigated the effects of ketogenic diet on individuals with schizophrenia and bipolar disorder, participants with schizophrenia showed a 32% reduction in Brief Psychiatric Rating Scale scores and overall Clinical Global Impression severity improved by an average of 31% (Sethi et al. 2024). These values might be used as expected effect sizes if the new study population was similar. Moreover, in a meta-analysis (Patikorn et al. 2023), types of low carbohydrate diet showed different effects on health outcomes with changes more than the minimal clinically important difference thresholds in different populations. Children and adolescents with refractory epilepsy who followed ketogenic or modified Atkins diets for 3–16 months were five times more likely to achieve a reduction in seizure frequency ≥ 50% from baseline compared with those who followed a regular diet (RR, 5.11; 95% CI, 3.18 to 8.21). If the expected effect size is large—Cohen's d = 0.80 or, for example, 44% recovery in active treatment versus 10% recovery in placebo—this enables a more powerful study with fewer participants. For example, a study with 52 participants (26 in each arm) would provide a large enough sample with 80% power to detect this large effect size. However, a smaller expected effect size would require a larger sample size to achieve equivalent power, disproportionately increasing as the expected effect size decreases.

When trying to determine or justify the expected effect size of an intervention, some researchers may be tempted to use small pilot study data. However, it is well documented that relying on results from pilot studies to justify power computations for larger clinical trials could be problematic due to the imprecision of the pilot study estimates, which stems from their small sample sizes (Kraemer et al. 2006; Leon et al. 2011; Westlund and Stuart 2017). Researchers should exercise caution when using overly optimistic or unrealistic pilot studies results for power computations. If the effect size derived from a pilot study is overly large and unrealistic, the subsequent clinical trial is likely to be underpowered due to an insufficient number of subjects, leading to inconclusive results.

Also included in the calculation of power is the *level of statistical significance* required, typically preselected to be 5%, which is the predetermined boundary for

Table 12.1 Examples of power changes due to unbalanced random allocations for given sample size, expected true effect size (i.e., dichotomous outcome success prevalence in treatment vs. placebo arm), and fixed level of significance of 5%.

40% Expected Success in Treatment Group and 20% in Placebo Group		**20% Expected Success in Treatment Group and 10% in Placebo Group**	
Randomization ratio (sample size)	Power	Randomization ratio (sample size)	Power
1:1 (n = 164; 82 in each arm)	80.3%	1:1 (n = 398; 199 in each arm)	80.0%
2:1 (164)	77.1%	2:1 (398)	77.1%
1:2 (164)	74.6%	1:2 (398)	73.4%
3:1 (164)	71.1%	3:1 (398)	71.5%
1:3 (164)	66.0%	1:3 (398)	64.0%

type I error (or "alpha"), that the study will erroneously identify an ineffective intervention as significant. Allowing a more lenient level of statistical significance will increase power; however, most granting agencies and scientific journals expect this type of error to be small (i.e., less than 5%). When testing multiple hypotheses together (e.g., if there are multiple primary outcomes or multiple comparisons between groups), each specific hypothesis must be assessed at a more stringent (i.e., smaller) level of significance to maintain the overall, so-called family-wise type I error rate. This adjustment is crucial for maintaining the integrity of statistical findings, where the likelihood of erroneously identifying an ineffective intervention as significant increases with the number of tests performed.

An aspect of the design which can affect power and have practical and ethical implications is the randomization ratio. An equal distribution across study arms (i.e., 1:1 allocation in the case of two arms), is the most straightforward randomization that maximizes the power of the study for a given sample size and facilitates the most direct assessment of treatment effect. However, if the treatment is expensive, limited, or hard to provide, a ratio 1:2 can be employed when for each participant allocated to the treatment arm, there are two participants allocated to the control group. While this can be a cost- and resources-effective approach, it does reduce the statistical power and may require larger total sample size to reach the same power as 1:1 allocation achieves. Conversely, if participant enrollment rates decrease due to the fear of being randomized to the placebo group, a 2:1 allocation favoring intervention can mitigate these concerns by increasing the participant's chances of receiving the active treatment. This can enhance a participant's willingness to enroll and thus achieve a higher total sample size needed to achieve the same statistical power as 1:1 allocation provides (Table 12.1).

Additionally, a randomization ratio that heavily prefers the treatment group over the comparator group (X:1 allocations) gives researchers more opportunities for secondary and exploratory analyses of the treatment effect and mechanisms within subjects allocated to active treatment. Such allocation will provide higher power for secondary and exploratory aims that focus on subjects in the treatment group and allow for more reliable treatment effect detection within subgroup analyses or mediation effect in mediational mechanism analyses.

Finally, *variability* in the measure used to assess the neuropsychiatric outcome affects the power of the study. The larger the variability among subjects and time points observed within the study, the greater the uncertainty in estimating overall typical neuropsychiatric measures, which will consequently decrease the power of the study. Many measures do not provide researchers with ample options on how to decrease their variability. Variability of one measure among multiple time points can be curbed by selecting the most precise measuring scales and devices. Variability among subjects can be decreased by judiciously choosing inclusion/exclusion criteria to study as homogeneous a population as possible. For instance, heart rate variability, which tracks the autonomic nervous system's response to stress, can show wide variability in healthy populations, depending on age and gender (Voss et al. 2015). It can also reveal variability when measured from visit to visit compared to variability from beat to beat (Shaffer and Ginsberg 2017).

5 Analyzing the Data from the Randomized Clinical Trial

Since it is not possible here to review all the various statistical considerations needed when analyzing data from a clinical trial, we strongly recommend that a qualified statistician be included as a key member on the study team throughout the planning, conducting, and, of course, analysis stages. It is also important to note that part of the preregistering trial process mentioned earlier involves the detailed specification of a statistical analysis plan, which lays out exactly what analyses will be performed. Here we describe a few topics that often arise during the analysis stage and some points that may be considered.

5.1 Intention-to-Treat Analysis

This principle ensures that all randomized participants are included in the analysis according to their original random group assignment, regardless of whether they completed the intervention according to the study protocol. This approach is crucial for maintaining the benefits of randomization and avoiding bias introduced by nonrandom attrition. Indeed, the only causal question unbiasedly answered with an RCT is the effect of the intervention when it is offered. Recall, the benefit of randomization is its ability to create balance between the groups on background characteristics. By analyzing participants based on their original random group assignment, intention-to-treat analysis (ITT) maintains this balance, minimizing bias that could arise from post-randomization exclusions. It is also important to recognize that the ITT treatment effect often provides a more conservative estimate of the treatment effect, as it dilutes the difference between the treatment and control groups by including participants who may not have fully adhered to the treatment protocol.

5.2 Accounting for Adherence to Treatment

Monitoring and analyzing treatment adherence is important, as nonadherence can significantly impact the efficacy of the intervention and the validity of the trial results. The ways in which adherence is measured (e.g., self-report, blood samples, attendance, check-ins, pill-caps) depend on the intervention, but it should be reported in all clinical trials. Because of the crucial importance of diet adherence in a long-term diet intervention study, it is especially important to examine participant-reported dietary changes as well as body weight (diet or otherwise related) over time (Greenland 2019) using food diaries and other methods of quantifying dietary intake. For drug- or supplement-based interventions, the International Society for Medication Adherence (ESPACOMP) provides medication adherence reporting guidelines (EMERGE) on minimum reporting criteria as well as recommendations for how to assess and report adherence (De Geest et al. 2018). Given knowledge of who is and is not adherent, "per-protocol" or "as-treated" analyses can be performed. The analysis is limited to those participants who followed the expected procedures; otherwise, participants are analyzed based on the treatment they actually received, rather than the treatment to which they were randomized. While this sub-setting or reallocation approach can provide insight into the effectiveness of the treatment, it risks confounding and bias, as the reasons for adherence or changes in treatment may be associated with outcomes (Heitjan 1999; Shiovitz et al. 2016). Thus, ITT is typically considered the primary analysis in clinical trials.

5.3 Handling Missing Data

Given the chronic nature of many neuropsychiatric conditions, participant dropout can lead to substantial missing data. Methods such as analyzing only complete cases or using the last observation carried forward (LOCF) are no longer considered good practice and should not be used. Instead, the following statistical methods consider all available data: full information maximum likelihood (e.g., within generalized linear mixed modeling frameworks), multiple imputation, or inverse probability weighting. They also offer robust and more reliability results under a missing at random assumption. In addition, methods such as pattern-mixture models can be considered when there is strong evidence that data are not missing at random.

5.4 Heterogeneity of Treatment Effects

Investigating the differential impact of treatment across various subgroups (e.g., based on age, sex, severity of disease) can provide insights into personalized medicine approaches and identify those who benefit most from the intervention. The heterogeneity of effects can be identified by subgroup analysis (e.g., perform the analysis for the trial subset only to women and then again subset only to men) or through using interaction tests of randomized group by the characteristic of interest. We strongly recommend, however, that these analyses be prespecified and interpreted cautiously to avoid spurious findings. When investigating treatment effect heterogeneity, it is important to account for multiple testing and to interpret findings in the context of the biological or clinical plausibility. Overreliance on statistical significance, without considering clinical relevance, can lead to misleading conclusions.

In the context of metabolic interventions in mental health, when studies are limited by small sample sizes, conducting stratified analyses can additionally reduce statistical power because of the even smaller subgroups that are being compared. This reduction in power makes it more difficult to detect true treatment effects within subgroups and increases the likelihood of type II errors (i.e., failing to detect a true effect). Furthermore, it is often unclear which factors should be used for stratification at early stages of research, as the understanding of key moderators may be limited. Therefore, exploratory stratified analyses should be interpreted with caution. An emphasis should be placed on replication and further research to validate potential findings in larger, more diverse samples.

An important consideration in metabolic psychiatry is how best to stratify patients for therapeutic interventions. Biological markers, such as insulin resistance and lipid profiles, show promise in guiding treatment selection by reflecting the underlying metabolic dysfunctions common in psychiatric disorders. Their clinical use, however, is still being validated. Combining these markers with clinical characteristics, such as symptom severity and comorbidities, may offer a more effective stratification approach. Continued research is needed to determine which markers are most associated with treatment response to be appropriate for stratification.

6 Conclusion

Addressing the points outlined in this chapter can significantly enhance the scientific rigor and clinical relevance of clinical trials focusing on metabolic interventions for neuropsychiatric outcomes, thereby improving patient care and treatment approaches. Many of the standards promulgated in the field of clinical trials were necessary to prevent the biases and confounds that plague our field and it is critical to adhere to best practices to generate high quality evidence. Finally, it is important to consider the clinical relevance of a trial's findings in addition to statistical significance. While statistical significance indicates the probability that an observed effect is not due to chance, clinical relevance evaluates the real-world impact and practical significance of the treatment effect. This distinction is especially important in neuropsychiatric trials, where the clinical implications of the results are vital for determining their overall impact.

References

Ames BN (2018) Prolonging Healthy Aging: Longevity Vitamins and Proteins. PNAS 115 (43):10836–10844. https://doi.org/10.1073/pnas.1809045115

Amin NB, Darekar A, Anstee QM, et al. (2022) Efficacy and Safety of an Orally Administered DGAT2 Inhibitor Alone or Coadministered with a Liver-Targeted ACC Inhibitor in Adults with Non-Alcoholic Steatohepatitis (NASH): Rationale and Design of the Phase II, Dose-Ranging, Dose-Finding, Randomised, Placebo-Controlled MIRNA (Metabolic Interventions to Resolve NASH with Fibrosis) Study. BMJ Open 12 (3):e056159. http://dx.doi.org/10.1136/bmjopen-2021-056159

Astrazeneca (2023) A Study to Assess the Safety, Tolerability, Pharmacokinetics, and Pharmacodynamics of AZD6234 after Repeat Dose Administration in Participants Who Are Overweight or Obese. https://www.astrazenecaclinicaltrials.com/study/D8750C00002/

Boehringer Ingelheim (2023) Phase III Studies Survodutide Obesity and Overweight. . https://www.boehringer-ingelheim.com/human-health/metabolic-diseases/phase-3-studies-survodutide-obesity-and-overweight

Brody T (2016) Clinical Trials: Study Design, Endpoints and Biomarkers, Drug Safety, and FDA and ICH Guidelines. Academic Press, New York

Chakhtoura M, Haber R, Ghezzawi M, et al. (2023) Pharmacotherapy of Obesity: an Update on the Available Medications and Drugs under Investigation. EClinicalMedicine 58:101882. https://doi.org/10.1016/j.eclinm.2023.101882

Cohen J (1988) Statistical Power Analysis for the Behavioral Sciences. Lawrence Erlbaum, Mahwah, NJ

Colagiuri B, and Boakes RA (2010) Perceived Treatment, Feedback, and Placebo Effects in Double-Blind RCTs: An Experimental Analysis. Psychopharmacology 208 (3):433–441. https://doi.org/10.1007/s00213-009-1743-9

CRAN, Champely S, Ekstrom C, et al. (2020) PWR Package: Basic Functions for Power Analysis. https://github.com/heliosdrm/pwr

Davis RL (2020) Mechanism of Action and Target Identification: A Matter of Timing in Drug Discovery. iScience 23 (9):101487. https://doi.org/10.1016/j.isci.2020.101487

De Geest S, Zullig LL, Dunbar-Jacob J, et al. (2018) Espacomp Medication Adherence Reporting Guideline (EMERGE). Ann Intern Med 169 (1):30–35. https://doi.org/10.7326/m18-0543

Eichler HG, Bloechl-Daum B, Bauer P, et al. (2016) "Threshold-Crossing": A Useful Way to Establish the Counterfactual in Clinical Trials? Clin Pharmacol Ther 100 (6):699–712. https://doi.org/10.1002/cpt.515

Evans SR (2010a) Clinical Trial Structures. J Exp Stroke Transl Med 3 (1):8–18. https://pmc.ncbi.nlm.nih.gov/articles/PMC3059315/

——— (2010b) Fundamentals of Clinical Trial Design. J Exp Stroke Transl Med 3 (1):19–27. https://pmc.ncbi.nlm.nih.gov/articles/PMC3083073/

Genentech (2023) First-Ever Clinical Trial Exclusively in Black and Hispanic / Latinx People Living with Multiple Sclerosis Shows Genentech's Ocrevus Effectively Manages Disease Activity. https://www.gene.com/media/press-releases/15005/2023-10-10/first-ever-clinical-trial-exclusively-in

Greenland P (2019) Dietary Adherence in a Clinical Trial of a Nutritional and Behavioral Intervention. JAMA 322 (15):1500. https://doi.org/10.1001/jama.2019.15855

Heitjan DF (1999) Causal Inference in a Clinical Trial: A Comparative Example. Control Clin Trials 20 (4):309–318. https://doi.org/10.1016/s0197-2456(99)00012-4

Huberts-Bosch A, Bierens M, Ly V, et al. (2024) Short-Term Effects of an Elimination Diet and Healthy Diet in Children with Attention-Deficit/Hyperactivity Disorder: A Randomized-Controlled Trial. Eur Child Adolesc Psychiatry 33 (5):1503–1516. https://doi.org/10.1007/s00787-023-02256-y

ICH GCP (2024) Neurocognitive and Neurobiological Improvement in ADHD Children by Modification of Dietary Protein (BrainProtein). https://ichgcp.net/clinical-trials-registry/NCT01856686

Insel T (2013) NIMH's New Focus in Clinical Trials. https://www.nimh.nih.gov/funding/grant-writing-and-application-process/concept-clearances/2013/nimhs-new-focus-in-clinical-trials

Kraemer HC, Mintz J, Noda A, Tinklenberg J, and Yesavage JA (2006) Caution Regarding the Use of Pilot Studies to Guide Power Calculations for Study Proposals. Arch Gen Psychiatry 63 (5):484–489. https://doi.org/10.1001/archpsyc.63.5.484

Krogh HB, Storebø OJ, Faltinsen E, et al. (2019) Methodological Advantages and Disadvantages of Parallel and Crossover Randomised Clinical Trials on Methylphenidate for Attention Deficit Hyperactivity Disorder: A Systematic Review and Meta-Analyses. BMJ Open 9 (3):e026478. https://doi.org/10.1136/bmjopen-2018-026478

Laferton JAC, Kube T, Salzmann S, Auer CJ, and Shedden-Mora MC (2017) Patients' Expectations Regarding Medical Treatment: A Critical Review of Concepts and Their Assessment. Front Psychol 8:233. https://doi.org/10.3389/fpsyg.2017.00233

Le Roux C, Steen O, Lucas KJ, et al. (2023) 51-Or: A Phase 2, Randomized, Double-Blind, Placebo-Controlled, Dose-Finding Study of Bi 456906 in People with Overweight/Obesity. Diabetes 72 (Supplement_1):51–OR. https://doi.org/10.2337/db23-51-OR

Leon AC, Davis LL, and Kraemer HC (2011) The Role and Interpretation of Pilot Studies in Clinical Research. J Psychiatr Res 45 (5):626–629. https://doi.org/10.1016/j.jpsychires.2010.10.008

Lin D, Xiao H, Yang K, et al. (2024) Safety, Tolerability, Pharmacokinetics, and Pharmacodynamics of Tg103 Injection in Participants Who Are Overweight or Obese: A Randomized, Double-Blind, Placebo-Controlled, Multiple-Dose Phase 1b Study. BMC Med 22 (1):209. https://doi.org/10.1186/s12916-024-03394-z

McGurk SR, Otto MW, Fulford D, et al. (2021) A Randomized Controlled Trial of Exercise on Augmenting the Effects of Cognitive Remediation in Persons with Severe Mental Illness. J Psychiatr Res 139:38–46. https://doi.org/10.1016/j.jpsychires.2021.04.033

Meinert CL (2011) An Insider's Guide to Clinical Trials. Oxford Univ. Press, New York. https://doi.org/10.1093/acprof:oso/9780199742967.001.0001

Mukherjee P, Roy S, Ghosh D, and Nandi SK (2022) Role of Animal Models in Biomedical Research: A Review. Lab Anim Res 38 (1):18. https://doi.org/10.1186/s42826-022-00128-1

Nair B (2019) Clinical Trial Designs. Indian Dermatol Online J 10 (2):193–201. https://doi.org/10.4103/idoj.idoj_475_18

Patikorn C, Saidoung P, Pham T, et al. (2023) Effects of Ketogenic Diet on Health Outcomes: An Umbrella Review of Meta-Analyses of Randomized Clinical Trials. BMC Med 21 (1):196. https://doi.org/10.1186/s12916-023-02874-y

Sackett DL (2007) Commentary: Measuring the Success of Blinding in RCTs: Don't, Must, Can't or Needn't? Int J Epidemiol 36 (3):664–665. https://doi.org/10.1093/ije/dym088

Schulz KF, Altman DG, and Moher D (2010) Consort 2010 Statement: Updated Guidelines for Reporting Parallel Group Randomised Trials. BMJ 340:c332. https://doi.org/10.1136/bmj.c332

Sethi S, Wakeham D, Ketter T, et al. (2024) Ketogenic Diet Intervention on Metabolic and Psychiatric Health in Bipolar and Schizophrenia: A Pilot Trial. Psychiatry Res 335:115866. https://doi.org/10.1016/j.psychres.2024.115866

Shaffer F, and Ginsberg JP (2017) An Overview of Heart Rate Variability Metrics and Norms. Front Public Health 5:258. https://doi.org/10.3389/fpubh.2017.00258

Shamseer L, Hopewell S, Altman DG, Moher D, and Schulz KF (2016) Update on the Endorsement of Consort by High Impact Factor Journals: A Survey of Journal "Instructions to Authors" in 2014. Trials 17 (1):301. https://doi.org/10.1186/s13063-016-1408-z

Sherman KJ (2020) The Trials and Tribulations of Selecting Comparison Groups in Randomized Trials of Nonpharmacological Complementary and Integrative Health Interventions. J Altern Complement Med 26 (6):449–455. https://doi.org/10.1089/acm.2019.0460

Shiovitz TM, Bain EE, McCann DJ, et al. (2016) Mitigating the Effects of Nonadherence in Clinical Trials. J Clin Pharmacol 56 (9):1151–1164. https://doi.org/10.1002/jcph.689

Soufizadeh P, Mansouri V, and Ahmadbeigi N (2024) A Review of Animal Models Utilized in Preclinical Studies of Approved Gene Therapy Products: Trends and Insights. Lab Anim Res 40 (1):17. https://doi.org/10.1186/s42826-024-00195-6

Stanford Medicine Health Care (2025) Impact of a Ketogenic Diet on Metabolic and Psychiatric Health in Patients with Bipolar or Schizophrenia Illness. https://stanford-healthcare.org/trials/i/NCT03935854.html

Tchang B, Aras M, Kumar R, and Aronne L (2000) Pharmacologic Treatment of Overweight and Obesity in Adults. Endotext. MDText.com, Inc, South Dartmouth, MA

Voss A, Schroeder R, Heitmann A, Peters A, and Perz S (2015) Short-Term Heart Rate Variability: Influence of Gender and Age in Healthy Subjects. PLoS One 10 (3):e0118308. https://doi.org/10.1371/journal.pone.0118308

Wadden TA, Chao AM, Machineni S, et al. (2023) Tirzepatide after Intensive Lifestyle Intervention in Adults with Overweight or Obesity: The Surmount-3 Phase 3 Trial. Nat Med 29 (11):2909–2918. https://doi.org/10.1038/s41591-023-02597-w

Westlund E, and Stuart EA (2017) The Nonuse, Misuse, and Proper Use of Pilot Studies in Experimental Evaluation Research. Am J Eval 38 (2):246–261. https://doi.org/10.1177/1098214016651489

Whiteley P, Haracopos D, Knivsberg AM, et al. (2010) The ScanBrit Randomised, Controlled, Single-Blind Study of a Gluten- and Casein-Free Dietary Intervention for Children with Autism Spectrum Disorders. Nutr Neurosci 13 (2):87–100. https://doi.org/10.1179/147683010x12611460763922

Open Access This chapter is licensed under the terms of the Creative Commons Attribution-NonCommercial-NoDerivatives 4.0 International License (http://creativecommons.org/licenses/by-nc-nd/4.0/), which permits any noncommercial use, sharing, distribution and reproduction in any medium or format, as long as you give appropriate credit to the original author(s) and the source, provide a link to the Creative Commons license and indicate if you modified the licensed material. You do not have permission under this license to share adapted material derived from this chapter or parts of it.

The images or other third party material in this chapter are included in the chapter's Creative Commons license, unless indicated otherwise in a credit line to the material. If material is not included in the chapter's Creative Commons license and your intended use is not permitted by statutory regulation or exceeds the permitted use, you will need to obtain permission directly from the copyright holder.

13

Disassociating Causality in Complex Interacting Systems

David Hofmann, Simon Carter, Helmut H. Strey, and Lilianne R. Mujica-Parodi

Abstract Understanding causal interactions between brain activity and metabolic processes is a central challenge in systems neuroscience and psychiatric research. This chapter reviews current data-driven approaches for inferring causal relationships in complex interacting systems, with a particular focus on the metabolism-brain nexus in psychopathology. Widely used methods (e.g., Granger causality, transfer entropy, dynamic causal modeling) are critically assessed and their theoretical underpinnings, limitations, and applicability to neuronal and metabolic data are discussed. Special attention is given to recent methodological developments including structural causal models, simulation-based inference, and models tailored to systems with feedback loops, such as the Bicycle framework. Using major depressive disorder as a case study, the need for integrated modeling frameworks is highlighted that combine biophysical realism with statistical rigor. It is argued that mechanistic computational models, when combined with experimental interventions, hold the

David Hofmann (✉)
Dept. of Biomedical Engineering and Laufer Center for Physical and Quantitative Biology, Renaissance School of Medicine, State University of New York at Stony Brook, Stony Brook, NY 11794, USA
Email: david@lcneuro.org

Simon Carter
Applied Mathematics and Statistics, State University of New York at Stony Brook, Stony Brook, NY 11794, USA

Helmut H. Strey
Dept. of Biomedical Engineering and Laufer Center for Physical and Quantitative Biology, Renaissance School of Medicine, State University of New York at Stony Brook, NY 11794, USA

Lilianne R. Mujica-Parodi
Dept. of Biomedical Engineering and Laufer Center for Physical and Quantitative Biology, Renaissance School of Medicine, State University of New York at Stony Brook; Martinos Center for Biomedical Imaging, Massachusetts General Hospital, Boston, MA 02129, USA

© Frankfurt Institute for Advanced Studies (FIAS) 2025
D. Öngür, J. M. Ford (eds.), *Metabolic Neuropsychiatry*, Strüngmann Forum Reports,
https://doi.org/10.1007/978-3-032-05630-6_13

greatest promise for uncovering true causal mechanisms in complex biological systems.

Keywords Causal inference, dynamic causal modeling, transfer entropy, Granger causality, structural causal models, feedback loops, metabolism-brain interaction, psychopathology, major depressive disorder, systems neuroscience, computational modeling

1 Introduction

Complex systems are ubiquitous in nature. The brain and metabolic networks are prime examples of these systems and their interaction poses a significant challenge for modeling. Identifying causal relationships within such complex systems remains a long-standing research endeavor. While direct experimental interventions have traditionally been the gold standard for discerning causal relationships, some parts of a complex system may not lend themselves to direct manipulation or the experiments might be prohibitively expensive. However, where measurements are feasible, extracting causal relationships through algorithmic inference methods becomes possible. Approaches to extracting causal inferences from data continues to evolve across various disciplines, such as earth system sciences (Runge et al. 2019), ecology (Sugihara et al. 2012), neuroscience (Vicente et al. 2011; Bielczyk et al. 2019), and molecular biology (Prill et al. 2010; Meinshausen et al. 2016; Hill et al. 2016).

In psychopathology, as with other medical fields, both diagnostic tools and medical interventions can significantly benefit from a deeper understanding of the causal relationships in our body. As research increasingly demonstrates the impact of metabolic processes on psychopathology (Campbell and Campbell 2024; Rojas et al. 2024), it becomes evident that examining causal relationships solely among neuronal populations or exclusively within metabolic circuits is likely insufficient. Instead, dissecting and studying the causal interactions between these two complex systems is crucial.

Information between neurons is transmitted via action potentials that lead to neurotransmitter release into the synaptic cleft, resulting in postsynaptic potentials. This process requires maintaining ionic gradients across the cell membrane and producing, transporting, and releasing neurotransmitters at synapses, all of which consume metabolic energy. Unsurprisingly, dysfunction of the energy metabolism can disrupt neuronal information processing and lead to psychopathologies (Campbell and Campbell 2024; Rojas et al. 2024; for a discussion of potentially implicated circuits, see also Chapter 5 and Chapter 6). Thus, impaired energy metabolism could cause psychopathologies, but also psychopathologies could cause dysfunctional energy metabolism, or both could result from a shared underlying cause—a third variable. Hence, it is not clear which direction the causal relation between metabolism and pathological neuronal activity takes, and it may indeed vary between patients.

We examine major depressive disorder (MDD) as an instructive case study. While the fatigue commonly reported during depressive episodes might intuitively suggest metabolic involvement (Campbell and Campbell 2024), the connection becomes less apparent when other characteristic symptoms are considered, such as negative cognitive bias (Hindash and Amir 2012), hopelessness (Haeffel 2011; Iacoviello et al. 2010), and helplessness behaviors observed in animal models (Amat et al. 2005). Nevertheless, emerging research provides compelling evidence for the involvement of energy metabolism in MDD (Gu et al. 2021; Zhou et al. 2019). The precise nature of this relationship—whether metabolic dysfunction precedes MDD, develops as a consequence of the disorder, operates within a bidirectional feedback loop, or involves a third unknown variable (e.g., the causal motifs shown in Figure 13.3)—remains an open question and may vary among individual patients (Zhou et al. 2019).

A well-studied hypothesis of the cause of MDD is an insufficiency of monoamine transmitters such as serotonin, norepinephrine, and dopamine as shown in experimental studies (Gu et al. 2021; Huys et al. 2015; Köhler et al. 2016). Theoretical work, however, in neuroscience based on Bayesian Decision Theory suggests that depression could be caused by erroneous prediction of expected rewards due to dysfunctional neuronal information processing in areas such as the caudate, the dorsolateral, and the medial prefrontal cortex (Huys et al. 2015). It is not possible for a human or an animal to consider the entire decision tree of potential future consequences of an action; thus, pruning of less-promising paths becomes necessary. Pruning, interpreted as halting the internal exploration of decision trees when faced with significant losses, is associated with subclinical symptoms of depression (Huys et al. 2012). Dayan and Huys (2008) have proposed that serotonin (5-HT) may be involved in the pruning process, which appears to be associated with specific brain regions, namely the subgenual and adjacent pregenual anterior cingulate cortices (sgACC and pgACC) (Lally et al. 2017).

Figure 13.1 depicts a potential circuit that links 5-HT synthesis to activity in the ACC and expands it to include the regulation of energy metabolism via the HPA axis. This model suggests that disrupted 5-HT synthesis, for instance, driven by dysfunction of TPH1 or TPH2 enzymes (Correia and Vale 2022), could cascade down to affect glucose, insulin, and other components of energy metabolism through a dysfunctional glucocorticoid regulation (Frigerio 2023), illustrating a common pathway potentially underlying both MDD and metabolic dysfunction. Alternatively, the dysfunction might originate in the energy metabolism, affecting information processing in the ACC (Zheng et al. 2015) or other regions implicated in MDD. Hyperactivity of the HPA axis, however, has also been argued to be a potential cause for MDD rather than its consequence (Pariante and Lightman 2008).

While there are numerous possible scenarios well beyond the hypotheses discussed here, this chapter does not seek to provide an exhaustive list. Instead, we aim to provide an overview of potential data-driven, rather than interventional, methods to disambiguate between these different scenarios. For an in-depth discussion of randomized controlled trials used in experimental interventions to disassociate causality, see Wall and Pavlicova (Chapter 12). Another noteworthy line of research that uses the concept of instrumental variables, which are variables (instruments) that influence treatment, is the Local Average Treatment Effect developed by Angrist and Imbens (1995). For this and related work on causal inference, they

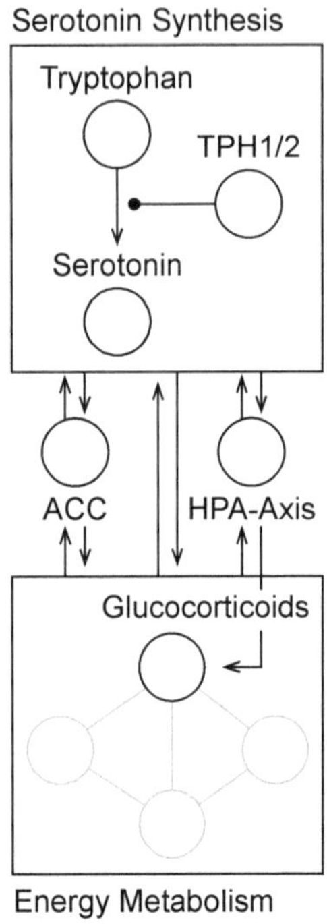

Figure 13.1 Serotonin circuit and major depressive disorder. Potential pathway starting at serotonin synthesis in the body (e.g., gastrointestinal tract, enterochromaffin cells) and in the brain (Raphe nuclei). Serotonin projections from Raphe nuclei affect, among other regions, the anterior cingulate cortex, known to be involved in ruminating thoughts in MDD. Serotonin also affects the activity of the HPA axis, which in turn influences energy metabolism, in particular by controlling glucocorticoids, and can influence energy metabolism directly.

were awarded the Nobel Prize in Economics. We refer the reader to a comprehensive book by Imbens and Rubin (2015) on causal inference, which provides in-depth discussions of experimental techniques and observational methods that build on them. To discuss the relationship between this line of research and the equally foundational structural causal models (SCM) discussed here, we point the reader to a book by Peters et al. (2017).

We begin with a review of state-of-the-art methods for purely data-driven discovery of causal relationships. Although these methods are increasingly used and hold considerable appeal, they are no panacea and must be applied rigorously and interpreted cautiously to avoid erroneous conclusions. For this purpose, we deem it essential to get familiar with the rigorous formalism for causality provided by Judea Pearl and colleagues (Pearl 2009; Peters et al. 2017). We conclude with a commentary on which methods seem promising for identifying the causal influences within the metabolism-brain nexus in psychopathology—an essential step toward uncovering mechanisms and formulating therapeutic targets (Bastiaanssen and Cryan 2021).

2 General Thoughts on Inference of Causality

In research, comprehending the fundamental representation of causality is crucial for a complete understanding of the system. This requires understanding concepts that are distinct from conventional statistical metrics.

As detailed by Pearl (2009), metrics fall into two categories: causal and statistical domains. Statistical domains include familiar concepts such as correlation, regression, conditional independence, association, likelihood, and Granger causality. In contrast, causal domains involve elements such as randomization, influence, effect, confounding, exogeneity, ignorability, and intervention (Pearl 2009). The key difference is that causal metrics, by nature, elude determination solely through observational data and necessitate either intervention or prior assumptions by the researcher.

Effective articulation of causal systems necessitates a medium to represent these relationships, and causal models are invaluable for this purpose. These relationships are primarily defined through two paradigms: causal Bayesian networks and functional models (Pearl 2009; Peters et al. 2017). Bayesian networks offer a clear medium to represent causality, assuming probabilistic system decomposition. For instance, the joint probability of variables $P(x_1,\ldots,x_n)=\prod_{k=1}^{n} p(x_i|pa_i)$ can be broken down into a series of conditional distributions where pa_i denotes a minimal set of predecessors that makes X_i independent of its other predecessors. As an example, Figure 13.2, a Directed Acyclical Graph (DAG), represents the set of conditional independencies $P(X_1)P(X_3|X_1)P(X_2|X_1)P(X_4|X_3,X_2)P(X_5|X_4)$. These conditional distributions represent the directionality of a Bayesian graph.

Furthermore, merely representing causal relationships through joint probability distributions is insufficient, as this does not capture all necessary information. Functional models, inspired by the view of causal determinism where every event is the consequence of previous events, enable the depiction of counterfactuals and a broader range of causal relationships, including feedback loops. In these models, each directional relationship is functionally defined as $x_i = f_i(pa_i, u_i)$, where each

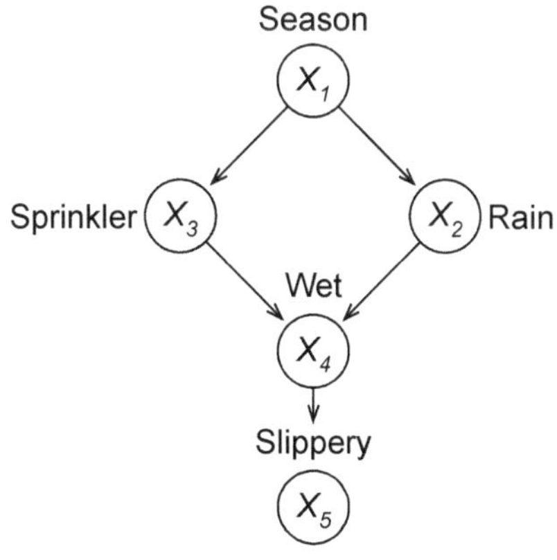

Figure 13.2 A simple example illustrating how a Bayesian net can express causal relations. *X*s denote different stochastic variables and arrows the causal direction. Figure from Pearl (2009), used with permission of Cambridge University Press through PLSclear.

variable X is a function of its parent nodes pa_i and a probabilistic error term U that represents unmodeled factors affecting X. This framework allows for a comprehensive representation using functional models, including structural equation modeling, which is a linear form of these relationships. The formalism for causal inference, including the conditions under which it is possible, has been rigorously defined using the *do*-calculus, forming the foundation for a mathematically rigorous understanding of causal relationships (Pearl 2009; Peters et al. 2017).

A causal representation not only defines a system of probabilistic models but also enhances model resilience to new information by detailing the underlying mechanisms and predicting its behavior under various conditions. As an example, we can examine the graph in Figure 13.2. If we were to learn the season, this would alter the conditional distribution between X_3 and X_4, but it would not alter their causal relationship. Even beyond this, if we change the mechanism by which the season affects sprinkler use, the fact that the sprinkler does not cause rain or that it makes the ground wet would remain invariant. The only potential change in the causal relationship from this would be the complete removal of the sprinkler's dependence on the season. In this regard, probabilistic models are insufficient for representing all the information in a causal relationship, and more robust causal models are needed to describe the system fully.

The language of *do*-calculus facilitates interaction with these causal models, succinctly encapsulating the act of intervention in mathematical terms. It is essential to differentiate intervention from observation; the former allows for analysis of the underlying causal structure, whereas the latter provides mere passive observation. In a causal system, observation of a variable (e.g., noticing the sprinkler is on) and the action of intervention (turning on the sprinkler) are distinct, with the intervention severing the dependence of the parents. An example of this difference can be seen when observing the relationship between X_3 and X_1. Observing that the sprinkler is on gives us information about the season because the DAG tells us the two are not independent. Turning on the sprinkler, however, renders this state independent of the season, i.e., $p\left(X_1 \middle| do\left(X_3 = on\right)\right) = p\left(X_1\right)$, and renders X_4 independent of X_2, thus changing the structure of the DAG. For DAGs, this *do* formalism is complete. Given a graphical structure, a set of rules as outlined by Pearl is known to describe both when a causal interaction can be inferred and what intervention is needed to efficiently facilitate this, under the assumption of no feedback; that is, we can represent the system as a DAG.

Despite the robustness of the theory behind causal links in DAGs, particularly within metabolic and neuronal networks, feedback loops are ubiquitous; thus the question arises of how this formalism can be extended to cyclic causal relationships. Two strategies can be pursued. One requires measuring time series with a sampling rate that exceeds the feedback loop's characteristic timescale. Then, unrolling the model in time effectively removes the feedback loop. Such a strategy is pursued with vector autoregressive models or, more generally, dynamic Bayesian networks (Ghahramani 1998; Rajapakse and Zhou 2007) and with dynamical system models. The second strategy is based on the assumption that the system is in equilibrium and the data have been gathered from the respective equilibrium distribution, which is then used to conclude the underlying dynamical system (Mooij et al. 2011). We will discuss examples for both below. Indeed, due to the nature of experimental data that are typically available within the respective fields of neuroscience and metabolomics, the former strategy is usually employed for causal inference in neuroscience, whereas the latter is pursued in genetic and metabolic network inference. Combining them seems a promising venue for tackling both metabolic and neuronal circuits and dissociating their causal relationships.

3 Methods for Time-Series Data

In neuroscience, the most prevalent approaches for inferring directed graphs are Granger causality and dynamic causal modeling (DCM). Granger causality is extensively used due to its mechanism-agnostic nature and computational and conceptual simplicity. On the other hand, DCM hinges upon a plausible definition of a mechanistic model that accurately reflects the underlying neuronal and measurement processes. Initially developed for analyzing fMRI data (Friston et al. 2003), DCM has since been expanded to other modalities of measuring neuronal activity (David et al. 2006; Friston et al. 2015; Moran et al. 2011). We will devote considerable attention to these methods but also expand on related techniques, such as *transfer entropy*, which generalizes Granger causality, or present approaches, such as *simulation-based inference*, that can potentially help improve existing techniques such as DCM.

3.1 Granger Causality

Clive Granger formulated a statistical definition of causality based on the premise that (a) a cause precedes its effect and (b) knowledge of a cause improves the prediction of its effect (Granger 1969). He later refined these premises to discuss dependencies between stochastic processes. Specifically, for two jointly-distributed stochastic processes X_t, Y_t in the context of an environment E_t (that excludes X_t and Y_t) at each time t, then Y does not Granger-cause X at time t if

$$P\left(X_t \middle| X_{t^-}, Y_{t^-}, \varepsilon_{t^-}\right) = P\left(X_t \middle| X_{t^-}, \varepsilon_{t^-}\right) \tag{13.1}$$

where $P(\cdot|\cdot)$ denotes conditional distributions and superscript "–" denotes history up to (but not including) time t.

Granger causality of a time series x_t over another y_t is established if including x_t's history improves the prediction of y_t compared to using only y_t's history. This improvement is quantified by the difference in prediction error variances from one-step linear predictors under two conditions: including versus excluding the history of the putatively causal time series. Mathematically, this is expressed as:

$$var\left(y_t - \hat{y}_t \mid y_{0:t-1}\right) > var\left(y_t - \hat{y}_t \mid y_{0:t-1}, x_{0:t-1}\right). \tag{13.2}$$

In practice, this is tested using finite-order vector autoregression (VAR) models, assessing the prediction error variances from separate VAR models—a full model including all components and a reduced model excluding the putatively causal series. Since the reduced model essentially forms an autoregression moving average (ARMA) model, a finite VAR cannot properly approximate and thus erroneous estimations can result. Instead, a formulation based on state space models or from the autocovariance sequence via Whittle's spectral factorization algorithm (Barnett et al. 2018; Seth et al. 2013) is robust and should thus be the default methods to estimate Granger causality (Barnett and Seth 2015; Solo 2016).

Despite the wide adoption of Granger causality, we point out that Granger causality performs poorly in reconstructing nonlinear-directed graphs, such as those modeled by Kuramoto oscillators, which are commonly used to simulate neuronal and, less frequently, metabolic networks. Lusch et al. (2016) found that Granger causality is unstable—small changes in input data can lead to large fluctuations in the estimated network—and that it systematically underestimates or overestimates connections depending on network sparsity and synchronization. These findings highlight significant limitations of Granger causality in inferring causality in systems with nonlinear and dynamic interactions.

3.2 Information Flow

Various methodologies have been proposed as measures of information flow (Ay and Polani 2008; Liang and Kleeman 2005; Lizier and Prokopenko 2010). Here we focus specifically on the approach initially defined by Liang and Kleeman (2005) and later extended to a multivariate method by Liang in subsequent studies (Liang 2018, 2021). Unlike pairwise inference methods, this multivariate approach aptly identifies common drive motifs (see Figure 13.3) if all nodes in the motif are actually measurable. Notably, this method has demonstrated robustness in scenarios involving the synchrony of (chaotic) oscillators—a situation where Granger causality is known to falter. This robustness makes it a significant tool for analyzing complex systems where traditional methods like Granger causality may not provide reliable results (Liang 2021).

3.3 Transfer Entropy

Transfer entropy (TE), introduced independently by Thomas Schreiber (2000) and Milan Paluš et al. (2001), has become a widely used method for inferring the directionality of information flow, and thereby constructing directed graphs. TE is based on the conditional mutual information between a target variable and the past of a source variable, conditioned on the part of the target. Like other information theoretic approaches, TE suffers from the curse of dimensionality, as it requires the estimation of full probability distributions (for discrete data) or densities (for continuous data). Nearest neighbor estimators are among the most commonly employed

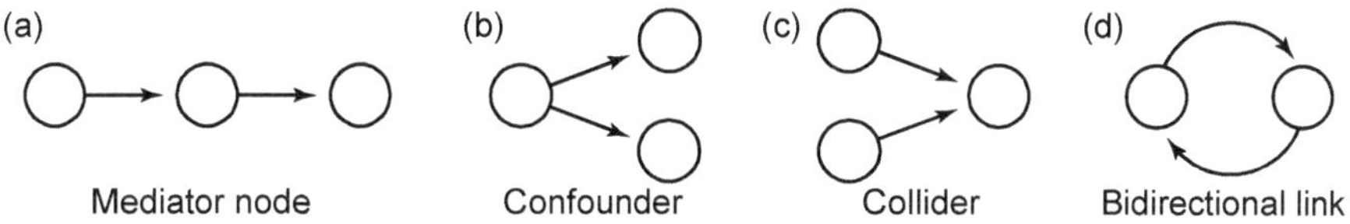

Figure 13.3 Simple motifs of directed relations. This list is not intended to be comprehensive; it addresses some simple yet important scenarios. (a) Causal graph motif with three nodes where the node in the middle is mediating the causal influence from the left to the right node and is thus called a mediator. A mediator could also be inaccessible to measurements and therefore alter the (indirect) causal relationship between the left and right node as a hidden variable. (b) A confounder variable (left node) is influencing two other variables. This creates a spurious causal relationship between the two affected nodes. In particular, when the confounder node is not accessible to measurements, inference methods fail to correctly infer the absence of a causal interaction. Note that such a motif forms what is named a backdoor path. More importantly, it is an open backdoor path. By eliminating all open backdoor paths in a graph, it is possible to isolate direct causal relationships and thus avoid spurious relationships. Blocking of an open backdoor can be done by experimental intervention or is given when a motif (c) closes the backdoor. In (c) the causal relationship is reversed and a confounder becomes a collider. This forms a closed backdoor path. For more detailed explanations, see Cunningham (2021). (d) directed links can also form cycles. Most methods of causal inference are suited for the estimation of directed acyclic graphs and it is considered a more challenging task to correctly infer cyclic relationships. In the main text, we discuss some methods to approach this challenge.

methods in practice to handle these challenges (Vicente et al. 2011). Additionally, the estimation of delay time helps to omit unnecessary historical data from the source variable, thus reducing the dimensionality to some extent. However, Runge et al. (2012) defined a method based on Markov chains to properly alleviate the curse of dimensionality and make TE also applicable to larger dimensional problems. For an in-depth discussion of the challenges and limitations when estimating TE from data and the implications for causality conclusions, we refer the reader to Natale et al. (2018) and references therein.

We note that for linear and Gaussian systems, the equivalence between TE and Granger causality can be shown analytically, thereby proving TE to be a natural nonlinear extension of Granger causality (Barnett et al. 2009).

Finally, we caution the reader about a common misconception that measuring the transfer of information is equivalent to getting causal directionality. This concept is fundamentally linked as a correlational measure, and information transfer can be used as the causal version, where we use interventional probability distributions rather than those purely observational (Lizier and Prokopenko 2010). Observational data can be used to infer causality but only under particular circumstances, such as strong knowledge of the underlying mechanics, or through special circumstances, such as those dictated by the backdoor criteria (Pearl 2009). For a detailed discussion see Lizier and Prokopenko (2010).

3.3.1 Analytical Solution for Networks of Coupled Oscillators

As discussed above (Section 3.1), simple models of coupled oscillators are employed as models for neuronal activity as well as metabolic networks, and, as an extension, could be used to model their interaction. The larger a system gets, the more challenging its computational treatment becomes along with its susceptibility to the curse of dimensionality. Here, analytical approaches can help alleviate these hurdles. An analytic treatment of information transfer in networked systems of coupled oscillators was provided by Kirst et al. (2016). They employed small Gaussian noise expansions on some collective dynamical state to get analytic expressions for conditional probability densities and, thus, time-delayed mutual information and TE. This allows the study of information routing, where information is encoded in the "noise" process while routing is determined by the collective behavior encoded by the underlying dynamical state. The control of information routing pairs can be achieved by controlling the collective network dynamics; that is, the reference state (of the noise approximation). Switching can be induced by larger amplitude external stimuli or by activity of another part of the network, thus providing a potential mechanism for self-organization of neuronal networks. It would be an interesting scientific endeavor, one that to our knowledge has not yet been pursued, to study switching dynamics in coupled metabolic and neuronal networks.

3.4 Dynamic Causal Modeling

DCM is a widely adopted method in neuroscience for extracting directed graphs from data that finds application in fMRI (Cha et al. 2016; Rupprechter et al. 2020; Seghier and Friston 2013; Sladky et al. 2015; Tik et al. 2018), EEG (David et al. 2006), and studies based on local field potentials (LFP) (Friston et al. 2015; Moran et al. 2011). However, aside from some exploration into the field of epidemiology (Parr et al. 2021), DCM has found application only in neuroscience. Essentially, DCM represents a procedure that combines Bayesian parameter estimation with Bayesian model selection, where the models are dynamical systems. The crucial parameters for inferring directed graphs are the elements of the adjacency matrix that link different brain regions, with the dynamics of each region described by ordinary differential equations. These models capture the underlying neuronal dynamics typically at the meso or macro scale, along with the measurement process (e.g., BOLD, EEG, LFP). A variational Bayes method is employed to estimate the evidence lower bound (ELBO)[1] and the posterior distribution of parameters. Model comparison involves contrasting the ELBO of different models. By varying the prior distribution, researchers can assess the presence or absence of connections between regions.

The effectiveness of DCM hinges on accurate model assumptions and a large enough sample size (Silchenko et al. 2023). Incorrect assumptions can lead to flawed conclusions about causal relationships. For a critical perspective on the method, particularly regarding issues with Bayesian model selection and its performance in high-dimensional problems, refer to the debate between Lohmann et al. (2012, 2013) and (Friston et al. 2013). The method's heavy reliance on accurate prior information is a potential limitation, especially when modeling many regions. These robustness issues of DCM for causality assessment led to the development of variants such as spectral DCM, the canonical microcircuit DCM, or the stochastic

[1] In the DCM literature, this objective function is also called free energy due to its equivalent functional form with the free energy defined in physics. We prefer to use the terminology commonly used in machine learning.

DCM alleviating these issues (Friston et al. 2019). For a comparison of DCM with Granger causality, see Seth et al. (2015).

3.4.1 Simulation-Based Inference

The above conceptual approach is not limited to a specific parameter estimation method or model selection method. Another promising approach is simulation-based inference, a modern family of techniques for Bayesian inference where an analytic expression for the likelihood is not needed. With these methods, the posterior is estimated using a neural network to learn the generally infeasible likelihood, to which traditional Monte Carlo methods may then be applied, or to directly infer the posterior distribution from the joint distribution (Cranmer et al. 2020). The name comes from using the simulator directly, where the conditional distributions are estimated from the pairs of parameters and simulator outputs (θ, x_i) sampled from a prior. Different flavors of simulation-based inference exist, offering different benefits based on their use cases. Amortized versions and sequential methods exist, though the use case for each tends to differ. Amortized versions, where a single neural network can be used to estimate the posterior of multiple different datasets, are advantageous because they can be applied to the same model with different parameters. In contrast, sequential versions, which can be used for only a single dataset, generally converge faster and require fewer samples. The performance is heavily dependent on the prior and the underlying conditional density estimation used (the neural network architecture chosen). Masked autoregressive flows (Papamakarios et al. 2018) and splines (neural spline flows) are the most common neural density estimation techniques used, as they offer fast performance and a wide variety of distributions. Similar to DCM, this method is generally used for Bayesian parameter extraction and thus can be adapted similarly for causal extraction.

4 Methods for Stationary Data

4.1 Structural Causal Models

SCMs provide a formal framework to model causality with a minimum set of necessary assumptions. A simple example SCM can be defined as follows: An SCM $C(f_E, P_N)$ with graph $C \rightarrow E$ consists of two assignments where we call C the cause and E the effect:

$$\begin{aligned} C &:= N_C, \\ E &:= f_E(C, N_E) \end{aligned} \tag{13.3}$$

where $N_E \perp\!\!\!\perp N_C$, that is, N_E is independent of N_C and denote random variables with distributions P_{N_C} and P_{N_E}. Thus the SCM describes a data-generating process that transforms exogenous random variables N_C and N_E into observed endogenous random variables C and E and entails a joint distribution $P(C, E)$ (Peters et al. 2017).

We may expand this formalism to a set of observables $X_1, \ldots, X_n$. These are associated with the vertices of a DAG. We have a SCM $C(f, P_N)$ if each observable is the result of an assignment

$$X_i := f_i\left(X_{pa_i}, N_i\right) \tag{13.4}$$

with f_i denoting deterministic functions that represent the causal mechanisms that relate X_i to its causal parents X_{pa_i}. This framework allows to represent cause-effect interventions by using the *do* formalism defined by Pearl (Pearl 2009; Peters et al. 2017). We note that the intervened system induces another distribution, which usually differs from the observational distribution. Instead of studying these two systems together, it seems more appropriate to treat them as two separate systems, which motivates the idea that after an intervention only parts of the data-generating process change. For example, if we intervene in the system defined in the beginning of this chapter by setting E:=4, we do not affect the mechanism that generates C.

Given this formal framework and data, the approach now is to assess independence between observables, rather than prediction as in the previously defined methods. Several methods have not yet found appreciation within the fields of neuroscience or metabolomics. However, methods such as LiNGAMs (linear non-Gaussian acyclic model) (Shimizu et al. 2006) tie in with the formal framework of SCMs (Peters 2017) and have been applied to fMRI data and other modalities (Bielczyk 2019), in particular LiNGAMs have been extended to deal with hidden variables (see Figure 13.3) (Adams et al. 2021).

4.2 Structural Dynamical Causal Models

Despite the similarities in name, a structural dynamical causal model (SDCM) should not be confounded with DCMs. SDCMs provide a formal framework that relates SCMs to random dynamical systems. Indeed, as such a SDCM forms a general framework that, at least on a conceptual level, includes DCMs (Bongers et al. 2022). Early work ventured into expanding the SCM formalism to cyclic linear graphs (Spirtes 1995), and later, the domain of application was expanded to nonlinear models with cyclic noise models (Mooij et al. 2011), both without any formal relation to differential equations. However, the expansion of SCMs to the equilibrium state of first-order ordinary differential equations opened up an interesting and natural approach to address cyclical causal relationships (Mooij et al. 2013; Rubenstein et al. 2018) that was only recently expanded to include equilibrium states of random[2] differential equations (Bongers et al. 2022). This is an exciting ongoing line of research that has the potential to provide a proper formal framework for methods such as DCMs and the method discussed next.

[2] Note the difference between random and stochastic ordinary differential equations: random differential equations account for stochasticity in initial conditions or parameters. Thus, there is no process noise that would alter the temporal solution of the dynamical system. Stochastic differential equations, on the other hand, refer to stochasticity in the process itself, which implies that the solution can only be computed by the Ito or Stratonovich formalism (Higham 2001).

4.3 Bicycle—A Method Designed for Feedback-Loop Inference

Acyclic causal models are well-studied, but the assumption that a causal graph is free of feedback loops (see Figure 13.3b) is often not valid, particularly in neuroscience and metabolic networks where feedback loops are common. Bicycle is a method that addresses this often neglected problem in inference of directed graphs (Rohbeck et al. 2024). It is based on a recently published method called *Dictys,* designed to model dynamics of gene regulatory networks (Wang et al. 2023). Dictys is an inference method that models latent, cyclic regulatory dynamics using a multivariate Ornstein–Uhlenbeck process, which it couples to a modality-specific technical noise model. The authors of Bicycle augmented this method conceptually by introducing experimental perturbations into the inference procedure through the independent causal mechanism (ICM) principle (Schölkopf et al. 2021). This modification appends a strong inductive bias toward sparsity of interventional parameter shifting, justified by the ICM principle. In essence, this means that parameters assumed to be affected by the causal intervention will be able to be fitted, while all the other parameters of the model are kept fixed. We note that the objective function of the model inference procedure is ELBO. The approach closely resembles task-based stochastic DCM, with all the interactions strictly assumed to be linear. Unlike the usual DCM approach, experimental interventions are thought to influence certain parameters directly, and model fitting is performed with all the other parameters being held constant (as inferred from data produced in the non-intervention state). In DCM, however, all parameters are being fit to data from interventions and data without interventions. Afterward, model comparison, again similar to DCM and other methods, leads to the final insights about causal structures.

5 Summary and Outlook

Genomics and metabolomics research often correlate gene expression data with psychopathologies (Pawlak et al. 2017; Wigner et al. 2018), yet these approaches typically do not establish causal links. Conversely, several studies utilize fMRI data to analyze functional connectivity patterns between brain regions of patients in comparison to healthy controls (Greicius et al. 2007; Kaiser et al. 2015; Yan et al. 2019). However, functional connectivity patterns do not establish causal links between brain regions. Extracting causal relationships from measurements is difficult in both research domains; the complexity only increases if combined. One key challenge is to find approaches that can relate to very different measurement types. While neuronal measurements are typically time series, metabolomics are measured by assessing concentrations of molecules at a given time. Only a few studies have successfully measured both; such experiments have been achieved in small animal models, such as *Drosophila*, which are arguably too distinct from humans to generate insight on psychopathology (Mann et al. 2021). At the same time, biophysical computational models that attempt to integrate neuronal and metabolic processes are rare (Jolivet et al. 2015). Thus, significant research opportunities remain.

5.1 Identifying Challenges

Causal relationships between brain regions based solely on neuronal activity are difficult to achieve due to the potential presence of many hidden confounders (Figure 13.3, motifs c and e), such as metabolic factors like ATP levels. Integrating measurements of metabolic molecules (e.g., glucose) alongside neuronal dynamics, although complex, enhances the dataset by revealing otherwise concealed variables, thus reducing the risk of identifying spurious connections.

Measuring time series is critical to causal inference; however, relying on measurements alone is usually insufficient to disassociate causal relationships in complex systems correctly. Predictions based on time-series data are susceptible to spurious correlations due to numerous unmeasured confounders, particularly with techniques with lower temporal resolution like functional neuroimaging. While it is widely known that *correlation is not causation*, studies sometimes still rely on correlations alone to judge cause and effect relationships (Van Hul et al. 2020). Furthermore, methods such as Granger causality and DCM are often misconceived as providing robust causal inference, possibly because their names suggest more definitive conclusions than they actually deliver.

Here, developments such as SDCMs hold the promise of putting methods, such as DCM, on a more solid formal basis. This would allow a consistent integration of the data-driven method and experimental interventions for the assessment of causal relationships.

We conclude that while it is possible to extract causal relationships from data alone, algorithms must be applied in compliance with their assumptions, and we caution the reader of potential premature conclusions about causal relationships. Experimental interventions, such as direct brain stimulation experiments, are often necessary to correctly infer the causal mechanisms and data-driven methods can guide experimental hypothesis building (Siddiqi et al. 2022).

Feedback loops are a critical feature of neuronal and metabolic networks. Feedback loops present a unique challenge as most analytical methods are designed for directed acyclic graphs (Runge et al. 2019). Despite the prevalence of feedback loops in complex systems, adaptations of the widely used Granger causality for identifying these loops from data are few. They have not been broadly adopted, suggesting their inadequacy for such complex analyses (Dong et al. 2012). However, computational models such as systems of ordinary differential equations allow the definition of any graph structure in principle. An approach exemplified by the Bicycle method, which is similar to DCM as described above, has thus been proposed to address the extraction of feedback loops by relying on steady state solutions of ordinary differential equations and experimental interventions (Rohbeck et al. 2024). Thus, the use of model-based techniques seems to be a promising way to further the disassociation of causal mechanisms in complex systems.

It is important to note that formal relationships between SCM and dynamical system models have been constructed. However, these hold only at the fixed-point solution of the ordinary differential equation. So far, relations exist for the deterministic case (Mooij et al. 2013; Rubenstein et al. 2018) as well as for random differential equations (Bongers et al. 2022). The latter provides a sufficiently general framework to encompass DCMs (Bongers et al. 2022) and arguably provides a formal framework that allows for more rigorous inference of causal relations.

5.2 Model-Based Causal Inference is a Promising Venue

Statistics alone provide only a superficial insight into causal mechanisms since causal mechanisms produce statistical dependencies, but the reverse argument is not guaranteed: statistical dependencies do not imply causal structures. However, combining mechanistic models with experimental interventions establishes a more substantial method for identifying causal mechanisms. We advocate for utilizing computational models and, where feasible, experimental interventions. These allow for enhanced inference of causal relationships through the incorporation of prior knowledge and help inform and guide experimental interventions aimed at falsifying and refining the model structure.

The use of prior information is crucial because relying solely on data-driven inference is susceptible to errors. For example, evaluating the DREAM (Dialogue for Reverse Engineering Assessment and Methods) challenges underscores the importance of incorporating prior knowledge and biophysical constraints into causal inference (Hill et al. 2016). Thus, detailed model-based approaches are valuable, provided the underlying model structure is accurate. Methods such as DCM, commonly used in neuronal imaging studies, could theoretically be expanded to include metabolic dynamics as well.

Among model-based methods for estimating causal relationships that incorporate a higher degree of biophysical detail, DCM is the most frequently used in neuronal imaging. To our knowledge, it has not yet been applied to metabolic dynamics; however, as detailed above, the recently developed method Bicycle to study gene expression data follows a similar approach (Rohbeck et al. 2024). This illustrates the potential of the modeling approach well: while DCM is applied to time-series data, Bicycle relies on steady-state data. Although simultaneous measurements across both domains may present challenges, we emphasize that both methodologies are fundamentally grounded in differential equations, suggesting a natural pathway for theoretical integration. We propose that significant advances could emerge from establishing a unified modeling framework encompassing both metabolic and neuronal circuits. Such computational simulations would facilitate the generation of testable hypotheses for subsequent experimental investigation. Admittedly, temporal dynamics absent from equilibrium data would necessitate additional experimental validation. However, through an iterative approach, a comprehensive dynamical systems model could be developed that incorporates both causal relationships and mechanistic insights. The successful development of such computational models and elucidation of underlying causal mechanisms will depend critically on sustained collaboration between computational scientists and experimental researchers, as outlined in this prospective analysis.

References

Adams J, Hansen N, and Zhang K (2021) Identification of Partially Observed Linear Causal Models: Graphical Conditions for the Non-Gaussian and Heterogeneous Cases. In: Ranzato M, Beygelzimer A, Dauphin Y, Liang PS, Wortman Vaughan J (eds) Nips '21: Proc 35th Intl Conf on Neural Information Processing Systems. Advances in Neural Information Processing Systems. Curran Associates, Inc., Red Hook, NY, pp 22822–22833

Amat J, Baratta MV, Paul E, et al. (2005) Medial Prefrontal Cortex Determines How Stressor Controllability Affects Behavior and Dorsal Raphe Nucleus. Nat Neurosci 8 (3):365–371. https://doi.org/10.1038/nn1399

Angrist JD, and Imbens GW (eds) (1995) Identification and Estimation of Local Average Treatment Effects. NBER Working Paper No. t0118, Rochester, NY

Ay N, and Polani D (2008) Information Flows in Causal Networks. Adv Complex Syst 11 (1):17–41. https://doi.org/10.1142/S0219525908001465

Barnett L, Barrett AB, and Seth AK (2009) Granger Causality and Transfer Entropy Are Equivalent for Gaussian Variables. Phys Rev Lett 103 (23):238701. https://doi.org/10.1103/PhysRevLett.103.238701

——— (2018) Solved Problems for Granger Causality in Neuroscience: A Response to Stokes and Purdon. Neuroimage 178:744–748. https://doi.org/10.1016/j.neuroimage.2018.05.067

Barnett L, and Seth AK (2015) Granger Causality for State-Space Models. Phys Rev E 91 (4):040101. https://doi.org/10.1103/PhysRevE.91.040101

Bastiaanssen TFS, and Cryan JF (2021) The Microbiota-Gut-Brain Axis in Mental Health and Medication Response: Parsing Directionality and Causality. Int J Neuropsychopharmacol 24 (3):216–220. https://doi.org/10.1093/ijnp/pyaa088

Bielczyk NZ, Uithol S, van Mourik T, et al. (2019) Disentangling Causal Webs in the Brain Using Functional Magnetic Resonance Imaging: A Review of Current Approaches. Netw Neurosci 3 (2):237–273. https://doi.org/10.1162/netn_a_00062

Bongers S, Blom T, and Mooij JM (2022) Causal Modeling of Dynamical Systems. https://doi.org/10.48550/arXiv.1803.08784

Campbell IH, and Campbell H (2024) The Metabolic Overdrive Hypothesis: Hyperglycolysis and Glutaminolysis in Bipolar Mania. Mol Psychiatr 29 (5):1521–1527. https://doi.org/10.1038/s41380-024-02431-w

Cha J, Greenberg T, Song I, et al. (2016) Abnormal Hippocampal Structure and Function in Clinical Anxiety and Comorbid Depression. Hippocampus 26 (5):545–553. https://doi.org/10.1002/hipo.22566

Correia AS, and Vale N (2022) Tryptophan Metabolism in Depression: A Narrative Review with a Focus on Serotonin and Kynurenine Pathways. Int J Mol Sci 23 (15):8493. https://doi.org/10.3390/ijms23158493

Cranmer K, Brehmer J, and Louppe G (2020) The Frontier of Simulation-Based Inference. PNAS 117 (48):30055–30062. https://doi.org/10.1073/pnas.1912789117

Cunningham S (2021) Causal Inference. The Mixtape. Yale University Press, New Haven. doi:https://doi.org/10.12987/9780300255881

David O, Kiebel SJ, Harrison LM, et al. (2006) Dynamic Causal Modeling of Evoked Responses in EEG and MEG. Neuroimage 30 (4):1255–1272. https://doi.org/10.1016/j.neuroimage.2005.10.045

Dayan P, and Huys QJM (2008) Serotonin, Inhibition, and Negative Mood. PLoS Comput Biol 4 (2):e4. https://doi.org/10.1371/journal.pcbi.0040004

Dong C-Y, Shin D, Joo S, Nam Y, and Cho K-H (2012) Identification of Feedback Loops in Neural Networks Based on Multi-Step Granger Causality. Bioinformatics 28 (16):2146–2153. https://doi.org/10.1093/bioinformatics/bts354

Frigerio F (2023) The HPA Axis and the Regulation of Energy Balance. In: Luzi L (ed) Cellular Physiology and Metabolism of Physical Exercise: Topical Clinical Issues. Springer, Cham, pp 101–114. https://doi.org/10.1007/978-3-031-27192-2_1

Friston K, Daunizeau J, and Stephan KE (2013) Model Selection and Gobbledygook: Response to Lohmann et Al. Neuroimage 75:275–278. https://doi.org/10.1016/j.neuroimage.2011.11.064

Friston KJ, Bastos AM, Pinotsis D, and Litvak V (2015) LFP and Oscillations—What Do They Tell Us? Curr Opin Neurobiol 31:1–6. https://doi.org/10.1016/j.conb.2014.05.004

Friston KJ, Harrison L, and Penny W (2003) Dynamic Causal Modelling. Neuroimage 19 (4):1273–1302. https://doi.org/10.1016/S1053-8119(03)00202-7

Friston KJ, Preller KH, Mathys C, et al. (2019) Dynamic Causal Modelling Revisited. Neuroimage 199:730–744. https://doi.org/10.1016/j.neuroimage.2017.02.045

Ghahramani Z (1998) Learning Dynamic Bayesian Networks. In: Giles CL, Gori M (eds) Adaptive Processing of Sequences and Data Structures. NN 1997. Lecture Notes in Computer Science, vol 1387. Springer, Heidelberg, pp 168–197. https://doi.org/10.1007/BFb0053999

Granger CWJ (1969) Investigating Causal Relations by Econometric Models and Cross-Spectral Methods. Econometrica 37 (3):424–438. https://doi.org/10.2307/1912791

Greicius MD, Flores BH, Menon V, et al. (2007) Resting-State Functional Connectivity in Major Depression: Abnormally Increased Contributions from Subgenual Cingulate Cortex and Thalamus. Biol Psychiatry 62 (5):429–437. https://doi.org/10.1016/j.biopsych.2006.09.020

Gu X, Ke S, Wang Q, et al. (2021) Energy Metabolism in Major Depressive Disorder: Recent Advances from Omics Technologies and Imaging. Biomed Pharmacother 141:111869. https://doi.org/10.1016/j.biopha.2021.111869

Haeffel GJ (2011) After Further Deliberation: Cognitive Vulnerability Predicts Changes in Event-Specific Negative Inferences for a Poor Midterm Grade. Cognit Ther Res 35 (4):285–292. https://psycnet.apa.org/doi/10.1007/s10608-010-9298-y

Higham DJ (2001) An Algorithmic Introduction to Numerical Simulation of Stochastic Differential Equations. SIAM Rev Soc Ind Appl Math 43 (3):525–546. https://doi.org/10.1137/S0036144500378302

Hill SM, Heiser LM, Cokelaer T, et al. (2016) Inferring Causal Molecular Networks: Empirical Assessment through a Community-Based Effort. Nat Methods 13 (4):310–318. https://doi.org/10.1038/nmeth.3773

Hindash AHC, and Amir N (2012) Negative Interpretation Bias in Individuals with Depressive Symptoms. Cognit Ther Res 36:502–511. https://doi.org/10.1007/s10608-011-9397-4

Huys QJM, Daw ND, and Dayan P (2015) Depression: A Decision-Theoretic Analysis. Annu Rev Neurosci 38 (1):1–23. https://doi.org/10.1146/annurev-neuro-071714-033928

Huys QJM, Eshel N, O'Nions E, et al. (2012) Bonsai Trees in Your Head: How the Pavlovian System Sculpts Goal-Directed Choices by Pruning Decision Trees. PLoS Comput Biol 8 (3):e1002410. https://doi.org/10.1371/journal.pcbi.1002410

Iacoviello BM, Alloy LB, Abramson LY, and Choi JY (2010) The Early Course of Depression: A Longitudinal Investigation of Prodromal Symptoms and Their Relation to the Symptomatic Course of Depressive Episodes. J Abnorm Psychol 119 (3):459–467. https://psycnet.apa.org/doi/10.1037/a0020114

Imbens GW, and Rubin DB (2015) Causal Inference for Statistics, Social, and Biomedical Sciences: An Introduction. Cambridge Univ. Press, New York. https://doi.org/10.1017/CBO9781139025751

Jolivet R, Coggan JS, Allaman I, and Magistretti PJ (2015) Multi-Timescale Modeling of Activity-Dependent Metabolic Coupling in the Neuron-Glia-Vasculature Ensemble. PLoS Comput Biol 11 (2):e1004036. https://doi.org/10.1371/journal.pcbi.1004036

Kaiser RH, Andrews-Hanna JR, Wager TD, and Pizzagalli DA (2015) Large-Scale Network Dysfunction in Major Depressive Disorder: A Meta-Analysis of Resting-State Functional Connectivity. JAMA Psychiatry 72 (6):603–611. https://doi.org/10.1001/jamapsychiatry.2015.0071

Kirst C, Timme M, and Battaglia D (2016) Dynamic Information Routing in Complex Networks. Nat Commun 7:11061. https://doi.org/10.1038/ncomms11061

Köhler S, Cierpinsky K, Kronenberg G, and Adli M (2016) The Serotonergic System in the Neurobiology of Depression: Relevance for Novel Antidepressants. J Psychopharm 30 (1):13–22. https://doi.org/10.1177/0269881115609072

Lally N, Huys QJM, Eshel N, et al. (2017) The Neural Basis of Aversive Pavlovian Guidance during Planning. J Neurosci 37 (42):10215–10229. https://doi.org/10.1523/JNEUROSCI.0085-17.2017

Liang XS (2018) Causation and Information Flow with Respect to Relative Entropy. Chaos 28 (7):075311. https://doi.org/10.1063/1.5010253

——— (2021) Normalized Multivariate Time Series Causality Analysis and Causal Graph Reconstruction. Entropy 23 (6):679. https://doi.org/10.3390/e23060679

Liang XS, and Kleeman R (2005) Information Transfer between Dynamical System Components. Phys Rev Lett 95 (24):244101. https://doi.org/10.1103/PhysRevLett.95.244101

Lizier JT, and Prokopenko M (2010) Differentiating Information Transfer and Causal Effect. Eur Phys J B 73 (4):605–615. https://doi.org/10.1140/epjb/e2010-00034-5

Lohmann G, Erfurth K, Müller K, and Turner R (2012) Critical Comments on Dynamic Causal Modelling. Neuroimage 59 (3):2322–2329. https://doi.org/10.1016/j.neuroimage.2011.09.025

Lohmann G, Müller K, and Turner R (2013) Response to Commentaries on Our Paper: Critical Comments on Dynamic Causal Modelling. Neuroimage 75:279–281. https://doi.org/10.1016/j.neuroimage.2012.07.041

Mann K, Deny S, Ganguli S, and Clandinin TR (2021) Causal Coupling between Neural Activity, Metabolism, and Behavior across the Drosophila Brain. Nature 593:244–248. https://doi.org/10.1038/s41586-021-03497-0

Meinshausen N, Hauser A, Mooij JM, et al. (2016) Methods for Causal Inference from Gene Perturbation Experiments and Validation. PNAS 113 (27):7361–7368. https://doi.org/10.1073/pnas.1510493113

Mooij JM, Janzing D, Heskes T, and Schölkopf B (2011) On Causal Discovery with Cyclic Additive Noise Models. In: Shawe-Taylor J, Zemel RS, Bartlett PL, Pereira FCN, Weinberger KQ (eds) Advances in Neural Information Processing Systems, vol 24. Curran Associates, Inc., Red Hook, NY,

Mooij JM, Janzing D, and Schölkopf B (2013) From Ordinary Differential Equations to Structural Causal Models: The Deterministic Case. https://arxiv.org/abs/1304.7920

Moran RJ, Mallet N, Litvak V, et al. (2011) Alterations in Brain Connectivity Underlying Beta Oscillations in Parkinsonism. PLoS Comput Biol 7 (8):e1002124. https://doi.org/10.1371/journal.pcbi.1002124

Natale JL, Hofmann D, Hernández DG, and Nemenman I (2018) Reverse-Engineering Biological Networks from Large Data Sets. In: Munsky B, Hlavacek WS, Tsimring LS (eds) Quantitative Biology: Theory, Computational Methods, and Models. MIT Press, Cambridge, MA, pp 213–246

Papamakarios G, Pavlakou T, and Murray I (2018) Masked Autoregressive Flow for Density Estimation. https://doi.org/10.48550/arXiv.1705.07057

Pariante CM, and Lightman SL (2008) The HPA Axis in Major Depression: Classical Theories and New Developments. Trends in Neurosciences 31 (9):464–468. https://doi.org/10.1016/j.tins.2008.06.006

Parr T, Bhat A, Zeidman P, et al. (2021) Dynamic Causal Modelling of Immune Heterogeneity. Sci Rep 11 (1):11400. https://doi.org/10.1038/s41598-021-91011-x

Pawlak J, Dmitrzak-Węglarz M, Maciukiewicz M, et al. (2017) Personality Traits as an Endophenotype in Genetic Studies on Suicidality in Bipolar Disorder. Acta Neuropsychiatr 29 (2):115–121. https://doi.org/10.1017/neu.2016.43

Pearl J (2009) Causality: Models, Reasoning, and Inference. Cambridge Univ. Press, Cambridge

Peters J, Janzing D, and Schölkopf B (2017) Elements of Causal Inference: Foundations and Learning Algorithms. Adaptive Computation and Machine Learning. MIT Press, Cambridge, MA

Prill RJ, Marbach D, Saez-Rodriguez J, et al. (2010) Towards a Rigorous Assessment of Systems Biology Models: The Dream3 Challenges. PLoS One 5 (2):e9202. https://doi.org/10.1371/journal.pone.0009202

Rajapakse JC, and Zhou J (2007) Learning Effective Brain Connectivity with Dynamic Bayesian Networks. Neuroimage 37 (3):749–760. https://doi.org/10.1016/j.neuroimage.2007.06.003

Rohbeck M, Clarke B, Mikulik K, et al. (2024) Bicycle: Intervention-Based Causal Discovery with Cycles. In: Procs of Machine Learning Research: 3rd Conference on Causal Learning and Reasoning, vol 236. https://proceedings.mlr.press/v236/rohbeck24a/rohbeck24a.pdf

Rojas R, Griñán-Ferré C, Castellanos A, et al. (2024) Beta-Hydroxybutyrate Counteracts the Deleterious Effects of a Saturated High-Fat Diet on Synaptic AMPA Receptors and Cognitive Performance. https://doi.org/10.1101/2024.01.23.576931

Rubenstein PK, Bongers S, Schoelkopf B, and Mooij JM (2018) From Deterministic Odes to Dynamic Structural Causal Models. https://doi.org/10.48550/arXiv.1608.08028

Runge J, Bathiany S, Bollt E, et al. (2019) Inferring Causation from Time Series in Earth System Sciences. Nat Commun 10:2553. https://doi.org/10.1038/s41467-019-10105-3

Runge J, Heitzig J, Petoukhov V, and Kurths J (2012) Escaping the Curse of Dimensionality in Estimating Multivariate Transfer Entropy. Phys Rev Lett 108 (25):258701. https://doi.org/10.1103/physrevlett.108.258701

Rupprechter S, Romaniuk L, Series P, et al. (2020) Blunted Medial Prefrontal Cortico-Limbic Reward-Related Effective Connectivity and Depression. Brain 143 (6):1946–1956. https://doi.org/10.1093/brain/awaa106

Schölkopf B, Locatello F, Bauer S, et al. (2021) Toward Causal Representation Learning. Proc IEEE 109 (5):612–634. https://doi.org/10.1109/JPROC.2021.3058954

Seghier ML, and Friston KJ (2013) Network Discovery with Large DCMs. Neuroimage 68:181–191. https://doi.org/10.1016/j.neuroimage.2012.12.005

Seth AK, Barrett AB, and Barnett L (2015) Granger Causality Analysis in Neuroscience and Neuroimaging. J Neurosci 35 (8):3293–3297. https://doi.org/10.1523/JNEUROSCI.4399-14.2015

Seth AK, Chorley P, and Barnett LC (2013) Granger Causality Analysis of fMRI BOLD Signals Is Invariant to Hemodynamic Convolution but Not Downsampling. Neuroimage 65:540–555. https://doi.org/10.1016/j.neuroimage.2012.09.049

Shimizu S, Hoyer PO, Hyvarinen A, and Kerminen A (2006) A Linear Non-Gaussian Acyclic Model for Causal Discovery. J Mach Learn Res 7:2003–2030

Siddiqi SH, Kording KP, Parvizi J, and Fox MD (2022) Causal Mapping of Human Brain Function. Nat Rev Neurosci 23 (6):361–375. https://doi.org/10.1038/s41583-022-00583-8

Silchenko AN, Hoffstaedter F, and Eickhoff SB (2023) Impact of Sample Size and Regression of Tissue-Specific Signals on Effective Connectivity within the Core Default Mode Network. Hum Brain Mapp 1 (44):17. https://doi.org/10.1002/hbm.26481

Sladky R, Höflich A, Küblböck M, et al. (2015) Disrupted Effective Connectivity between the Amygdala and Orbitofrontal Cortex in Social Anxiety Disorder During Emotion Discrimination Revealed by Dynamic Causal Modeling for fMRI. Cereb Cortex 25 (4):895–903. https://doi.org/10.1093/cercor/bht279

Solo V (2016) State-Space Analysis of Granger-Geweke Causality Measures with Application to fMRI. Neural Comput 28 (5):914–949. https://doi.org/10.1162/NECO_a_00828

Spirtes PL (1995) Directed Cyclic Graphical Representations of Feedback Models. https://doi.org/10.48550/arXiv.1302.4982

Sugihara G, May R, Ye H, et al. (2012) Detecting Causality in Complex Ecosystems. Science 338 (6106):496–500. https://doi.org/10.1126/science.1227079

Tik M, Sladky R, Luft CDB, et al. (2018) Ultra-High-Field fMRI Insights on Insight: Neural Correlates of the Aha!-Moment. Hum Brain Mapp 39 (8):3241–3252. https://doi.org/10.1002/hbm.24073

Van Hul M, Le Roy T, Prifti E, et al. (2020) From Correlation to Causality: The Case of *Subdoligranulum*. Gut Microbes 12 (1):1–13. https://doi.org/10.1080/19490976.2020.1849998

Vicente R, Wibral M, Lindner M, and Pipa G (2011) Transfer Entropy–a Model-Free Measure of Effective Connectivity for the Neurosciences. J Comput Neurosci 30 (1):45–67. https://doi.org/10.1007/s10827-010-0262-3

Wang L, Trasanidis N, Wu T, et al. (2023) Dictys: Dynamic Gene Regulatory Network Dissects Developmental Continuum with Single-Cell Multiomics. Nat Methods 20 (9):1368–1378. https://doi.org/10.1038/s41592-023-01971-3

Wigner P, Czarny P, Synowiec E, et al. (2018) Association between Single Nucleotide Polymorphisms of Tph1 and Tph2 Genes, and Depressive Disorders. J Cell Mol Med 22 (3):1778–1791. https://doi.org/10.1111/jcmm.13459

Yan C-G, Chen X, Li L, et al. (2019) Reduced Default Mode Network Functional Connectivity in Patients with Recurrent Major Depressive Disorder. PNAS 116 (18):9078–9083. https://doi.org/10.1073/pnas.1900390116

Zheng H, Jia F, Guo G, et al. (2015) Abnormal Anterior Cingulate N-Acetylaspartate and Executive Functioning in Treatment-Resistant Depression after rTMS Therapy. Int J Neuropsychopharmacol 18 (11):pyv059. https://doi.org/10.1093/ijnp/pyv059

Zhou X, Liu L, Lan X, et al. (2019) Polyunsaturated Fatty Acids Metabolism, Purine Metabolism and Inosine as Potential Independent Diagnostic Biomarkers for Major Depressive Disorder in Children and Adolescents. Mol Psychiatr 24 (10):1478–1488. https://doi.org/10.1038/s41380-018-0047-z

Open Access This chapter is licensed under the terms of the Creative Commons Attribution-NonCommercial-NoDerivatives 4.0 International License (http://creativecommons.org/licenses/by-nc-nd/4.0/), which permits any noncommercial use, sharing, distribution and reproduction in any medium or format, as long as you give appropriate credit to the original author(s) and the source, provide a link to the Creative Commons license and indicate if you modified the licensed material. You do not have permission under this license to share adapted material derived from this chapter or parts of it.

The images or other third party material in this chapter are included in the chapter's Creative Commons license, unless indicated otherwise in a credit line to the material. If material is not included in the chapter's Creative Commons license and your intended use is not permitted by statutory regulation or exceeds the permitted use, you will need to obtain permission directly from the copyright holder.

14

Designing Innovative Interventions for Metabolic Neuropsychiatry

Corey Weistuch, Virginie-Anne Chouinard, Sharmili Edwin Thanarajah, Peter Falkai, David Hofmann, Dost Öngür, Martin P. Paulus, Melanie M. Wall, and Lilianne R. Mujica-Parodi

Abstract Interest in the role of metabolism in psychiatric disorders and their treatment is rapidly expanding, driven by emerging evidence that neurometabolic dysfunction contributes to the pathophysiology of conditions such as depression, bipolar disorder, and schizophrenia. This chapter provides an overview of strategies for designing mechanistically informed studies and clinical trials that evaluate metabolic interventions in psychiatric populations. Discussion includes the importance of integrating biomarkers, neuroimaging, and computational modeling to identify and stratify patient subgroups most likely to benefit from metabolic therapies. Key considerations for developing proof-of-concept and efficacy studies are presented, including optimal study design, statistical and mechanistic modeling, biomarker selection, and the alignment of mechanistic hypotheses with clinical endpoints. By adopting a translational framework that connects experimental data, clinical observations, and systems-level mathematical modeling, researchers can more effectively evaluate the therapeutic potential of metabolic interventions. This roadmap aims to accelerate the development of personalized and data-driven treatment strategies and establish a rigorous evidence base for incorporating metabolic targets into psychiatric care.

Keywords Clinical trial design, computational modeling, metabolic interventions, dynamical biomarkers, personalized mental health treatment

---

Corey Weistuch (✉)
Department of Medical Physics, Memorial Sloan Kettering Cancer Center, 1275 York Avenue, New York, NY 10065, USA
E-Mail: weistucc@mskcc.org

Affiliations for the coauthors are available in the List of Contributors.

© Frankfurt Institute for Advanced Studies (FIAS) 2025
D. Öngür, J. M. Ford (eds.), *Metabolic Neuropsychiatry*, Strüngmann Forum Reports,
https://doi.org/10.1007/978-3-032-05630-6_14

Group photos (top left to bottom right) Virginie-Anne Chouinard, Corey Weistuch, Sharmili Edwin Thanarajah, Dost Öngür, Lilianne Mujica-Parodi, Peter Falkai, Melanie Wall, David Hofmann, Martin Paulus, Peter Falkai, Lilianne Mujica-Parodi, Corey Weistuch, Martin Paulus, David Hofmann, Melanie Wall, Dost Öngür, Sharmili Edwin Thanarajah, Virginie-Anne Chouinard

1 Leveraging Research toward Improved Clinical Care

Interest in the role of metabolism in psychiatric disorders and treatments is rapidly expanding. As the existing database on neurometabolic abnormalities is reviewed in other chapters of this volume, we focus here on approaches that can be used to generate mechanistically informed proof-of-concept and efficacy trials to target such abnormalities. Since common metabolic interventions, such as the ketogenic diet, generate systemic physiological changes that vary over time, developing next-generation metabolic treatments requires a comprehensive framework to parse the multiple competing mechanisms by which such interventions might improve health. Here, we present a statistical (*mediator*) roadmap that integrates clinical and experimental research to evaluate competing mechanistic hypotheses, model the value of different types of experimental data, and inform study design.

Within the mediator framework, dynamic perturbations such as acute, repeatable hypoxic interventions (brief, controlled reductions in oxygen availability) serve to inform a statistical model of the multiple competing processes (the "mediators") by which a desired therapeutic intervention can act (Ehrenreich et al. 2023). Once learned, the model can distinguish between normal and disordered states and propose the minimal intervention required. Likewise, it can identify the most likely mechanisms of action of existing interventions. Since these functions also depend on context and environment (e.g., age, sex, and other "moderator" variables), future studies and modeling efforts need to consider patient heterogeneity and its impact on outcomes. These and other related ideas are developed below, leading to specific recommendations for choice of biomarkers, perturbations, and study design considerations.

2 Conceptual: What Is the System?

To probe the mechanistic pathways by which metabolic interventions impact neuropsychiatric disorders, we first need to identify the primary physiological control circuits that maintain homeostasis during health, and thus narrow the search space to a set of candidate critical points of failure. These circuits include the following:

- Energy sensing: AMP-activated protein kinase (AMPK) and mammalian target of rapamycin (mTOR) pathways
- Energy production: glycolysis, tricarboxylic acid (TCA) cycle, mitochondrial regulation
- Energy storage
- Insulin signaling: glucose-insulin-glucagon dynamics
- Oxidative stress: radicals and hypoxia
- Inflammation: cytokines
- Hormonal regulation: hypothalamic-pituitary-adrenal (HPA) axis, thyroid, and reproductive hormones

These components interact dynamically to maintain homeostasis and contribute to the multisystem dysregulatory phenotype associated with metabolic disorders. Disentangling the driving causes from their cascading effects is thus essential to treat the disease, rather than just its symptoms.

Crucially, cognitive and affective symptoms in neuropsychiatric disorders depend not only on which circuit is affected but on how it is affected. The basic structure of all control circuits includes several key components: *sensors*, *set points*, *excitatory* and *inhibitory components of negative feedback loops* (each of which includes both gains and lags), as well as various filters and other gating mechanisms (Alon 2019). The many ways the circuit can break can generally be described as the circuit's "points of failure." Different mechanisms may affect the same circuit in various ways by targeting different points of failure. An obvious example of this is the distinction between type 1 diabetes, type 2 diabetes, and reactive hypoglycemia. Although all three disorders implicate the same control circuit regulating glucose, they involve different points of failure. This distinction is most clearly revealed by perturbing the system (i.e., via a glucose bolus) and measuring the resulting glucose-insulin-glucagon dynamics as the system attempts to regain homeostasis (Figure 14.1).

Perturbing the metabolic system and measuring the resulting dynamics as it attempts to regain homeostasis can not only identify the dysregulation *type* (in terms of its point[s] of failure), but also quantify its *degree of severity*; namely, its *control error*. Control error refers to the difference between the system's optimal versus actual output over time. In the context of physiological homeostasis, control error measures how negative feedback loops can effectively maintain key variables (e.g., blood oxygen concentration and pH) within physiological limits. A well-regulated control circuit minimizes control error over time, while chronically taxing it leads to dysregulation, causing errors to propagate and further stress the system. If that dysregulation becomes sufficiently severe, the system effectively "breaks," a point

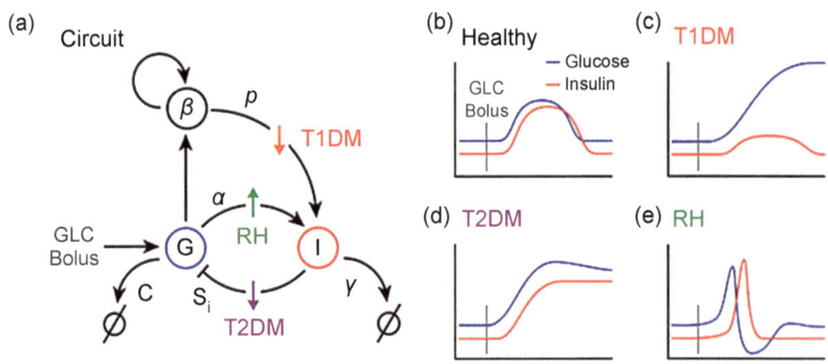

Figure 14.1 Physiological (including neural) control circuits may be compromised at distinct points of failure, leading to qualitatively different types of circuit dysregulation, output dynamics, and resulting clinical symptoms. (a) Schematic of a typical physiological control circuit (after Karin et al. 2016): regulation between blood glucose (G) and blood insulin (I) in response to exogenous glucose (GLC) bolus, with specific disorders (type 1 and type 2 diabetes mellitus, reactive hypoglycemia) associated with the breakdown of specific feedforward and feedback components. (b) Healthy homeostatic regulation (c) shows failure to downregulate glucose due to impaired insulin secretion versus (d) impaired insulin sensitivity. (e) Both contrast with the dynamics seen with excessive glucose-stimulated insulin secretion rate. Abbreviations: G: blood glucose, I: blood insulin, β: pancreatic β population, C: glucose removal rate at zero insulin, S_i: insulin sensitivity, p: insulin secretion per cell, α: insulin secretion rate due to glucose stimulation, γ: insulin removal rate.

that often coincides with clinical symptoms (Mackey and Glass 1977). The aim of early (prodromal) detection of metabolic dysregulation is that detecting the system's "bend" before its "break" may allow therapeutic intervention when the system is already stressed homeostatically but still amenable to recovery (Antal et al. 2025).

Control theory provides a way to model and predict how the behavior of physiological circuits can develop an "overshoot and collapse" response to increasing demand, exceeding a limited supply of biological precursors (hitting the system's *carrying capacity*) and then breaking (McEwen and Wingfield 2003). This phenomenon is common in many physiological disorders (e.g., the common conversion of type 2 to type 1 diabetes), and there are many examples in psychiatry. In post-traumatic stress disorder, for example, cortisol levels are initially abnormally high and then become abnormally low (Yehuda 2002; Bremner et al. 2003). Likewise, in schizophrenia, parenchymal glutamate (Glu) concentrations are elevated in young (especially first episode) schizophrenia patients but decline over time and become reduced in chronically ill patients, compared to age-matched healthy controls (Merritt et al. 2021). In major depressive disorder, magnetic resonance spectroscopy (MRS) Glu studies have reported that Glu levels are lower during episodes of illness (Yüksel and Öngür 2010) and improve back to normal levels with treatment (Boucherie et al. 2023). These observations of dynamic change in brain chemistry measures are not limited to Glu and can also be made for metabolic and redox measures (Dwir et al. 2023). Taken together, the literature suggests that "points of failure" in neurotransmission and neurometabolism are not reflected in abnormal steady-state metabolite levels but rather in their inability to remain within set ranges over time.

3 Experimental: How Do We Test Multiple Competing Hypotheses?

As described above, neuropsychiatric outcomes are mediated by multiple, closely linked causal factors. The mediator model provides a valuable framework for organizing these competing factors into a discrete set of mechanistic hypotheses (see Figure 14.2). Within this framework, mediators act as the intermediate variables through which an intervention (e.g., the ketogenic diet) influences the outcomes of interest. By building a statistical model that incorporates these fundamental elements and their interactions, we can assess how these processes influence brain function and test multiple hypotheses in parallel. This can be achieved through two parallel strategies: bottom-up and top-down modeling (Chesebro et al. 2025). The bottom-up strategy begins with a detailed representation of the individual components of the system and is particularly useful for connecting variables that cannot be measured simultaneously, such as the biophysical properties of neurons and fMRI dynamics (DiNuzzo et al. 2024; van Nieuwenhuizen et al. 2024). The top-down strategy involves measuring indicators of the mediators and outcomes under various conditions to learn the relationships from data and is crucial for integrating new biomarkers. This turns hypothesis testing into a parameter-fitting exercise: by examining the evidence for each mediator in accounting for the observed change in the outcomes of a study, one can rank the relative likelihood of alternative explanations for the data, thereby enabling the simultaneous testing of multiple competing

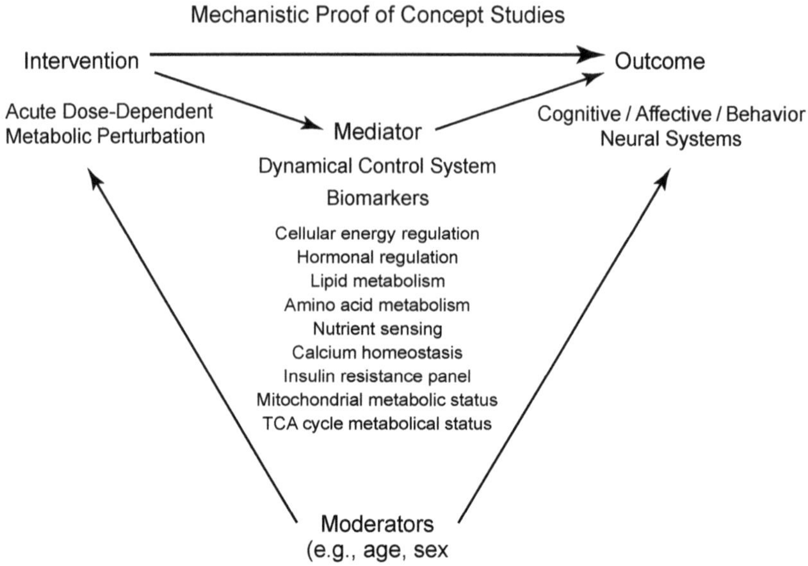

Figure 14.2 The mediator model illustrates the relationship between an intervention (the input), multiple mediators, and study outcomes. Factors such as sex and age can moderate the strength and direction of the relationship between the intervention and outcomes, influencing both the intervention-to-mediator and mediator-to-outcomes interactions. Two key points are worth noting. First, the outcome is measured over time to assess not only the magnitude of the measure but also its temporal stability and dynamics. Second, the intervention's effects can be nonlinearly influenced by dose, exposure, and specificity.

hypotheses (Antal et al. 2024; Stephan et al. 2010). This is critical for interpreting the study findings and revealing competing hypotheses that the current data cannot distinguish between.

Time series data are crucial for evaluating competing hypotheses in the metabolic-neuronal system, as they allow for distinguishing between different points of metabolic circuit failure (the mediators) and assessing if an outcome temporally follows its proposed cause (Mujica-Parodi and Strey 2020). This entails tracking potential mediating features (e.g., AMPK activity, insulin levels) and outcomes (e.g., functional neuroimaging, blood glucose levels) over time. Tools like dynamic causal modeling (DCM) can then be used, in conjunction with the mediator model, to help distinguish between driving and modulating mediators of the measured outcomes (Hofmann et al. 2024; Stephan et al. 2010). By adopting this temporal framework, researchers can elucidate multiphasic responses, highlight adaptation mechanisms, and account for state-dependent factors like sleep and diet. For example, the mediator effect of an acute hypoxic stressor may initially peak in measures such as AMPK activity and blood glucose levels before potentially declining with system adaptation (Burtscher et al. 2024). Meanwhile, the outcome effect, reflecting changes in brain function and cognitive performance, follows its own temporal dynamics, potentially stabilizing or improving post-intervention due to sustained adaptations. Establishing and refining this framework is essential for interpreting the complex dynamical signatures unique to different interventions within the system.

To utilize the mediator model to interpret and inform clinical studies, we first need to collect the appropriate data to train the mediator-outcome relationships. Phase I proof-of-concept studies are crucial for this process. Their primary goal is to test specific hypotheses about the mechanistic effects of interventions in humans while gathering preliminary data on safety, efficacy, and biological activity. Within the mediator framework, these interventions can also be seen as perturbations that alter the metabolic state and dynamic response of an individual. For example, one dynamic perturbation could be an acute, repeatable hypoxic intervention, characterized by brief periods of reduced oxygen availability, typically lasting from seconds to a few minutes (Ehrenreich et al. 2023). The primary aim is to induce a controlled hypoxic state that stimulates physiological and neurobiological responses without causing harm. These short, intermittent bouts of hypoxia, however, can also be used to generate a unique response function in mediating metabolic circuits, thus providing general constraints on the internal structure of the mediator model. The "outcomes" can therefore be both static and dynamic measures. Furthermore, the specificity of preclinical interventions, such as the precise regulation of oxygen levels, allows researchers to investigate dose-response relationships and the temporal dynamics of targeted mediator components. Thus, the outcome can be modeled using mediator-specific parameters such as the intensity of hypoxia (e.g., the percentage of oxygen in the inhaled air), duration of each episode, and the frequency of these episodes.

The mediator framework provides a conceptual map for assessing the relative influence of the seven core metabolic circuit classes on neuropsychiatric outcomes. However, these components are deeply co-regulated and cannot easily be isolated in experimental settings. Moreover, the outcomes and interactions between the mediators are influenced by nonspecific "moderator" variables such as sex and age. For instance, sex-specific hormonal differences can affect metabolic responses to hypoxia, with evidence suggesting that estrogen modulates oxidative stress and mitochondrial function differently in males and females (Pellegrino et al. 2022). Similarly, older adults may exhibit reduced mitochondrial biogenesis and antioxidant defenses, potentially diminishing the cognitive benefits of hypoxic interventions (Militello et al. 2024). To dissect these competing mediators and their co-regulatory factors, two additional considerations are critical. First, at least one representative component from each circuit class should be measured to assess which mediators could potentially provide alternative explanations for the effect of an intervention. Second, as will be discussed next, these variables should be measured dynamically when possible to provide evidence of an alternative causality that could be explored through future research.

4 Biomarkers: What Do We Measure?

A biomarker is a measurable indicator of some biological condition or disorder. Although the term is used throughout modern medical research and clinical care, searching for biomarkers has been particularly urgent in psychiatry, where biological indicators that do not rely on self-reports and symptoms are still notably lacking (Abi-Dargham et al. 2023). In metabolic neuropsychiatry, the biomarker search is more constrained and targeted to processes we understand better than in the rest of psychiatry. Therefore, using biomarkers to identify treatment targets and monitor treatment impacts is a realistic near-term goal.

We can identify three types of potential biomarkers: *diagnostic* (for clinical diagnoses), *prognostic* (to predict the future clinical trajectory of a diagnosed disease), and *theranostic* (to predict whether and/or how a particular therapeutic will alter the disease's clinical trajectory). In Appendix 14.1, we present an extensive list of possible biomarkers of potential use in metabolic neuropsychiatry. Here, we discuss broad metabolic systems that are of interest to the field. Then, within each system, we specify the measurable target biomarker, the tissue or source of the measurement, as well as the timescale over which the biomarker is expected to change. This table (Appendix 14.1) may serve as a useful tool for investigators designing studies in metabolic neuropsychiatry who wish to explore the full range of metabolic processes their work may impact.

Since metabolic processes consist of a complex network of pathways in constant mutual interaction, treatment research in metabolic neuropsychiatry will be well-served by avoiding a narrow focus on single biomarkers. Measuring multiple processes will not only provide convergent evidence for whether the desired changes are occurring, but also reveal unexpected changes in other parts of the complex network. This could include, for example, an unexpected change in certain biomarkers, no change in biomarkers is expected (perhaps because they are too important and therefore maintained within a narrow range), or change occurs in an unexpected direction in some biomarkers (because of overcompensation and nonlinear interactions).

In addition to measuring multiple biomarkers in intervention studies, it is imperative to collect time series data in these studies, as described earlier. To emphasize this point, we added a column in Appendix 14.1 to indicate the rate of change. In complex networks, the timescale for change can vary widely, depending on the relationships between different parameters and pathways, from seconds to years. By collecting time series data on the appropriate scale, investigators can make an informed decision about the impact of treatment and timing of outcome measurements.

Finally, there are many additional considerations that can constrain the choice of biomarker. These include specificity of the intervention being studied, scale of the study (e.g., sample size, number of measures), the trade-off between accuracy and measurement error in a specific study implementation, cost of required measurements, the need for specialized staff, equipment, as well as material. Appendix 14.1 does not address each of these issues but rather provides a starting point for investigators to consider their choice of biomarker.

5 Intervention: How Do We Perturb the System to Target Specific Points of Failure?

In this section we review potential avenues for perturbing and intervening on metabolic pathways to target specific points of failure. Candidate pharmaceuticals used in other metabolic disorders are well-positioned to be repurposed for metabolic interventions in psychiatry. Although metabolic systems and targets will continue to evolve in this rapidly expanding field, we review candidate interventions by metabolic targets, considering that interventions may have multiple effects on interconnected metabolic pathways. In Appendix 14.2, we list modulators of major metabolic circuits to provide potential candidates for modulation of metabolic pathways and their mechanism of action. Depending on known mechanisms of action, interventions may have more targeted or specific effects on the metabolic system. Candidate modulators include a range of pharmaceutical compounds, nutritional supplements, heat or cold therapy, environmental manipulations (hypoxia), and lifestyle interventions. By leveraging known metabolic interventions, most candidate interventions highlighted in Appendix 14.1 have a well-studied safety profile. However, known adverse effects must be considered for each intervention.

Several factors in the administration of candidate metabolic interventions will affect the design and interpretation of the intervention. Here, we consider the mode of administration, timing of the intervention, and the possibility of combining candidate interventions.

The mode of administration may have an impact on the onset of action, ease of use, and targeted tissues. Different modes of administration into metabolic systems include oral, intravenous, and intranasal administration. Interventions delivered intranasally, such as insulin, offer the possibility of probing CNS response to insulin (e.g., brain insulin resistance). Insulin administered in the nasal cavity is thought to bypass the blood-brain barrier, with delivery to the brain following olfactory and trigeminal pathways (Thorne et al. 2004). This allows for the measure of biomarkers and outcomes in response to a central manipulation. In addition to mode of administration, dosing of metabolic manipulations should be carefully considered with dose-response studies.

Perturbations may achieve similar targeted metabolic regulation by different means, which may alter the system effects of the intervention. Recently, there has been interest in targeting mitochondrial regulation by using alternative fuel sources for the brain, mainly ketogenic interventions (Bernard et al. 2025; Ozan et al. 2024). The metabolic state of ketosis can be achieved by administration of exogenous ketones through a ketogenic diet, or by fasting. However, dietary lifestyle changes and administration of nutritional supplements may exert different effects on the metabolic system and pathways.

Some metabolic interventions may be used for single intervention or repeated interventions in a short time frame to probe mechanistic effects, whereas others may be more useful for longer-term interventions that assess efficacy on psychiatric, cognitive, or functional outcomes. Examples of short-term interventions designed for rapid administration and repeated use include exogenous ketone bodies, oral glucose, and intranasal insulin. Limitations of short-term probes include their inability to predict longer-term outcomes. While lifestyle interventions, such as diet and exercise, can be useful acute interventions, they are best suited for longer-term trials to test efficacy of the intervention on psychiatric or cognitive outcomes. Of note, dietary interventions can manipulate macronutrients and micronutrients, caloric intake, and fasting periods. Similarly, exercise interventions can manipulate different factors, including low- or high-intensity exercise and resistance training, among others (Pajonk et al. 2010).

Finally, combining various interventions could enhance therapeutic outcomes by targeting multiple pathways simultaneously. This may involve combining interventions with different known mechanisms of action and different types of interventions, such as combining pharmacologic or nutritional supplementation interventions alongside lifestyle modifications (e.g., dietary changes or exercise). Such combinations would enable the quantification of potential synergistic effects, the evaluation of the relative efficacy of individual interventions (e.g., comparing ketogenic diets to exercise), and the exploration of alternative mechanisms through which these interventions may exert their effects, such as through simple weight loss. While the list of candidate metabolic interventions should continue to expand and evolve, considering the mechanisms of action and metabolic targets for an intervention sets the stage for developing study designs of metabolic treatments in neuropsychiatric disorders.

6 Extending the Impact: How to Design Better Clinical Trials in Metabolic Neuropsychiatry

The previous sections outlined a conceptual and quantitative framework for metabolic neuropsychiatry; what follows is a strategy for its practical implementation. Key issues essential to the design and success of clinical trials in this emerging field are identified. By addressing these issues, we aim to establish a roadmap for developing rigorous metabolic neuropsychiatry clinical trials that can effectively evaluate the impact of therapeutic interventions on mental health. This multidisciplinary approach, which incorporates elements ranging from mathematical modeling to clinical and experimental research, will accelerate the translation of findings into real-world applications, ultimately improving patient care and advancing the field.

6.1 Study Population and Stratification

The choice of study population is critical to the success of a clinical trial. Selecting appropriate inclusion and exclusion criteria ensures that the findings are both externally valid and generalizable. For metabolic neuropsychiatry trials, it is essential to decide whether to focus on broad patient populations (e.g., all individuals with major depressive disorder) or to narrow the focus to more specific subgroups (e.g., individuals with treatment-resistant depression or who exhibit insulin resistance). Enriching the study population with individuals who are more likely to benefit from the intervention can increase the likelihood of detecting meaningful effects. However, it is also important to be wary of self-selection bias, particularly in lifestyle intervention trials where participants may need to change behaviors significantly. Safety considerations must also be paramount, excluding individuals with conditions that could be exacerbated by the intervention, such as active eating disorders or severe metabolic imbalances. Stratified randomization should be employed to ensure balance between treatment groups, especially in smaller trials. Stratification can be based on prognostic factors known to affect outcomes, such as body mass index, age, sex, or baseline metabolic health. By ensuring these factors are evenly distributed, researchers can minimize confounding variables and better isolate the effects of the intervention. Stratification is also useful when there are specific subgroups of interest (e.g., patients with the target diagnosis but no comorbidities), facilitating adequate sample sizes within these groups for more detailed analysis.

6.2 Blinding and Control Conditions

Blinding is a crucial component of clinical trial design, particularly in neuropsychiatric trials where placebo effects can be substantial. While it is often challenging to blind participants in lifestyle intervention studies, blinding evaluators can help mitigate bias. For example, in dietary intervention studies, although participants know their diet changes, outcome assessors can be kept unaware of the group assignments to reduce bias in their assessments. Designing appropriate control conditions is equally important. Control interventions should be active and match the experimental intervention in key aspects, such as caloric intake or social interaction. For example, in a trial testing a ketogenic diet, the control diet should be similarly structured but without the specific macronutrient composition that induces ketosis. As metabolic interventions affect healthy individuals as well, it is imperative to include subjects with and without the intervention, as well as both with and without the targeted disease. This approach helps isolate the specific metabolic effects of complex interventions, such as the ketogenic diet, from other nonspecific factors.

6.3 Intervention Design

The design of the intervention must strike a balance between efficacy and feasibility. The duration and dose of the intervention should be optimized to achieve the best possible outcomes while minimizing the burden on participants. For dietary interventions, this might involve specifying macronutrient compositions, caloric restrictions, and fasting periods. It is crucial to provide flexibility and fallback options, such as alternative dietary recommendations, to accommodate participants who may struggle to adhere strictly to the intervention protocols. Pharmacological interventions should be chosen based on their mechanisms of action and potential to target specific metabolic pathways. For instance, intranasal insulin can be used to probe central insulin resistance, whereas GLP-1 receptor agonists may be repurposed from diabetes treatment to address metabolic dysregulation in psychiatric disorders (Hanssen et al. 2023). Combining different types of interventions, such as pharmacological and lifestyle changes, can also be explored to enhance therapeutic outcomes by targeting multiple pathways simultaneously.

6.4 Adherence Strategies

Ensuring adherence to lifestyle interventions is challenging but critical for the success of clinical trials. Regular consultations with clinicians and dietitians can help participants understand the importance of the intervention and provide continuous support. Offering specific food recommendations and easy-to-prepare recipes that cater to participants' preferences may increase compliance. Delivering food items directly to participants can further reduce barriers to adherence. Group settings and community support can also play a significant role in enhancing adherence. Participants may find motivation and encouragement in group settings where they can share experiences and support each other. For those randomized to control arms, offering a waitlist for the active intervention can maintain engagement and reduce dropout rates.

6.5 Monitoring and Outcomes

Continuous monitoring of relevant biomarkers is essential to ensure both adherence and safety in lifestyle intervention trials. Parameters such as blood glucose, lactate, and ketone bodies should be regularly measured. Weight and other anthropometric measures should also be tracked to monitor changes over time. Outcomes should include both mechanistic and clinical measures. Mechanistic outcomes might involve biomarkers directly impacted by the intervention, such as mitochondrial enzyme activities or oxidative stress markers. Clinical outcomes should assess the broader impacts on psychiatric symptoms, cognitive function, and overall health. Combining these outcome measures provides a comprehensive understanding of the intervention's effects and its potential mechanisms of action.

By considering these issues, a roadmap for the development of clinical trials in metabolic neuropsychiatry can be designed to investigate the impacts of various interventions on mental health effectively. This multidisciplinary approach utilizes elements from mathematical modeling to clinical and experimental research, and will enhance our ability to translate findings into real-world applications, ultimately improving patient care and advancing the field of metabolic neuropsychiatry.

7 Where Do We Go?

Integrating mathematical modeling and a control systems framework with clinical and experimental research from the outset provides a robust approach for understanding the complex mechanisms underlying metabolic interventions. It enables us to move beyond descriptive post hoc analyses, using models to dynamically guide data collection and test competing hypotheses. This is especially crucial for probing metabolic circuits, where therapeutic effects may be systemic and progress nonlinear, and be consistent with multiple equally plausible explanations. By elucidating the underlying mechanisms, we can intervene in a more informed and efficient manner to determine whether there are treatment benefits or drawbacks in targeting specific metabolic pathways versus more general interventions such as diet.

Using the example of a hypoxic intervention to examine brain function, the mediator model demonstrates how acute, repeatable hypoxic episodes influence cognitive performance and brain function through specific metabolic changes. Key factors (e.g., AMPK activity, blood glucose, lactate levels, mitochondrial enzyme activities, and ROS markers) offer insights into the biochemical and physiological responses to hypoxia. Tracking these factors over time helps delineate the pathways through which hypoxic stimuli exert their effects, revealing both immediate and long-term adaptations. This model underscores the importance of considering multiple timescales, state-dependent factors, and individual variability to fully capture the intervention's impact.

It is important, however, to emphasize that the field is new and evolving. This creates important challenges and opportunities based on the inherent complexity and heterogeneity of metabolic interventions. Different diseases and individual responses complicate the identification of universal biomarkers and intervention strategies. Adopting a dynamic systems perspective, however, allows for a more nuanced understanding of disease progression and intervention outcomes. In addition, more precise and effective interventions can be developed if collaboration is fostered among experts in mathematical modeling, metabolomics, clinical research, and other specialties. Such a multidisciplinary approach would advance experimental medicine and clinical trial design and enhance our ability to translate findings into real-world applications, with the ultimate goal of improving patient care in metabolic neuropsychiatry and beyond.

Appendix 14.1 Relevant analytes, their assessment of metabolic biomarkers and approximate time course. Based on general metabolic dynamics and response to physiological changes, rate of change is categorized as "Fast" (within minutes to hours), "Moderate" (hours to days), and "Slow" (days to weeks).

Metabolic System	Analyte	Tissue (Measure)	Rate of Change (in blood)
Cellular energy regulation	*mTOR/AMPK Pathway*		
	Phosphorylated mTOR (p-mTOR)		Moderate
	AMPK activity		Moderate
	Downstream targets (e.g., S6K, 4E-BP1)		Moderate
	Glycolytic Regulation		
	Glucose levels	Blood, CSF, brain (1H-MRS, PET)	Fast
	Lactate levels	Blood, CSF, brain (1H-MRS)	Fast
	Hexokinase and phosphofructokinase activity		Moderate

Metabolic System	Analyte	Tissue (Measure)	Rate of Change (in blood)
	Mitochondrial regulation		
	Mitochondrial DNA (mtDNA) copy number		Slow
	Mitochondrial enzyme activities (e.g., Complex I-IV)		Slow
	ATP levels		Fast
	TCA Cycle	Blood	
	Citrate, isocitrate, alpha-ketoglutarate, succinate		Moderate
	Fumarate, malate		
	Enzyme activities (e.g., citrate synthase, isocitrate dehydrogenase)		Moderate
	ROS (reactive oxygen species)		
	Malondialdehyde (MDA) and 4-hydroxynonenal (4-HNE)		Fast
	Superoxide dismutase (SOD) activity		Moderate
	Glutathione levels (GSH/GSSG ratio)		Moderate
Hormonal regulation	*Glucose/insulin*		
	Fasting blood glucose	Blood	Fast
	Insulin levels	Blood	Fast
	HbA1c	Blood	Slow
	Glucocorticoid		
	Cortisol levels	Blood, saliva, hair	Fast
	Corticosterone levels	Blood	Fast
	Thyroid		
	Thyroid-stimulating hormone (TSH)	Blood	Moderate
	Free T3 and free T4 levels	Blood	Moderate
Lipid metabolism regulation	*Lipid markers*		
	Triglycerides and cholesterol (Total, HDL, LDL)	Blood	Slow
	Free fatty acids	Blood	Moderate
	Peroxisome proliferator-activated receptors (PPARs)		
	Expression levels of PPARα, PPARγ		Slow
	PPAR target genes (e.g., ACOX1, CPT1)		Slow
Amino acid metabolism	*Regulation*		
	Plasma amino acid levels		Moderate
	Urea and ammonia levels		Moderate
Nutrient sensing	*Sirtuins*		
	SIRT1 activity or expression levels		Slow
	NAD+/NADH ratio		Moderate

Metabolic System	Analyte	Tissue (Measure)	Rate of Change (in blood)
	Insulin-like growth factor (IGF)		
	IGF-1 Levels		Moderate
Calcium homeostasis	*Calcium signaling molecules*		
	Serum calcium levels		Fast
	Parathyroid hormone (PTH) levels		Moderate
	Calcium-binding proteins (e.g., calmodulin)		Moderate
Insulin resistance panel	*Core biomarkers*		
	Fasting plasma glucose (FPG)		Fast
	Fasting insulin		Fast
	Hemoglobin A1c (HbA1c)		Slow
	Homeostasis model assessment of insulin resistance (HOMA-IR)		Moderate
	Additional biomarkers		
	C-Peptide		Moderate
	Adiponectin		Moderate
	Leptin		Moderate
	Resistin		Moderate
	Inflammatory Markers (CRP, TNF-α, IL-6)		Moderate
	Lipid Profile (Triglycerides, HDL, LDL)		Slow
	Emerging biomarkers		
	Fetuin-A		Slow
	Omentin		Slow
	Retinol-Binding Protein 4 (RBP4)		Moderate
	Visfatin		Moderate
Mitochondrial metabolic status	*Core biomarkers*		
	Lactate and pyruvate		Fast
	Creatine kinase (CK)		Fast
	Coenzyme Q10 (CoQ10)		Slow
	Carnitine and acylcarnitines		Moderate
	Adenosine triphosphate (ATP)		Fast
	Additional biomarkers		
	Fibroblast growth factor 21 (FGF21)		Slow
	Growth differentiation factor 15 (GDF15)		Slow
	NAD+		Moderate

Metabolic System	Analyte	Tissue (Measure)	Rate of Change (in blood)
	Inflammatory and oxidative stress markers		
	C-reactive protein (CRP)		Moderate
	8-Hydroxy-2'-deoxyguanosine (8-OHdG)		Moderate
	Malondialdehyde (MDA)		Fast
	SOD and GPx		Moderate
	Metabolomics and lipidomics markers		
	Amino acids, e.g., branched-chain amino acids (BCAAs)		Moderate
	Short-chain fatty acids (SCFAs)		Moderate
	Ceramides		Slow
	Genetic and epigenetic markers		
	mtDNA copy number		Slow
	mtDNA mutations and deletions		Slow
	Epigenetic modifications (e.g., DNA methylation of mitochondrial genes)		Slow
TCA cycle metabolic status	*Core biomarkers*		
	Citrate		Moderate
	α-Ketoglutarate (α-KG)		Moderate
	Succinate		Moderate
	Fumarate		Moderate
	Malate		Moderate
	Oxaloacetate		Moderate
	Enzymatic activity markers		
	Lactate Dehydrogenase (LDH)		Fast
	Isocitrate Dehydrogenase (IDH)		Moderate
	Malate Dehydrogenase (MDH)		Moderate
	Related metabolites		
	Pyruvate		Fast
	Acetyl-CoA		Moderate
	Inflammatory and oxidative stress markers		
	Reactive oxygen species (ROS)		Fast
	Antioxidant enzymes (e.g., SOD, GPx)		Moderate
	Hormonal and regulatory markers		
	Insulin		Fast
	Adiponectin		Moderate

Appendix 14.2 Interventions that are relatively system specific, acute experimental perturbations of the system; * denotes potential chronic perturbations.

System	Modulators	Mechanism of Action
mTOR/AMPK	Rapamycin	Inhibits mTORC1, reducing protein synthesis and cell growth
	Metformin*	Activates AMPK, inhibiting mTOR and enhancing insulin sensitivity
	PPAR*	Activates PPAR receptors, regulating gene expression involved in glucose and lipid metabolism
Glucose/insulin	Oral glucose	Increases blood glucose levels, stimulating insulin secretion
	Intranasal insulin	Directly delivers insulin to the brain
	Glucagon	Increases blood glucose levels by promoting glycogen breakdown in the liver
	Somatostatin	Inhibits the release of insulin and glucagon, regulating blood glucose levels
	Clamps (fix one probe)	Used to maintain a constant glucose level for precise insulin/glucose response measurement
Glycolysis	Hypoxia*	Induces anaerobic glycolysis by reducing oxygen availability, increasing lactate production
	Exercise* (high-intensity)	Enhances glucose uptake and utilization in muscles through increased AMPK activation
	Intensive cognitive training	May enhance brain glucose metabolism and neuroplasticity
Mitochondrial regulation/ TCA cycle	Ketones*	Provide an alternative energy source, reducing reliance on glucose
	Ketogenic diet*	Increases ketone production, enhancing mitochondrial function and energy efficiency
	MCT*	Provides a rapid source of ketones, bypassing traditional metabolic pathways
	Carb reduction	Reduces blood glucose levels, promoting ketone production and fat utilization
	Resveratrol	Activates sirtuins and AMPK, enhancing mitochondrial biogenesis and function
	CoQ10	Supports mitochondrial electron transport chain, improving ATP production
	GLP-1, PPAR, metformin*	GLP-1 enhances insulin secretion, PPAR regulates lipid metabolism, and Metformin activates AMPK
	Cold/heat	Activates thermogenic and stress response pathways, influencing mitochondrial function
	PDK inhibitors	Inhibit pyruvate dehydrogenase kinase, enhancing glucose oxidation and reducing lactate production
HPA/glucocorticoids	Social stress	Activates the HPA axis, increasing cortisol release
	Dexamethasone	Synthetic glucocorticoid that suppresses the HPA axis and reduces inflammation
	Metyrapone	Inhibits cortisol synthesis by blocking 11β-hydroxylase
Thyroid	T4*	Thyroxine (T4) is converted to the active form T3, regulating metabolism and energy expenditure

System	Modulators	Mechanism of Action
Radicals/redox	NR*	Nicotinamide riboside increases NAD+ levels, supporting mitochondrial function and reducing oxidative stress
	NAC*	N-acetylcysteine replenishes glutathione, reducing oxidative stress and improving redox balance
General	Anti-inflammatory (TNFα, IL-1β, IL-6, COX1,2)	Reduce inflammation by inhibiting pro-inflammatory cytokines and enzymes
	Sex hormones	Estrogen and testosterone modulate various metabolic and neuroprotective pathways
	Estrogen receptor modulators	Modulate estrogen receptors, influencing gene expression and metabolic processes
Autonomic nervous system (ANS)	ß-blockers	Block β-adrenergic receptors, reducing heart rate and blood pressure
	Mimetics	Mimic neurotransmitters to modulate sympathetic and parasympathetic nervous system activity

References

Abi-Dargham A, Moeller S, Ali F, et al. (2023) Candidate Biomarkers in Psychiatric Disorders: State of the Field. World Psychiatry 22 (2):236–262. https://doi.org/10.1002/wps.21078

Alon U (2019) An Introduction to Systems Biology: Design Principles of Biological Circuits. CRC Computational Biology, 2 edn. CRC Press, Boca Raton

Antal BB, Chesebro AG, Strey HH, Mujica-Parodi LR, and Weistuch C (2024) Achieving Occam's Razor: Deep Learning for Optimal Model Reduction. PLoS Comput Biol 20 (7):e1012283. https://doi.org/10.1371/journal.pcbi.1012283

Antal BB, van Nieuwenhuizen H, Chesebro AG, et al. (2025) Brain Aging Shows Nonlinear Transitions, Suggesting a Midlife "Critical Window" for Metabolic Intervention. PNAS 122 (10):e2416433122. https://doi.org/10.1073/pnas.2416433122

Bernard J, Tamouza R, Godin O, et al. (2025) Mitochondria at the Crossroad of Dysregulated Inflammatory and Metabolic Processes in Bipolar Disorders. Brain Behav Immun 123:456–465. https://doi.org/10.1016/j.bbi.2024.10.008

Boucherie D, Reneman L, Ruhé H, and Schrantee A (2023) Neurometabolite Changes in Response to Antidepressant Medication: A Systematic Review of ^{1}H-MRS Findings. Neuroimage Clin 40:103517. https://doi.org/10.1016/j.nicl.2023.103517

Bremner JD, Vythilingam M, Vermetten E, et al. (2003) Cortisol Response to a Cognitive Stress Challenge in Posttraumatic Stress Disorder (PTSD) Related to Childhood Abuse. Psychoneuroendocrinology 28 (6):733–750. https://doi.org/10.1016/S0306-4530(02)00067-7

Burtscher J, Citherlet T, Camacho-Cardenosa A, et al. (2024) Mechanisms Underlying the Health Benefits of Intermittent Hypoxia Conditioning. J Physiol 602 (21):5757–5783. https://doi.org/10.1113/jp285230

Chesebro AG, Antal BB, Weistuch C, and Mujica-Parodi LR (2025) Challenges and Frontiers in Computational Metabolic Psychiatry. Biol Psychiatry Cogn Neurosci Neuroimaging 10 (3):258–266. https://doi.org/10.1016/j.bpsc.2024.10.011

DiNuzzo M, Dienel GA, Behar KL, et al. (2024) Neurovascular Coupling Is Optimized to Compensate for the Increase in Proton Production from Nonoxidative Glycolysis and Glycogenolysis during Brain Activation and Maintain Homeostasis of pH, pCO_2, and pO_2. J Neurochem 168 (5):632–662. https://doi.org/10.1111/jnc.15839

Dwir D, Khadimallah I, Xin L, et al. (2023) Redox and Immune Signaling in Schizophrenia: New Therapeutic Potential. Int J Neuropsychopharmacol 26 (5):309–321. https://doi.org/10.1093/ijnp/pyad012

Ehrenreich H, Gassmann M, Poustka L, et al. (2023) Exploiting Moderate Hypoxia to Benefit Patients with Brain Disease: Molecular Mechanisms and Translational Research in Progress. Neuroprotection 1 (1):55–65. https://doi.org/10.1002/nep3.15

Hanssen R, Rigoux L, Kuzmanovic B, et al. (2023) Liraglutide Restores Impaired Associative Learning in Individuals with Obesity. Nat Metab 5 (8):1352–1363. https://doi.org/10.1038/s42255-023-00859-y

Hofmann D, Chesebro AG, Rackauckas C, et al. (2024) Increasing Spectral DCM Flexibility and Speed by Leveraging Julia's Modelingtoolkit and Automated Differentiation. https://doi.org/10.1101/2023.10.27.564407

Karin O, Swisa A, Glaser B, Dor Y, and Alon U (2016) Dynamical Compensation in Physiological Circuits. Mol Syst Biol 12 (11):886. https://doi.org/10.15252/msb.20167216

Mackey MC, and Glass L (1977) Oscillation and Chaos in Physiological Control Systems. Science 197 (4300):287–289. https://doi.org/10.1126/science.267326

McEwen BS, and Wingfield JC (2003) The Concept of Allostasis in Biology and Biomedicine. Horm Behav 43 (1):2–15. https://doi.org/10.1016/S0018-506X(02)00024-7

Merritt K, McGuire P, Egerton A, et al. (2021) Association of Age, Antipsychotic Medication, and Symptom Severity in Schizophrenia with Proton Magnetic Resonance Spectroscopy Brain Glutamate Level: A Mega-Analysis of Individual Participant-Level Data. JAMA Psychiatry 78 (6):667–681. https://doi.org/10.1001/jamapsychiatry.2021.0380

Militello R, Luti S, Gamberi T, et al. (2024) Physical Activity and Oxidative Stress in Aging. Antioxidants 13 (5):557. https://doi.org/10.3390/antiox13050557

Mujica-Parodi LR, and Strey HH (2020) Making Sense of Computational Psychiatry. Int J Neuropsychopharmacol 23 (5):339–347. https://doi.org/10.1093/ijnp/pyaa013

Ozan E, Chouinard V-A, and Palmer CM (2024) The Ketogenic Diet as a Treatment for Mood Disorders. Curr Treat Options Psychiatry 11 (3):163–176. https://doi.org/10.1007/s40501-024-00322-z

Pajonk F-G, Wobrock T, Gruber O, et al. (2010) Hippocampal Plasticity in Response to Exercise in Schizophrenia. Arch Gen Psychiatry 67 (2):133–143. https://doi.org/10.1001/archgenpsychiatry.2009.193

Pellegrino A, Tiidus PM, and Vandenboom R (2022) Mechanisms of Estrogen Influence on Skeletal Muscle: Mass, Regeneration, and Mitochondrial Function. Sports Med 52:2853–2869. https://doi.org/10.1007/s40279-022-01733-9

Stephan KE, Penny WD, Moran RJ, et al. (2010) Ten Simple Rules for Dynamic Causal Modeling. Neuroimage 49 (4):3099–3109. https://doi.org/10.1016/j.neuroimage.2009.11.015

Thorne RG, Pronk GJ, Padmanabhan V, and Frey WH, 2nd (2004) Delivery of Insulin-Like Growth Factor-I to the Rat Brain and Spinal Cord Along Olfactory and Trigeminal Pathways Following Intranasal Administration. Neuroscience 127 (2):481–496. https://doi.org/10.1016/j.neuroscience.2004.05.029

van Nieuwenhuizen H, Chesebro AG, Polizu C, et al. (2024) Ketosis Regulates K^+ Ion Channels, Strengthening Brain-Wide Signaling Disrupted by Age. Imaging Neurosci 2:10.1162. https://doi.org/10.1162/imag_a_00163

Yehuda R (2002) Post-Traumatic Stress Disorder. N Engl J Med 346 (2):108–114. https://doi.org/10.1056/nejmra012941

Yüksel C, and Öngür D (2010) Magnetic Resonance Spectroscopy Studies of Glutamate-Related Abnormalities in Mood Disorders. Biol Psychiatry 68 (9):785–794. https://doi.org/10.1016/j.biopsych.2010.06.016

Open Access This chapter is licensed under the terms of the Creative Commons Attribution-NonCommercial-NoDerivatives 4.0 International License (http://creativecommons.org/licenses/by-nc-nd/4.0/), which permits any noncommercial use, sharing, distribution and reproduction in any medium or format, as long as you give appropriate credit to the original author(s) and the source, provide a link to the Creative Commons license and indicate if you modified the licensed material. You do not have permission under this license to share adapted material derived from this chapter or parts of it.

The images or other third party material in this chapter are included in the chapter's Creative Commons license, unless indicated otherwise in a credit line to the material. If material is not included in the chapter's Creative Commons license and your intended use is not permitted by statutory regulation or exceeds the permitted use, you will need to obtain permission directly from the copyright holder.

Subject Index

A
acidosis 38
adenosine triphosphate (ATP) 34–36, 41, 60–62, 67, 71, 83–85, 98, 100, 113–116, 143, 160, 226
  production 64, 88, 91–94, 172
  synthase 68, 140, 141, 166, 234
  synthesis 3, 168
adolescence 47, 57, 63, 70, 71, 231
aerobic glycolysis 21, 22, 47, 64, 160
aging 34, 57, 70, 121, 123, 124, 138, 142, 144, 239
allostatic load 31–34, 45, 46, 57–64, 164
  chronic high 65–66
  maternal 227
Alzheimer disease (AD) 21, 39, 127, 157, 174, 239
  genetic risk 20, 136, 141, 237
  hypometabolism 234, 238
  imaging 18, 22
  ketogenic diet 235
  mitochondrial dysfunction 135–153
  Seattle Alzheimer's Disease Cell Atlas 21
AMP-activated protein kinase (AMPK) 113–116, 128, 249, 307, 317, 321
  signaling 39, 111
amyloid-beta (Aβ) plaques 20, 135, 142, 145, 237
amyloid precursor protein (APP) 136–138, 142
  mouse model 139
anhedonia 100, 174, 196, 198, 203, 206–208, 211, 212, 245, 253
anorexia nervosa 68, 69, 235
anxiety 67, 157, 162, 173, 174, 196, 213, 251, 265
  hormonal disturbance 229–231
  Netherlands Study of Depression and Anxiety 200
  obesity 161
APOE4 136–141, 237
apoptosis 64, 67, 226, 229, 232, 250
astrocyte-neuron lactate shuttle (ANLS) 93, 94, 116, 128, 172
attention-deficit/hyperactivity disorder 162, 264
autism 3, 86, 235, 251, 264
Avon Longitudinal Study 204

B
Bicycle framework 285, 297, 299
bifurcations 122–124, 129
big data neuroscience 15–28
bioenergetic dysfunction 2–5, 64, 83, 89, 91, 98–100, 135, 144
bioenergetic function 83, 87, 90, 97–98, 143
biomarkers 1, 99, 124, 127, 166–168, 174, 196, 228, 233, 265, 280, 305
  assessing 272, 311, 312, 317
  chronic inflammation 201–202
  developing 96, 239
  fluid 20, 22
  immuno-metabolic 199, 203, 207
bipolar disorder 63, 83–86, 97, 99, 157, 165, 196, 200, 212, 234, 245, 247, 265, 305
  bioenergetic dysfunction 3–4
  dysglycemia 248
  genetic risk 164, 165
  ketogenic diet 235, 276
  mitochondrial dysfunction 84, 88, 89, 160
blood-brain barrier (BBB) 4, 33, 36, 37, 39, 40, 64, 65, 168, 206, 232, 234, 313–322
blood cell bioenergetic profiling 142
blood-oxygen-level-dependent (BOLD) 2, 33, 294
brain energy supply. *See* energy budget
brain pH 35–38, 60–63, 71, 84, 93–95, 166
BrainSpan dataset 21, 22

C
cable theory 122, 123
cancer 4, 69, 112, 145, 157, 226
cardiovascular disease 1, 65, 69, 155–158, 196, 212, 226, 231, 233
  in schizophrenia 65, 66
  risk 202, 236, 269
causality 200, 206, 253, 266, 285–304, 311
cerebral blood flow (CBF) 16, 17, 37, 38, 40, 45–47, 62, 234
cerebral metabolic rate of oxygen ($CMRO_2$) 35–38, 40, 45–47
cholesterol 21, 91, 158, 160, 172, 200, 203, 237

© Frankfurt Institute for Advanced Studies (FIAS) 2025

D. Öngür, J. M. Ford (eds.), *Metabolic Neuropsychiatry*, Strüngmann Forum Reports,
https://doi.org/10.1007/978-3-032-05630-6

clinical trial design 314–316. *See also* randomized control trials
CONSORT guide 267
cognitive decline 127, 135, 144, 173, 174, 225, 238, 245, 253
ketogenic diet 264
computational modeling 122–128, 285, 297–299, 305
control circuits 5, 111–116, 119, 124, 128, 197, 307–309
cortisol 63–65, 91, 115, 251, 309
C-reactive protein 199, 203, 211, 320
criticality 111, 120–123, 127, 129
curse of dimensionality 292–294

D

dementia 33, 112, 113, 175, 236–239. *See also* Alzheimer disease
depression 115, 171–173, 176, 201–206, 226, 234, 265, 305. *See also* major depressive disorder
Australian Genetics Depression Study 204
bipolar 209, 210, 248
genetic basis 199–201
hormonal disturbance 229–231
inflammation 195–214, 234
ketogenic diet 235, 264
Netherlands Study of Depression and Anxiety (NESDA) 200
obesity 161
unipolar 60, 61
development 34, 47, 57, 63, 70, 71, 223, 231
adolescence 47, 57, 63, 70, 231
critical periods 45–47, 59, 63, 69–71, 225–227, 230–233
early-life 162, 164, 171, 198, 226, 227, 234, 253
diabetes 40, 119, 157, 158, 163, 196, 201, 210, 212, 213, 236, 264, 308
risk for 87–88
type 2 69, 99, 208, 213, 224, 226, 236, 238, 239, 248, 269
diet 16, 69, 85, 91, 144, 147, 162, 174, 176, 224–228, 251, 264, 265, 270, 272, 279, 310
ketogenic 1, 4, 7, 23, 99, 121, 235, 236, 251, 263–265, 271, 272, 276, 313–323
Directed Acyclical Graph (DAG) 289–291
dopamine 65, 68, 69, 116, 224, 229, 233, 249, 287
dynamic causal modeling (DCM) 285, 291, 294–298, 310
dysglycemia 1, 92, 159, 245, 248
dyslipidemia 158, 205, 210, 230, 253

E

early-life adversity 162, 164, 171, 198, 226, 227, 234, 253
eating disorders 68, 157, 161, 173, 238, 314–323
risk 226–228
electron transport chain (ETC) 64, 67, 136–141, 166, 168–171
complex I 67, 88, 138–140, 165, 166, 213
complex II 139, 142
complex III 139, 140
complex IV 136, 140, 142, 170
complex V 140, 141, 166
energy budget 3–11, 31–36, 41, 45–47, 60–62, 67–69, 162, 197
model 32
energy metabolism 2–4, 6, 21, 33, 68, 167, 247, 286
impaired 286
measuring 19, 70
regulating 63, 115, 287, 288
whole-body 6, 29, 31, 33, 45, 155, 157
environmental stressors 70, 162, 171, 226–228, 251. *See also* allostatic load
epilepsy 100, 112, 113, 121
ketogenic diet 7, 272, 276
Estonian Biobank 204
estrogen 64, 65, 116, 228, 229, 311, 322
deficiency 231, 232, 250
euglycemia 37, 115
excitatory-inhibitory balance 59, 115, 127
exercise 34, 35, 66–69, 91, 116, 144, 147, 228, 264, 265, 272, 313–323
extracellular vesicles 142, 145–146

F

fatigue 162, 171, 174, 198, 203–206, 208, 210
fatty acids 85, 90, 95–98, 141, 200, 202, 234, 237, 264, 318, 320
fibroblasts 87, 143
fluid biomarkers 20, 22
functional models 289, 290

future research 2, 7–11, 22, 98–100, 125–128, 172, 211, 227–228, 232, 235, 236, 238–240, 253–255, 266, 298–300, 307, 311

G

genetic risk 20, 84, 86–88, 98, 100, 136, 141, 163–167, 173, 199, 205, 237
genomic structural equation models (GenomicSEM) 204, 205
glial cells 2, 3, 21, 36, 41, 115, 116
GLP-1 agonists 23, 161, 197–198, 206, 213, 239, 247, 252, 269, 315
glucagon 113, 115, 161, 197, 224, 234, 239, 247, 269, 321
gluconeogenesis 90, 115
glucose 2, 16–19, 29, 33, 34, 47, 71, 83, 113–117, 224, 249–251, 321
  extraction 36–40
  fasting levels 87, 91, 159, 200
  fates of 85, 91–93
  hypometabolism 234
  intolerance 229, 248
  metabolism 43, 61, 68, 84, 90, 93, 98–100, 127, 136, 155, 159–161, 164, 166, 172, 234, 248
  regulating 92, 308, 316, 317
  transport 37–39, 45, 60, 63, 64, 115, 119, 129, 163, 233
glutaminolysis 63, 100, 167
glycerol-3-phosphate dehydrogenase (GpDH) 141–142
glycocholic acid 146–147
glycolysis 3, 19–22, 37, 90–93, 137, 159, 160, 164, 167, 172, 307
  aerobic 21, 22, 47, 64, 120, 160
  hyper- 63, 100, 167
  non-oxidative 34, 38, 43, 63
Granger causality 285, 289, 291–292, 295, 298
growth differentiation factor 15 (GDF15) 162, 252, 319
gut microbiome dysbiosis 157, 174

H

heterochronic parabiosis 144
homeostasis 29, 41–45, 47, 57–61, 65, 66, 115, 139, 195–198, 208, 211
  control circuits 112, 113, 119, 124, 307–309
  glucose 33, 91, 98, 113, 248
  glucose-insulin 223, 230, 232–236, 240
  oxygen 33, 36
  regulation 112, 113, 203, 206–208, 308
  restoring 67–68, 121, 163, 308
HPA axis 68, 162, 230, 234, 251, 287, 288, 307, 321
hyperglycemia 40, 162, 233, 237, 252
hyperphagia 204, 206, 207
hypoglycemia 37, 91, 115, 248
  postprandial 163, 308
hypokalemia 45, 66

I

immune system 7, 31, 155, 157, 162, 163, 173, 195, 197, 209, 223, 227, 233–235, 240, 245, 248, 250
inflammation 65, 98, 139, 161–164, 174, 195–214, 224, 229, 250, 253, 265, 307, 321
  hyperinsulinemia 233, 234
  maternal 226
  neuro- 7, 166–168
insulin 63, 163, 197, 206, 207, 224, 248–250, 252, 320, 321
  coma therapy 91
  fasting 159, 251, 319
  intranasal 315
  resistance 65, 87, 91, 113, 119, 123, 158, 174, 198–201, 205, 225–229, 234–237, 248, 253–255, 280, 319
  sensitivity 68, 87, 113, 115, 199, 230, 231, 308, 321
  signaling 7, 83–85, 90–91, 98, 159, 232, 248, 307
interleukin-6 (IL-6) 91, 197–200, 207, 319, 322
interventions 314–316. *See also* exercise, ketogenic diet
  designing 70, 268–270, 305, 315
  dietary 85, 174, 236, 264, 272, 273, 279, 313
  personalized 5, 98, 112, 128, 155, 161, 175, 252, 255, 264, 279
  points of failure 312–314

K

ketamine 61, 127
ketogenesis 127, 235, 238

ketogenic diet 1, 4, 7, 23, 99, 121, 235, 236, 251, 263–265, 271, 272, 276, 307, 313–323
ketone bodies 37, 64, 129, 235, 313–323
Kir channels 43–45

L
lactate 34, 37, 44, 60–62, 64, 71, 84–86, 92–95, 98, 100, 116–118, 160, 319
  dehydrogenase B (LDHB) 20, 160
  homeostasis 63, 66
  monitoring 316, 317
  production 60, 120, 172, 321
  in schizophrenia 62, 63, 94, 166
Leigh syndrome 169
linoleic acid 146–147
lipids 85, 95–98, 199, 202, 234, 254, 280, 310, 318–321
  dysregulation 158, 174, 205, 210, 230, 237–239, 253
  extracellular 146–147
  synthesis 160
  TREM2 141, 237
Local Average Treatment Effect 287

M
major depressive disorder (MDD) 83, 84–86, 97, 99, 157, 166, 173, 196, 200, 202–210, 285, 287, 288
  in adolescents 96
  bioenergy abnormalities 3–4
  genetic risk 164
  Glu levels 309
  obesity in 162
  role of trauma 163
mania 63, 115, 167
mediator model 307, 309–311, 317
menopause 228–229, 231
metabolic control circuits. *See* control circuits
metabolic dysregulation 111, 112, 118–120, 128, 155, 158, 163, 237, 250, 252, 255, 315
  early detection 236, 309
  measuring 124–125
metabolic syndrome 4, 5, 29, 157, 173, 175, 199, 231
mitochondrial biogenesis 113, 116, 129, 235, 311
mitochondrial cascade hypothesis 136, 137, 146
mitochondrial dysfunction 92, 98, 115, 158, 166, 169, 170, 172, 174, 200, 214, 234, 236, 237
  Alzheimer disease 135–153
  bipolar disorder 84, 88, 89, 160
  schizophrenia 89
mitochondrial function 2, 21, 65, 83, 87, 111, 113–115, 155, 158, 169, 202, 228, 240, 265, 321, 322
  Alzheimer disease 136–138, 143
monocarboxylate transporters (MCT) 37, 39, 93, 321
mood disorders 4, 83–86, 100, 173, 213, 230, 251, 252
mTOR 98, 129, 249, 307, 317, 321
  AMPK regulation 111, 113–115
multiple sclerosis 145, 165, 239, 270

N
neural mass models 123, 125, 126, 129
neural synchrony 121, 123–125, 167, 171, 292
neurogenesis 3, 64, 144
neuroinflammation 7, 166–168
neurometabolic coupling (NMC) 37, 38, 41, 60–62
neuronal signaling 31–33, 35, 41, 43, 45, 66
neuroplasticity 3, 90, 99, 100, 321
neurovascular coupling (NVC) 33, 37, 38, 40, 41, 62
non-oxidative glycolysis 34, 38, 43, 63
noradrenaline 224, 233
norepinephrine 287

O
obesity 158, 161, 162, 198, 201, 204–207, 210, 212, 213, 229, 239, 264, 269
  global prevalence 69
  maternal 226–228
obsessive-compulsive disorder 157
osteoporosis 231
oxidative phosphorylation (OXPHOS) 3, 19, 63, 64, 67, 83, 88, 92, 93, 100, 137, 142, 166–171, 227
oxidative stress 40, 88, 92, 139, 140, 145, 155, 166–168, 171, 229, 234, 237, 307, 320, 322

role of estrogen 250, 311
oxygen 2, 29, 33, 38, 47, 65, 247
extraction 34–37
hyperoxia 40
measuring 17–19, 71
reduced 62, 307, 311
transport 39–41, 45, 46, 60

P

Parkinson disease 68, 127, 145, 157, 169, 213
parvalbumin-positive interneurons (PVIs) 167, 171
peripheral blood mononuclear cells (PBMCs) 86, 98, 159–160, 166
personality disorders 70, 157, 173
PET imaging 17–20, 22, 62, 93, 100, 127, 138, 142, 238
PGC-1α 116, 129
polycystic ovary syndrome 229, 232
post-traumatic stress disorder (PTSD) 61, 157, 163, 164, 173, 309
pregnancy 70, 226–232
progesterone 228, 231
Psychiatric Genomics Consortium (PGC) 164, 199, 201, 204, 205
psychosis 4, 64, 91, 100, 157, 165, 167, 173, 174, 265
anhedonia 100, 174, 196, 198, 203, 206–208, 211, 212, 245, 253
glucose dysregulation 92, 159–161, 248
hormonal disturbance 229–231
insulin resistance 65, 234, 254
risk for 168, 226–228
pyruvate 39, 62, 64, 87, 90, 92, 94, 95, 160, 202, 319

Q

quality of life 1, 85, 97, 165, 266, 271

R

radiotracers 17–19, 22
randomized controlled trials (RCTs) 96, 99, 235, 263, 287, 316
avoiding bias 265, 266, 272–274
data analysis 278–280
designing 268–278
rare genetic mitochondrial diseases 165
reactive oxygen species (ROS) 64, 88, 136–142, 164, 171, 226, 235, 318, 320
redox 2, 3, 20, 39, 66, 198, 322
dysregulation 40, 65, 167–170, 200, 212
resting-state neuronal activity 3, 33, 207, 208

S

schizophrenia 3–4, 83–89, 99, 127, 157, 166–171, 245, 247, 265, 305, 309
biomarkers 196, 200
brain pH 85, 95
cardiovascular disease 65, 66
genetic risk 86–89, 164–167, 248
glucose metabolism 93, 159, 234
imaging 19–20, 94, 95
impact of hormones 65, 230
insulin resistance 91
ketogenic diet 235, 264, 276
lactate levels 62, 63, 66, 94, 166
obesity 161, 162
treating 68, 97
Scottish Family Health Study 86, 204
Seattle Alzheimer's Disease Cell Atlas 21
selfish brain 3
serotonin 224, 227, 229, 287, 288
simulation-based inference 126, 285, 291, 295
sleep 35, 58, 59, 69, 115, 176, 197, 198, 203–210, 224, 252–254, 265, 310
stress 57–60, 71, 115, 116, 138, 162–164, 174, 197, 198, 245, 247, 250–253
early-life 171, 198, 226, 227, 234
energy budget model 31–33
oxidative 40, 88, 92, 139, 140, 145, 155, 166–168, 171, 229, 234, 237, 307, 311, 316, 320
PTSD 61, 157, 173, 309
role of estrogen 64–65
stroke 16, 18, 69
structural causal models (SCM) 285, 288, 295, 296
structural dynamical causal model (SDCM) 296, 298
synaptic metabolic homeostasis 41–45
synaptosomes 17

T

testosterone 228–229, 322
transfer entropy (TE) 285, 291–294

trauma 57, 157, 163, 164, 171, 173, 176
  early-life 197, 198, 226–228, 250
  PTSD 61, 157, 163, 164, 173, 309
tricarboxylic acid (TCA) cycle 61, 90, 92, 93, 115, 142, 159–161, 307, 318–321
type 2 diabetes 69, 99, 208, 213, 224, 226, 236, 238, 239, 248, 269

U
UK Biobank 200, 202–205

W
Warburg effect 120, 160
whole-body energy metabolism 6, 29, 31, 33, 45, 155, 157

Strüngmann Forum Reports

Series Editor: Julia R. Lupp
Editorial Assistance: Aimée Gessner, Catherine Stephen
Lektorat: BerlinScienceWorks

Sex and Gender: Toward Transforming Scientific Practice
Edited by L. Zachary DuBois, Anelis Kaiser Trujillo, Margaret M. McCarthy
Springer Cham
DOI: https://doi.org/10.1007/978-3-031-91371-6
ISBN: 978-3-031-91370-9

The Nature and Dynamics of Collaboration
Edited by Paul F. M. J. Verschure
DOI: https://doi.org/10.7551/mitpress/15533.001.0001
ISBN: 9780262548144

The Frontal Cortex: Organization, Networks, and Function
Edited by Marie T. Banich, Suzanne N. Haber and Trevor W. Robbins
DOI: https://doi.org/10.7551/mitpress/15679.001.0001
ISBN: 9780262549530

Digital Ethology: Human Behavior in Geospatial Context
Edited by Tomáš Paus and Hye-Chung Kum
DOI: https://doi.org/10.7551/mitpress/15532.001.0001
ISBN: 9780262548137

Migration Stigma: Understanding Prejudice, Discrimination, and Exclusion
Edited by Lawrence H. Yang, Maureen A. Eger and Bruce G. Link
DOI: https://doi.org/10.7551/mitpress/15529.001.0001
ISBN: 9780262548120

Exploring and Exploiting Genetic Risk for Psychiatric Disorders
Edited by Joshua A. Gordon and Elisabeth Binder
DOI: https://doi.org/10.7551/mitpress/15380.001.0001
ISBN electronic: 9780262377423

Intrusive Thinking: From Molecules to Free Will
Edited by Peter W. Kalivas and Martin P. Paulus
ISBN: 9780262542371

Deliberate Ignorance: Choosing Not to Know
Edited by Ralph Hertwig and Christoph Engel
ISBN 9780262045599

Youth Mental Health: A Paradigm for Prevention and Early Intervention
Edited by Peter J. Uhlhaas and Stephen J. Wood
ISBN: 9780262043977

The Neocortex
Edited by Wolf Singer, Terrence J. Sejnowski and Pasko Rakic
ISBN: 9780262043243

© Frankfurt Institute for Advanced Studies (FIAS) 2025

D. Öngür, J. M. Ford (eds.), *Metabolic Neuropsychiatry*, Strüngmann Forum Reports,
https://doi.org/10.1007/978-3-032-05630-6

Interactive Task Learning: Humans, Robots, and Agents Acquiring New Tasks through Natural Interactions
Edited by Kevin A. Gluck and John E. Laird
ISBN: 9780262038829

Agrobiodiversity: Integrating Knowledge for a Sustainable Future
Edited by Karl S. Zimmerer and Stef de Haan
ISBN: 9780262038683

Rethinking Environmentalism: Linking Justice, Sustainability, and Diversity
Edited by Sharachchandra Lele, Eduardo S. Brondizio, John Byrne, Georgina M. Mace and Joan Martinez-Alier
ISBN: 9780262038966

Emergent Brain Dynamics: Prebirth to Adolescence
Edited by April A. Benasich and Urs Ribary
ISBN: 9780262038638

The Cultural Nature of Attachment: Contextualizing Relationships and Development
Edited by Heidi Keller and Kim A. Bard
ISBN (Hardcover): 9780262036900 ISBN (ebook): 9780262342865
Winner of the Ursula Gielen Global Psychology Book Award

Investors and Exploiters in Ecology and Economics: Principles and Applications
Edited by Luc-Alain Giraldeau, Philipp Heeb and Michael Kosfeld
ISBN (Hardcover): 9780262036122 ISBN (eBook): 9780262339797

Computational Psychiatry: New Perspectives on Mental Illness
Edited by A. David Redish and Joshua A. Gordon
ISBN: 9780262035422

Complexity and Evolution: Toward a New Synthesis for Economics
Edited by David S. Wilson and Alan Kirman
ISBN: 9780262035385

The Pragmatic Turn: Toward Action-Oriented Views in Cognitive Science
Edited by Andreas K. Engel, Karl J. Friston and Danica Kragic
ISBN: 9780262034326

Translational Neuroscience: Toward New Therapies
Edited by Karoly Nikolich and Steven E. Hyman
ISBN: 9780262029865

Trace Metals and Infectious Diseases
Edited by Jerome O. Nriagu and Eric P. Skaar
ISBN 9780262029193

Pathways to Peace: The Transformative Power of Children and Families
Edited by James F. Leckman, Catherine Panter-Brick and Rima Salah,
ISBN 9780262027984

Rethinking Global Land Use in an Urban Era
Edited by Karen C. Seto and Anette Reenberg
ISBN 9780262026901

Schizophrenia: Evolution and Synthesis
Edited by Steven M. Silverstein, Bita Moghaddam and Til Wykes,
ISBN 9780262019620

Cultural Evolution: Society, Technology, Language, and Religion
Edited by Peter J. Richerson and Morten H. Christiansen,
ISBN 9780262019750

Language, Music, and the Brain: A Mysterious Relationship
Edited by Michael A. Arbib
ISBN 9780262019620

Evolution and the Mechanisms of Decision Making
Edited by Peter Hammerstein and Jeffrey R. Stevens
ISBN 9780262018081

Cognitive Search: Evolution, Algorithms, and the Brain
Edited by Peter M. Todd, Thomas T. Hills and Trevor W. Robbins,
ISBN 9780262018098

Animal Thinking: Contemporary Issues in Comparative Cognition
Edited by Randolf Menzel and Julia Fischer
ISBN 9780262016636

Disease Eradication in the 21st Century: Implications for Global Health
Edited by Stephen L. Cochi and Walter R. Dowdle
ISBN 9780262016735

Better Doctors, Better Patients, Better Decisions: Envisioning Health Care 2020
Edited by Gerd Gigerenzer and J. A. Muir Gray
ISBN 9780262016032

Dynamic Coordination in the Brain: From Neurons to Mind
Edited by Christoph von der Malsburg, William A. Phillips and Wolf Singer,
ISBN 9780262014717

Linkages of Sustainability
Edited by Thomas E. Graedel and Ester van der Voet
ISBN 9780262013581

Biological Foundations and Origin of Syntax
Edited by Derek Bickerton and Eörs Szathmáry
ISBN 9780262013567

Clouds in the Perturbed Climate System: Their Relationship to Energy Balance, Atmospheric Dynamics, and Precipitation
Edited by Jost Heintzenberg and Robert J. Charlson
ISBN 9780262012874
Winner of the Atmospheric Science Librarians International Choice Award

Better Than Conscious? Decision Making, the Human Mind, and Implications For Institutions
Edited by Christoph Engel and Wolf Singer
ISBN 978-0-262-19580-5

MIX
Papier aus verantwortungsvollen Quellen
Paper from responsible sources
FSC® C105338

If you have any concerns about our products,
you can contact us on
ProductSafety@springernature.com

In case Publisher is established outside the EU,
the EU authorized representative is:
Springer Nature Customer Service Center GmbH
Europaplatz 3, 69115 Heidelberg, Germany

Printed by Libri Plureos GmbH
in Hamburg, Germany